A HISTORY OF NAUTICAL ASTRONOMY

BY THE SAME AUTHOR

The Elements of Navigation
The Principles and Practice of Radio Direction Finding
The Master and His Ship
The Apprentice and His Ship
The Complete Coastal Navigator
The Physical Geography of the Oceans
The Astronomical and Mathematical Foundations of Geography

A History of Nautical Astronomy

CHARLES H. COTTER
EX C BSC F INST NAV

*Senior Lecturer in the Department of Maritime Studies
at the University of Wales Institute of Science and Technology*

'. . . but touching the Theoricke, by observation of
the Heavens, and guid-starres there, to finde
a passage in the vast Ocean where no paths are,
none is more necessarie for their vse, then the
perfect knowledge of this Trigonometrie.'

*From the Epistle Dedicatorie in Ralph Handson's
'Trigonometry: or THE DOCTRINE OF TRIANGLES.'
First written in Latine by Bartholmevs Pitiscvs
of Grunberg in Silesia.*

NEW YORK
AMERICAN ELSEVIER PUBLISHING COMPANY INC.

© Charles H. Cotter 1968

First published in the United States 1968

AMERICAN ELSEVIER PUBLISHING COMPANY, INC.
52 Vanderbilt Avenue
New York, New York 10017

LIBRARY OF CONGRESS CATALOG CARD NUMBER 68-12049

Made and printed in Great Britain by
William Clowes and Sons, Limited, London and Beccles

Contents

Preface by Alton B. Moody, ix
Author's Foreword, xi

I. *The Development of Nautical Astronomy*, 1
   1. Introductory, 1
   2. Babylonians and Phoenicians, 2
   3. The Greeks, 7
   4. Hipparchus, 13
   5. Ptolemy, 16
   6. The Arabs, 19
   7. Early Renaissance Scholars, 20
   8. Copernicus, Tycho Brahe, 22
   9. Kepler, Galileo, Newton, 25
   10. The Dawn of Modern Nautical Astronomy, 28

II. *Astronomical Methods of Time-measuring at Sea*, 32
   1. The Units of Time, 32
   2. The Calendar, 35
   3. The Gnomon, 36
   4. The Divisions of the Day, 37
   5. The Nocturnal, 39
   6. Sun Time and the Ring Dial, 41
   7. Mechanical Clocks, 42
   8. Arithmetical Navigation, 43
   9. The Azimuth Compass, 47
   10. Nautical Tables for Determining Time, 47
   11. The *Nautical Almanac*, 48
   12. Computing Local Time, 51
   13. The Marine Chronometer, 52

III. *The Altitude-measuring Instruments of Navigation*, 57
   1. Introductory, 57
   2. The Seaman's Quadrant, 57

3. The Astrolabe, 60
4. The Cross-staff, 64
5. The Kamal, 69
6. The Back-staff, 70
7. The Reflecting Quadrant, 74
8. The Hadley Quadrant, 77
9. The Reflecting Circle, 83
10. Perfected Altitude-measuring Instruments, 87
11. The Artificial Horizon, 91

IV. *The Altitude Corrections*, 97
1. Introductory, 97
2. Refraction, 97
3. Depression or Dip of the Sea Horizon, 111
4. The Sun's Semi-diameter, 117
5. The Moon's Semi-diameter, 118
6. Parallax, 119
7. Irradiation, 121
8. Personal Error or Equation, 122

V. *Methods of Finding Latitude*, 123
1. Introductory, 123
2. Latitude by the Pole Star, 130
3. Latitude by Meridian Altitude of the Sun, 137
4. Latitude by Meridian Altitude of a Star, 139
5. Latitude by the Southern Cross, 140
6. Declination Tables, 141
7. The Double-Altitude Problem, 143
8. Meridian and Maximum Altitudes, 162
9. Latitude by Ex-meridian Altitude, 165

VI. *Methods of Finding Longitude*, 180
1. Introductory, 180
2. Longitude from Eclipse Observations, 182
3. Longitude from Observations of Jupiter's Satellites, 184
4. Longitude from Observations of Moon Occultations, 189
5. Longitude by Lunar Transit Observation, 191
6. Rewards for Discovering the Longitude, 192

7. Longitude by Lunar Distance, 195
   (a) Historical Survey, 195
   (b) Maskelyne and the *Nautical Almanac*, 203
   (c) Principles and Practice, 205
   (d) Methods for Clearing the Distance, 208
8. Finding G.M.T. from a Lunar Observation, 237
9. Longitude by Chronometer, 243
   (a) Methods, 243
   (b) Observations for Checking the Chronometer, 254
   (c) Absolute-altitude and Equal-altitudes Observations, 257

VII. *Position-line Navigation*, 268
1. Introductory, 268
2. Historical Development of Position-line Navigation, 271
3. Sumner's Discovery, 275
4. Azimuth Tables in Position-line Navigation, 284
   (a) Burdwood and Davis, 284
   (b) Heath and A, B and C Tables, 284
5. Marcq St Hilaire and the New Navigation, 293

VIII. *Navigation Tables*, 309
1. Early Astronomical and Mathematical Tables, 309
2. The *Nautical Almanac*, 313
3. Short-method and Inspection Tables, 317
4. Graphical Solutions of the PZX Triangle, 334
5. Mechanical Aids to Calculation, 340

Appendix 1. Spherical Astronomy, 343
Appendix 2. Spherical Trigonometry, 349
Bibliography, 357
Index, 373

List of Plates

Between pages 92 and 93

1. Rev. Nevil Maskelyne D.D.
2. Mariner's Astrolabe. Probably Spanish, c. 1585.
3. Nocturnal in Boxwood. English, c. 1646.
4. Azimuth Compass. English, c. 1720.
5. Mariner's Quadrant. c. 1600.
6. Back-staff or Davis Quadrant. By John Gilbert, c. 1740.

Between pages 236 and 237

7. Hadley Octant. By Benjamin Martin, c. 1760.
8. Reflecting Circle. By Edward Troughton, c. 1800.
9. Sextant. By Kelvin Hughes, 1967.
10. Facsimile of Plate 3 of Sumner's Pamphlet.
11. Facing Pages of 1797 *Nautical Almanac*.
12. Facing Pages 1897 *Nautical Almanac*.

Plates 1 to 8 are reproduced by courtesy of the Trustees of the National Maritime Museum, Greenwich. Plate 9 is reproduced by courtesy of Messrs. Kelvin Hughes.

Preface

ALTON B. MOODY,
former President of the United States Institute of Navigation

In this fast-moving age, characterized by a rapid increase in knowledge sometimes referred to as an 'information explosion', it is easy to succumb to the temptation to limit one's reading to new developments in one's own specialty. With the plethora of books and technical periodicals that come to the attention of the professional, it becomes a problem even to keep informed of these. As a result, a certain superficiality pervades much of the thinking of those who exhibit interest in professional subjects.

Navigation is particularly susceptible to this danger because of the broad scope of disciplines it encompasses. As a result, a great amount of effort is expended by talented individuals who lack perspective regarding the problems they seek to solve, as attested by the many solutions that look good on paper but do not find a ready acceptance by those who would seem to be beneficiaries of the work.

The history of nautical astronomy is a fascinating subject involving the hopes, fears, superstitions and thoughtful observations of many individuals over a very long period of time. Early man sensed the value of celestial observations as a means of providing guidance at sea, where no landmarks were available and electronic signals were unknown, but lacked the knowledge and instruments needed fully to utilize this source of guidance. There was something frightening about putting to sea under these conditions, and a widespread fear of what might happen if one reached the physical boundary of the earth offered little comfort to those with sufficient curiosity to set forth into the unknown. As a result, only the more intrepid adventurers deliberately attempted long voyages out of sight of land.

But there were hardy individuals in various periods who sought to widen the horizon of man's knowledge. Little by little nature grudgingly yielded to these attempts, and the story of this struggle is the story of the progress of man. Certainly there is no more captivating story than that relating to man's attempt to 'discover' the longitude at sea, and many there were who despaired of a practical solution ever being found. As late as 1594, nearly two centuries before the problem was solved but many centuries after man had ventured beyond the sight of land, Davis wrote: 'Now there be some that are very inquisitive to have a way to get the longitude, but that is too tedious for seamen, since it requireth the deep knowledge of astronomy, wherefore I would not have any man think that the longitude is to be found at sea by any instrument, so let no seamen trouble themselves with any such rule, but let them keep a perfect account and reckoning of the way of their ship.'

This book is a history of nautical astronomy. But it is more than that. Captain Charles H. Cotter has done more than trace the sequence of events leading to man's present extensive knowledge of celestial navigation. With discerning care he has delved into representative solutions of various stages of man's developing knowledge, and gives the reader the mathematical as well as philosophical explanations of methods which are now generally known only by name—or not at all by many members of a rising generation who have been too absorbed in finding solutions to today's problems to take time to learn of those of other ages.

What then is the value of this book? As a reference, it brings together a wealth of information that would require extensive searching to find, and puts it in perspective. As a source of inspiration to those who may be discouraged at the difficulty of conquest over the unknown, it stands as a beacon on a hill. As a repository of cultural information on man's emergence as the master of his environment, it is a worthy addition to any gentleman's library.

Washington, D.C., 1967   ALTON B. MOODY

Author's Foreword

The several factors, including wind and hidden current, which operate during a voyage to set a vessel from the desired path make it imperative for a navigator frequently to check the progress of his ship. Without accurate knowledge of his latitude and longitude a navigator is unable to rectify the course of his ship with confidence. In this present work I have attempted to trace the fascinating story of the development of the astronomical methods used for finding a ship's position at sea when out of sight of land.

Few would deny that astronomical navigation, or nautical astronomy to give the subject its time-honoured name, is an obsolescent craft. The perfected astronomical methods used at present mark the culmination of an evolutionary process which began even before the first of the Phoenician sea-traders navigated their craft in Eastern Mediterranean waters, using the heavenly bodies to guide them, some three thousand years ago. The ancient craft of the nautical astronomer is now rapidly being replaced by sophisticated navigational systems born of that advanced technology which is the distinctive feature of the marvellous age in which we live.

My principal aim in writing this history has been to present in broad outline an historical account of the diverse problems of nautical astronomy and the ways in which they were solved. I have been conscious of a strong desire to associate with these problems the philosophers, scientists and navigators, who prepared and sometimes found the paths which led to their solutions. These men occupied prominent positions in a series of actions, events and purposes which, considered collectively, possess a remarkable dramatic unity.

The history of nautical astronomy spans a long era of many millennia. It had its beginnings when sea-venturers first learnt the rudimentary use of the stars to guide them in their exploratory and commercial voyages. The era is ending in our own times when the applications of radio and electronics are superseding the relatively simple techniques of the nautical astronomer.

My own intense interest in the history of navigation was sparked off during my formative years as a pupil at Smith Junior Nautical School, Cardiff. From my wise and inspiring teachers I learnt that mere technical mastery of navigation and nautical astronomy, however important this may be to an officer on the bridge of a ship at sea, does not in itself make a complete and educated navigator. This indeed is a valuable lesson to learn, and I too am firm in the belief that no practitioner can have complete respect for the science or craft he practises without having some acquaintance with the historical development or evolution of his subject. Moreover I hold the view that men of general culture can have little regard for a science they know all too little about. With these sentiments it is my sincere desire that this work will serve usefully to fill a gap in the literature of the history of the science of navigation.

I owe a debt of gratitude to several assistants, past and present, at the libraries of the British Museum and the National Maritime Museum in London, and the Public Library of Newcastle upon Tyne, for their kind help during the many years during which I was engaged in making researches. I wish also to record that I have been appreciative of the friendly interest shown by some of my colleagues in the progress of this work.

It is a very real pleasure for me to place on record my thanks to the staff of the publishers for much advice and many valuable suggestions which have helped considerably in clarifying and otherwise improving my original text.

Cardiff 1967   CHARLES H. COTTER

CHAPTER I

The development of nautical astronomy

1. INTRODUCTORY

John Seller, Hydrographer to the King during the late 17th century, declared, in his popular work on *Practical Navigation*, that the part of navigation

'...which may properly bare the name and principally deserves to be entituled the art of Navigation, is that part which guides the ship in her Course through the Immense Ocean to any part of the known World; which cannot be done unless it be determined in what place the Ship is at all times, both in respect of Latitude and Longitude: this being the principal care of a navigator and the Masterpiece of Nautical Science.'

'To the Commendable Accomplishment of this knowledge,' Seller added, 'four things are subordinate Requisites. Viz:
Arithmetic.
Geometry.
Trigonometry.
The Doctrine of the Spheres.'

The Doctrine of the Spheres covered the necessary spherical or mathematical astronomy, a knowledge of which was essential for guiding a ship across the pathless oceans, for finding her position, and for estimating her progress towards her destination.

The history of Ocean Navigation, or *la Navigation Grande* of the old French navigators, began when astronomy became scientific, that is to say, when men first began to reason about, and speculate upon, the nature of the celestial bodies and their movements.

Astronomy, which literally means the law of the stars or 'star distribution,' is a branch of knowledge probably as old as mankind itself; and our most ancient ancestors must have, as we have, gazed upon the firmament, rejoiced in its splendour, and pondered about its nature.

The knowledge, acquired by painstaking observations, of the relative movements of the Sun, Moon, Planets and stars was, little by little, built into a tradition which provided early man with the means for finding direction, time and season. The earliest civilizations found it essential to have compass, clock and calendar; and the orderly movements of the heavenly bodies, relative to the Earth, provided for their needs.

There is reason to suppose that the flourishing civilizations of at least three thousand years ago possessed good practical knowledge of astronomy. The earliest astronomical observations, and the most primitive cosmogonal ideas, belong to an extended period of astronomical prehistory for which no written records exist. During this period the stars were grouped to form the constellations, eclipses were observed, and the apparent paths of the Sun, Moon and planets across the backcloth of fixed stars were delineated.

The Greek philosophers of the 5th century before the beginning of the Christian Era appear to have been the first to enquire into the causes of celestial phenomena. They are, therefore, to be honoured as being the founders of scientific astronomy.

The astronomical science of the Ancient Greeks of the period between the 5th and 3rd centuries BC, was based upon observations made by earlier philosophers, notable amongst whom were those of Babylon and Egypt.

2. BABYLONIANS AND PHOENICIANS

The Babylonians, who occupied the seaboard of Syria, formed a branch of the Semitic race who cultivated a love for the sea. Some historians have argued that the Phoenicians—as these people are called—originated on the eastern shore of the Red Sea in the 'Land of Edom'; and certain it is, as we shall see, that Phoenician seamen voyaged in the Red Sea as well as in the Mediterranean.

The notable sea ports of the Phoenicians included Tyre and Sidon. Numerous references are made in the *Old Testament* to Tyrians and Sidonians, and to Phoenicia and its seamen, and to their nautical—as well as their commercial—skill.

The earliest biblical account of a long sea voyage appears in the

THE DEVELOPMENT OF NAUTICAL ASTRONOMY 3

First Book of Kings, where we learn that King Solomon made a navy of ships in the land of Edom; and that Hiram, king of Tyre, sent shipmen who had knowledge of the sea to join the servants of Solomon. And the navy, we are told, voyaged to Ophir—believed to be Ceylon—the voyage occupying three years. And from Ophir were brought gold, silver and great plenty of almug trees, as well as apes, peacocks and precious stones.

There is every reason to believe that the ancient Phoenician seamen used astronomical methods for navigating their vessels. Reference to Homer's *Odyssey*, the epic poem which describes the adventures of the mythical hero Odysseus, reveals that the stars were used for navigational purposes before the days of Homer. The period during which Homer flourished is not known with any degree of certainty, but classical scholars date it between the 12th and 7th centuries BC.

In Book Five of *The Odyssey*, which describes how Calypso helps Odysseus to build a craft and gives him sailing directions for his voyage, it is related of Odysseus in Pope's translation that

> 'Placed at the helm he sate, and mark'd the skies,
> Nor closed in sleep his ever watchful eyes.
> There viewed the Pleiads, and the Northern Team,
> And great Orion's more refulgent beam;
> To which, around the axle of the sky,
> The Bear revolving, points his golden eye,
> Who shines exalted on th' ethereal plain,
> Nor bathes his blazing forehead in the main.'

It was the Bear—the Great Bear—which the fair Calypso bade Odysseus to keep on his port side as he traversed the sea.

From the time of Solomon, who lived about ten centuries before the birth of Christ, there is little recorded in respect of navigation until the year 610 BC when, as we are informed by Herodotus, some Phoenician ships, by order of Necho, king of Egypt, sailed down the Red Sea and, after rounding the African continent, entered the Mediterranean through the Pillars of Hercules after a voyage lasting three years. It was during this voyage that the Phoenicians, it is thought, discovered the Canary Islands.

From the time of Necho, navigation on the western coast of the

African continent was neglected until the rise of the Portuguese in the early 15th century under the sponsorship of Prince Henry the Navigator.

That the Phoenicians were great seamen there can be no doubt; and for many centuries the ships and mariners of Tyre and Sidon were indispensable to the great powers of the Eastern Mediterranean, including Persia, Greece and Rome.

Modern historical research has revealed that Phoenician sea power in the Eastern Mediterranean sprang from an earlier Cretan or 'Minoan' sea power during the 12th century BC. The rise of the Phoenicians, and the decline of Cretan sea power, coincided with considerable disorder in the region due principally to invasions of people from the north, and the weakening of Egyptian power in the lands on the seaboard of the Eastern Mediterranean.

The wide extent of the commercial relations of the Phoenicians during their ascendancy, by both land and sea, may be appreciated from what is written in Chapter 27 of the *Book of the Prophet Ezekiel*. Colonies of Phoenicians were planted in commercially strategic points in the Mediterranean littoral, as well as on islands within the Mediterranean. Colonies of Phoenicians were to be found in North Africa, Spain, Cyprus, Malta and Sicily. Trade with Spain was of the greatest importance, because it was from here that silver and lead, and other important metals, were obtained. There is no doubt that Phoenician ships traded in the Atlantic, and the shipmen of Tyre and Sidon certainly voyaged to North-west Spain and possibly to Southern Britain in their quest for sea trade. It was the important trade with South-east Spain, however, that doubtless led to the colonization of the Western Mediterranean littoral by Phoenicians as early as the 9th century BC.

With the rise of Greek culture in the Aegean Sea, the power of Tyre and Sidon declined. The Phoenician colonies, therefore, turned for their protection to Carthage. Carthage, the great Phoenician trading emporium in Tunisia, established during the 8th century BC, emerged as a major power in the 6th century BC.

Let us now consider briefly the observational astronomy of the Babylonians. The astronomers of Babylon recorded lunar eclipses from as far back as the 8th century BC. Systematic observations of the Moon's apparent movement against the background of

THE DEVELOPMENT OF NAUTICAL ASTRONOMY

fixed stars reveal the Moon's periodic motions. The same applies to the planets. Moreover, observations of the fixed stars which lie on or near to the Sun's annual apparent revolution on the celestial sphere enable the determination of the length of the year.

The appearance of a fixed star for the first time in the eastern sky after sunset is known as the *achronychal rising* of the star; and its setting at the time of sunset is known as its *achronychal setting*. A star rising or setting respectively at the time of sunrise or sunset is said to rise or set *cosmically*. When a star first becomes visible in the morning before sunrise, or in the evening after sunset, it is said to rise or set *heliacally*. Observations of the achronychal, cosmical and heliacal risings and settings of selected stars, or star groups, were obvious ones to have been made by an ancient astronomer who studied the stars systematically.

Systematic observations, made by the early astronomers, were related to the practical problems of timekeeping, so necessary for agriculturist and administrator alike. Records of such observations enabled astronomers to predict astronomical events. Nowadays the astronomical measurement of time is not related to the rising or setting of celestial objects, but to their meridian passages.

For convenience of civil life a method of fitting days into periods such as months and years to form a calendar is necessary. The earliest use to which astronomical observations were put was related to the forming of calendars. Three astronomical periods were of importance in ancient calendar-making. These were: the diurnal rotation of the Earth; the monthly motion of the Moon, in which a complete sequence of phases from New Moon to the next New Moon is exhibited; and the orbital motion of the Earth around the Sun, a reflection of which is the apparent annual motion of the Sun around the celestial sphere. It is the incommensurable nature of these three periods that made the problem of devising a satisfactory calendar one of great complexity to the Ancients.

The most important astronomical periodic cycle is that of the Earth's revolution around the Sun, this regulating, as it does, the seasons and, therefore, the times for sowing and harvesting. The monthly cycle of the Moon is extremely erratic, so that the Moon is unsuitable for calendar-making despite the fact that the organization of much of the religious and civil life of the ancient peoples, especially that of the Egyptians, was related to the Moon and her phases.

The astronomical observations made by the Chaldeans—the priestly caste of Babylon—led to the discovery of an important eclipse cycle known as the Saros. The discovery of the Saros, which is a period of 223 lunations, made it possible to predict the times of eclipses.

The observations which enabled astronomers to predict astronomical events, particularly those observations related to the planets and the Moon, resulted in astronomers being regarded as having magical powers. It is not surprising, therefore, that the earliest astronomers, in taking advantage of their knowledge and skill, practised astrology. Events on Earth were regarded as being related to such astronomical events as occultations, eclipses, conjunctions and oppositions. The forecasting of rain and wind—phenomena which are related to the seasons (and these are clearly foretold by astronomical methods), as well as the prophesying of such events as victory or defeat in battle, illness or prosperity (which are unrelated to astronomical events), became the business—and a very lucrative one no doubt—of the astronomer-turned-astrologer.

The grouping of the stars into *asterisms* or *constellations* is of great antiquity, and the present names of many of the constellations suggest that the Babylonians were responsible for forming and naming them.

The star groups of the Ancient Greeks were manifestly borrowed from the Phoenicians. The figures or shapes of the constellations were, doubtless, initially simple, and were derived from commonplace things; but the Greek poets metamorphosed these simple figures so that they became hardly recognizable in their original forms. The figure of the constellation *Chimah*, for instance, which is mentioned in the *Old Testament* books of *Job* and *Amos*, is that of an armed man; and this immediately directs our attention to *Orion* the hunter, so prominent a figure in Greek mythology. The figure of the constellation *Aish*, referred to in the *Book of Job*, signifies a cluster; and it appears obvious that it is associated with the *Pleiades* of the Greeks. There are many similar examples of Babylonian constellations having been remodelled by the Ancient Greeks to link their fabulous history with the stars.

The term *Mazaroth* refers to a broad zone of the celestial sphere on which the tracks of the Sun and Moon are traced. As different star groups within this zone rise heliacally on different months, we may comprehend what is meant by the statement 'bringing forth Mazaroth in its season.' which appears in Chapter 9 of the *Book of Job*.

Observations of the fixed stars laid the foundations of astronomy, because, by these alone is it possible to determine the lengths of the year and month and the periods of the planets, as well as other astronomical quantities such as the rate of the precession of the equinoxes*; the celestial positions of the stars, Sun, Moon and planets; and the irregularities in the apparent motions of these bodies.

We are expressly told by Herodotus that the Greeks borrowed from the Phoenicians the gnomon and the method of dividing the day into twelve parts. The *gnomon*, which in its simplest form is merely a rod planted vertically in the ground, served to mark the passage of time during the daytime by the movement and length of the shadow it cast. It also served to mark the succession of the seasons, from the varying length of the shadow cast by successive noonday Suns. The Babylonians are credited with the invention of the so-called *sexagesimal system* of measuring angles, in which the circle is divided into 360 parts.

In the year 330 BC Babylon was conquered by Alexander the Great, the founder of the great Egyptian city and centre of learning, Alexandria. It was a direct result of Alexander's conquests, after which the lands of the entire eastern part of the Mediterranean were welded together into one great political unit, that Babylonian influence upon Greek science became possible.

3. THE GREEKS

The ascendancy of Greek science may be regarded as having coincided with the time of Thales, who flourished during the end of the 7th and the early part of the 6th centuries BC. Thales of Miletus travelled to Egypt and learnt much from the Egyptians. He learnt from the priests secret information such as the length of the year, the signs of the Zodiac and the positions of the solstices and equinoxes. Thales is said to have been the inventor of the theorem which is usually known as Pythagoras' Theorem,

* See p. 123, Chapter V.

and to have ascertained the heights of the pyramids by measuring the lengths of their shadows when the Sun's altitude was 45°.

Thales, according to the poet Callimachus, is said to have formed the constellation of the Lesser Bear; but this constellation was undoubtedly used by Phoenician navigators before the time of Thales. It is very likely, however, that Thales introduced this constellation to the Greeks. Although Thales left nothing in writing, it is believed that he explained the correct causes of eclipses and the cause of the phases of the Moon.

The suggestion that the Earth has the form of a sphere, appears to have been first advanced in the 5th century BC by Parmenides. The idea that the Earth is spherical may have suggested that the firmament of heaven is also spherical; and this may have led to the explanation of celestial phenomena by circular motion.

Some twenty or so years after the death of Thales, during mid-6th century BC, the famous Pythagoras was born. His name is closely linked with the study of geometry; and he is said to have been the inventor of many of the propositions which form the first book of Euclid. Pythagoras and his followers invented a celestial system in which the Earth was regarded as revolving around a central fire called *anlichthon*, which was believed to be located at the centre of the universe.

The cosmology of Pythagoras was based on fantastic principles. The Pythagoreans were convinced that the total number of moving objects in the heavens must be ten—the perfect number, as they thought. The Sun, Moon, Earth, and the five planets, Mercury, Venus, Mars, Jupiter and Saturn, together with the sphere of the fixed stars made nine. The tenth moving body was a supposed counter-Earth which revolved around the central fire. Because the central fire and the counter-Earth were not visible from the Earth, the Pythagoreans supposed the part of the Earth on which they lived to be directed away from both. When sea voyages were extended, and observers failed to see either the central fire or the counter-Earth, the hypothesis fell out of favour.

The deductive methods of the early Greek philosophers, in which ideas of the universe were formulated from general principles rather than from observations and knowledge, are exemplified in the writings of Aristotle (384–322 BC) and Plato (b. 429

BC). Plato's ideas on the universe were based on what was thought to be appropriate. The universe he believed to be modelled on a perfect plan and, therefore, because the most perfect shape is a sphere and the most perfect curve is a circle, the universe must be spherical, and the motions of the heavenly bodies must be circular about the Earth which he believed to be fixed at the centre of the universe.

Aristotle pictured the universe as comprising a number of concentric spheres with the Earth reposing at the centre. Surrounding the sphere of the Earth were the spheres of water, air and fire. Water, air and fire, together with earth, were the four *elements*. Surrounding the sphere of fire were the spheres of the Moon, Sun, and each of the five ancient planets; and beyond these was the sphere of the fixed stars. Aristotle believed that a force was in operation which kept the spheres of the planets, stars, Sun and Moon moving, each at its own allotted speed. The force necessary to do this was thought to reside in an additional sphere outside the sphere of the stars. This was the *Primum Mobile* which was identified with the Creator of the universe.

Eudoxus of Cnidus (408–355 BC), in reply to Plato's postulate that a set of circular movements would explain the observed planetary motions, devised what is regarded as being the first mathematical theory of planetary motion. To fit his theory with observation Eudoxus regarded the Earth as being at the centre of the universe, Surrounding the Earth he postulated a number of revolving spheres the outermost one corresponding to the sphere of the fixed stars introduced by Aristotle. Secondary spheres were regarded as revolving around points on the inner spheres; and it was these secondary spheres which carried the Sun, Moon and planets. The rotation periods of the many spheres which belonged to the system were made to fit the observations of the movements of the heavenly bodies. The planetary theory of Eudoxus was a geometrical conception designed to facilitate the compilation of tables of eclipses: there is no suggestion that he thought the universe was constructed in this way. The irregularities of the motions of the heavenly bodies, particularly those of the Moon, which were manifested when observations of astronomical events did not coincide with predictions, resulted in more spheres being added to the system of Eudoxus in an attempt to improve the mathematical laws of prediction. Callippus, a follower of Eudoxus, used thirty-four spheres, compared with twenty-seven in the system of Eudoxus, in order to explain the motions of the Sun, Moon and planets.

The problems of calendar-making were tackled by the Ancient Greeks, and many calendar reforms were suggested by them. The common year of the Greeks consisted of 360 days. It is obvious that had this period been adopted without correction the months and seasons would have fallen out of step with one another. To prevent this from happening it was necessary to intercalate days. The astronomer Meton, who flourished about 430 BC, introduced a calendar reform based on a nineteen-year cycle of 235 lunations. The *Metonic Cycle* is a period at the end of which the Sun and Moon occupy the same positions in the celestial sphere, relative to the fixed stars, as they did at the commencement of the cycle. The Metonic Cycle is still used for establishing the date of Easter in the ecclesiastic calendar, the Golden Number of the prayer book being the number in the cycle used for fixing the date of Easter Sunday.

Contemporary with Plato and Eudoxus was Philolaus, who asserted that the Earth revolves around the Sun once in a year. It is not known, however, by what arguments or observations he made this assertion.

A Syracusan named Nicetas, who lived at about the same time as Philolaus, believed the Earth to rotate once per day—a hypothesis put forward, it is supposed, to overcome the difficulty arising from the common belief which required the celestial sphere to rotate diurnally around the Earth at a fantastic speed.

Immediately following the vast conquests of Alexander the Great, the principal centre of learning in the Mediterranean region was established in Alexandria; and it was the Alexandrian Greeks who became the foremost scientists, and who were to occupy the focal point of learning for nigh on a millennium, until the year AD 642, when the famous city was sacked by the Arabs, and its splendid library destroyed.

Before entering upon a discussion on the improvements in astronomical knowledge made by the Alexandrian school, mention must be made of Pytheas of Marsala who, at about the time of Alexander the Great, determined the lengths of the midday shadows cast by the gnomon at the times of the solstices, and found, in effect, that the latitudes of Marsala and Byzantium were roughly the same. Although the latitudes of these places differ by about two degrees, Pytheas's observations are interesting and original.

Pytheas is credited with being the first to distinguish climates by the varying lengths of day and night. He is remembered mainly on account of his voyage in the Atlantic, and for his 'discovery' of Iceland. It was during this voyage that Pytheas found that the Sun was just circumpolar on the day of the summer solstice at Thule, located at the northern limit of his Arctic voyage. It seems clear from this that a change of latitude was recognized by a change in the so-called Arctic Circle—the circle centred at the celestial pole, or axis of the heavens, within which the celestial bodies were circumpolar and did not, therefore, rise or set.

On his death in 323 BC Alexander's great kingdom was divided. The western part, including Egypt, fell to Ptolemy—one of Alexander's generals. Ptolemy chose for his capital the city founded by Alexander and not yet completed at the death of its founder. The first Ptolemy was no less ambitious than Alexander in making Alexandria a great centre of commerce and a great seat of learning and, during the Ptolemaic dynasty—which ended with the death of Cleopatra in 30 BC when the Romans defeated the Egyptians—Alexandria was the cultural centre of the world.

The first amongst the Alexandrian Greeks who applied themselves to the study of astronomy were Timocharus and Aristyllus. Instruments were set up by Timocharus and Aristyllus, who fixed the positions of the zodiacal stars relative to the ecliptic. This marked a great stride forward in the development of astronomical observations. Hitherto, stars were 'fixed' by determining their heliacal risings and settings, and the determinations of these gave rise to considerable mathematical difficulties.

The division of the celestial sphere into two hemispheres, using the ecliptic, facilitated the determination of planetary motions. The visible planets could now be fixed relative to the ecliptic and to certain fixed stars near their paths.

To the poet Aratus, who flourished about 270 BC, we are indebted for the description of the constellations in elegant verse. We are reminded by Aratus that the Ancient Greeks used the Great Bear to determine direction in their voyaging, whereas the more skilful Phoenicians used the Lesser Bear. Although Helice

—the Great Bear—is bright and conspicuous, the Lesser Bear is: 'better for sailors, for the whole of it turns in a lesser circuit, and by it the men of Sidon steer the straighest course.'

It was in the 6th century BC that Thales is believed to have recommended the Phoenician practice of using the Lesser Bear to the ancient Greek navigators.

Navigation is related closely to geography; and the foundation of geography is based on astronomy. One of the earliest attempts to determine position on the Earth's surface consisted in observing, by means of a gnomon, the lengths of the longest and shortest days. We have had occasion to mention the observations made by Pytheas by means of which he determined the ratio between the length of the midday shadow at Marsala, and the length of the gnomon, on the day of the summer solstice. Pytheas is credited with being the first to establish the latitude of Marsala by using this method.

It is Eratosthenes (276–196 BC) who is credited with being the first to reduce the problem of terrestrial position-finding to a regular system. He based his system on the gnomon, and imagined a line linking places at which the longest day had the same length. Such a line is a parallel of latitude; and the parallel that was delineated by Eratosthenes passed through the island of Rhodes. This line was always used as a basis for ancient maps. Other parallels were traced, one through Alexandria and another through Syene in southern Egypt. Eratosthenes also traced a meridian line which he regarded as passing through Rhodes, Alexandria and Syene.

Eratosthenes, in pursuing his geographical studies, sought to determine the size of the Earth by using a measured length of the arc of a meridian between Alexandria and Syene. He noticed that on the day of the summer solstice the midday sun was at the zenith at Syene, whereas on the same day, at noon, the gnomon cast a shadow at Alexandria. He argued that the remote Sun's rays at the two places were parallel, and that the angle at the centre of the spherical Earth between radii terminating at the two places was equal to the zenith distance of the noon Sun at Alexandria. Knowing this angle and the distance between Alexandria and Syene, he was able to estimate the Earth's circumference. How near to the truth was the determination of the Earth's circumference by Eratosthenes is not known, owing to uncertainty as to the length of the unit of distance—the stadium —which he used.

Eratosthenes is also credited with making a very accurate estimation of the obliquity of the ecliptic, that is, the angle of inclination of the Sun's apparent annual path with the plane of the equator. His value for this angle is 11/166 of a circle, which is 23° 51′—a matter of 5′ too large for the true value at the time.

4. HIPPARCHUS

We come now to the prince of ancient astronomers—the great Hipparchus—who flourished about 160 BC. Hipparchus undertook the arduous task of making a star catalogue, having been prompted to do so by the appearance of a new star or *nova*. This event, according to Pliny, led Hipparchus to wonder if the stars were fixed, and whether or not they had motions peculiar to themselves.

> 'Wherefore,' as Pliny says, 'he attempted the task of numbering the stars for posterity and the reduction of the stars to a rule, so that by the help of instruments the particular place of each one may be exactly designed, and whereby men might discern, not only whether they disappear or newly appear, but also whether they change their stations; as likewise whether their magnitudes increase or diminish; leaving heaven for an inheritance for the wits of succeeding ages, if any were found acute and industrious enough to comprehend the mysterious order thereof.'

Pliny remarked further that this was the first time that the fixed stars were catalogued according to their latitudes and longitudes.

The ends of the axis around which the celestial sphere performs its diurnal rotations are the poles of the equinoctial. The *equinoctial* is a great circle* on the celestial sphere which is coplanar with the Earth's equator. A semi-great circle extending between the celestial poles is a celestial meridian; and the arc of a celestial meridian between the equinoctial and a star is a measure of the declination of the star. The points of intersection of the

---

* A great circle of a sphere is a circle on the sphere's surface, on whose plane the centre of the sphere lies.

equinoctial and ecliptic are known as the *spring* and *autumnal equinoxes* respectively. When the Sun is at the spring equinox it occupies a point in the sky which, at the time of Hipparchus, was occupied by the zodiacal constellation of Aries. At the time of the spring equinox the Sun entered this constellation, when he is said to be at the *First Point of Aries*. At this time the Sun's declination changed from southerly to northerly. The arc of the equinoctial, or the angle at the celestial pole between the celestial meridian of the First Point of Aries and the meridian of the star, is known as the star's *Right Ascension*.

Declination and Right Ascension on the celestial sphere correspond to latitude and longitude on the terrestrial sphere; and it was declination and Right Ascension that Pliny meant when he mentioned that Hipparchus was the first to make a star catalogue based on latitude and longitude.

It was upon this principle of arranging the stars that rested the great improvements which Hipparchus introduced to the problem of terrestrial position-finding. The rule for defining terrestrial positions was the same as that for defining celestial positions; geography and astronomy were henceforth firmly linked, and this marked an important event in the history of astronomical navigation.

The determination of latitude is a relatively simple matter, and it has been possible to do so since long before the time of Hipparchus. The problem of finding longitude, however, was one of exceptional difficulty to the Ancients. This problem is related to that of finding time, the difference of longitude between two places being a measure of their difference in local times, reckoning 15° of difference of longitude to one hour difference in local times.

Hipparchus was the first to suggest a method of determining longitude by eclipses of the Moon. If the times of eclipses are predicted for a particular meridian, the difference between a predicted time and the time of the eclipse at some meridian different from the one for which the predictions apply will give the difference of longitude between the two meridians.

Eclipses are relatively rare occurrences and it is small wonder that the longitudes of only a few places were determined by the method suggested by Hipparchus. The longitudes of places were determined largely from reports of travellers; and the difficulties of estimating distances resulted in large discrepancies in all ancient maps.

Hipparchus determined the obliquity of the ecliptic, and agreed with the figure for this deduced by Eratosthenes. He determined the length of the *tropical year*, which is the interval between successive instants when the Sun is at the First Point of Aries, to be 365 days 5 hours 53 minutes. This is about four minutes short of its true length. The tropical year is slightly shorter than the time taken for the Sun to make one apparent revolution around the Earth relative to a fixed star in its path—a period known as a *sidereal year*. The difference between the lengths of the tropical and sidereal years arises from a westward, or retrograde, movement of the equinoxes known as the *precession of the equinoxes*. The discovery of the precession of the equinoxes belongs to Hipparchus, who compared his star positions with those of Timocharus and Aristyllus, which had been determined some 150 years before the date of the catalogue of Hipparchus. The discovery of the precession of the equinoxes was necessary for the progress of accurate astronomical observations.

Another great feat of Hipparchus was the discovery of the irregular apparent motion of the Sun and the measurement of the equation of time. To comply with Plato's demand for uniform circular motion, he supposed the circular orbit of the Sun to be centred at some distance from the Earth. The line from the Earth to the point about which he imagined the Sun to revolve in a circular path he called the *apse line*; and the point, the *ex-centric*. A circle centred at the ex-centric and radius of length equal to the apse line he called the *equant*; and he supposed that the radius from the ex-centric to the Sun sweeps out equal areas in equal intervals of time. On this basis he computed tables for predicting the celestial positions of the Sun.

The plane of the Moon's orbit around the Earth makes an angle of about $5\frac{1}{4}°$ with the plane of the ecliptic. The points of intersection of the two planes are known as the *nodes*. Hipparchus is credited with discovering the retrograde motion of the nodes—a motion similar to that of the precession of the equinoxes. He also observed that the Moon's motion is irregular, and accounted for this by inventing an ex-centric and apse line and an equant for the Moon, with the aid of which he was able to compute tables of the Moon's motion.

In addition to the remarkable catalogue of discoveries made by the prince of astronomers, Hipparchus is credited with being the inventor of trigonometry—the mathematics of triangles. He constructed a table of chords, and is believed to have been the first to express the most important theorem of trigonometry:

$$\sin(A+B) = \sin A \cos B + \cos A \sin B$$

This theorem is usually known as Ptolemy's theorem.

Hipparchus is credited with being the inventor of spherical trigonometry as well as of plane trigonometry; and the solution of spherical triangles was to play a most important role in the practice of astronomical navigation of a later age. Hipparchus occupies a prominent place amongst those who were responsible for bringing the science of navigation to its state of excellence.

Hipparchus died about 120 BC. For a space of two centuries or more after this date we find no record of a philosopher of importance emanating from the Alexandrian school. The great wealth of knowledge and discovery which was assembled in Alexandria during the reigns of the first Ptolemy's was not to be repeated in human history for about fifteen hundred years. The decline in Alexandrian learning was not due to any one cause; but the fact that the professors were appointed by the Pharaohs and paid by the State meant that when the Pharaohs lost interest in the progress of science the professors and scholars did likewise, and the spirit of enquiry, so necessary for the advancement of learning, became stifled. A feature of the Alexandrian school of learning was the wide gap between scholar and artisan. Much of the knowledge acquired by the philosophers was never put to practical use. The researches and discoveries of the scholars were, however, recorded; and the vast library of Alexandria became a storehouse of the world's knowledge. This great assemblage of knowledge was not to bear fruit for many centuries: the world of the practical man, including the seaman, went on its way without knowing that the seeds of science and technology had been sown.

5. PTOLEMY

Following the period of relative inactivity which began soon after the death of Hipparchus, the first philosopher of note whom we encounter is the famous Claudius Ptolemy. Ptolemy, who must not be confused with the Pharaohs of the same name, flourished during the middle part of the 2nd century AD. We are indebted to Ptolemy, not so much for the part he played in the advancement of science, but for systematizing the astronomical and geographical knowledge of his time. His most well-known and important book is the *Syntaxis* or *General Composition of Astronomy*, commonly called by its Arabic name the *Almagest*. This work is a veritable cyclopaedia of astronomy.

In addition to the *Almagest* Ptolemy wrote an important treatise on optics, in which he made a study of atmospheric refraction. He knew that light passing from one substance to another of different optical density was bent at the common surface of the two substances. He rightly assumed that light from celestial bodies, on passing downwards through the atmosphere, would be refracted in the same way. Ptolemy is credited for introducing a law of refraction of light in air which, although not true, gives fairly good results for small zenith distances.

Ptolemy was the first to introduce a form of astrolabe and, as we shall see later, this instrument, in a modified form, was adapted for the use of seamen. In the hands of an astronomer the astrolabe was a valuable instrument for determining time, as well as altitudes, azimuths and amplitudes, of heavenly bodies.

Although the name *astrolabe* (from Greek ἀστρολαβον meaning star-taker) has been used for a variety of astronomical instruments, the astrolabe described in Ptolemy's *Almagest* is of a type known as an armillary astrolabe. Ptolemy's astrolabe consisted of a series of concentric rings, the innermost one carrying a pair of sights. The outer rings were designed so that when a heavenly body was observed in the sights of the inner ring, the celestial latitude and longitude of the body could be read off, thus saving the considerable mathematical labour of converting altitude and azimuth into ecliptic coordinates. The so-called plane astrolabe, or *planisphere*, employs the stereographic projection for solving astronomical problems, and is believed to have been invented by Hipparchus.

Ptolemy discussed the principles of map-making; and, in his monumental *Geographia*, which was to have a marked influence on seamen during the Great Age of Discovery, a long list of latitudes and longitudes of places, for the purpose of constructing maps of the world, or *mappa mundi* as these maps were called, was presented. Ptolemy repeated the advice of Hipparchus in respect of the advisability of fixing terrestrial positions by astronomical methods, and pointed out, as did Hipparchus before him, how eclipse observations could be used for the determination of longitude.

We owe a great debt of gratitude to Ptolemy for it is through him that we learn much about the work of Hipparchus, most of whose writings are lost. Certain it is that the star catalogue found in Ptolemy's *Almagest* is that devised by Hipparchus.

It is the system of the universe bearing his name for which Ptolemy is widely known. The Ptolemaic system of the universe consists of a fixed Earth at the centre of the system, with the plane of reference for the supposed circular orbits of the Sun, Moon and planets, being the ecliptic. Ptolemy replaced the spheres of Eudoxus and Callippus by a system of circles. In his system, the Moon and Sun were regarded as moving in circular orbits around the Earth. The orbits of the planets were regarded as comprising a system of deferents and epicycles. The *deferent* of a planet is a circular orbit which carries the so-called fictitious planet, the real planet being regarded as moving in a circular orbit, known as the *epicycle* of the planet, centred at the fictitious planet. The centres of the epicycles of Mercury and Venus—the inferior planets—were supposed to lie on a straight line joining the Earth and the Sun; and these planets were supposed to revolve in their epicycles in their own periodic times and to revolve in their deferents around the Earth in a year. This was in contrast to the superior planets, Mars, Jupiter and Saturn, which were regarded as revolving in their deferents in their periodic times and in their epicycles once in a year.

Ptolemy's scheme, although it was wildly erroneous, provided a suitable means for predicting astronomical events. The cumbrous system of deferents and epicycles survived for no less than fifteen hundred years and, until the time of Tycho Brahe and Kepler, there seemed little for astronomers to do except to observe and endeavour to improve the knowledge of the periods of the Moon, Sun and planets, and the diameters of their deferents in relation to those of their epicycles. The principal object of Ptolemy's geometrical system was to facilitate the preparation of tables for predicting the places of the Moon and planets. He viewed the problem of astronomical prediction, not as a problem of mechanics and a law of forces, but as a purely mathematical abstraction. His solution to the problem held the field until the time when the accurate observations of Tycho Brahe, in the hands of

THE DEVELOPMENT OF NAUTICAL ASTRONOMY 19

the illustrious Kepler, demanded a new approach to the problem of planetary motion.

6. THE ARABS

During the 6th century AD, a new, and what was to become a powerful, religious movement sprang up in Arabia. This led to the ascendancy of the Arab people, with their vision of world conquest for Islam. With incredible speed a great Arab Empire was formed by Mohammet, the self-proclaimed prophet of the one God. A holy war was preached and millions of Arabs became converted to the cause.

The Arab Empire extended from the boundaries of China in the east to Spain in the west, and it gave the world a new culture. Following the early stage of Arab conquest, when the *Koran* was considered to contain a complete code of conduct and an all-embracing philosophy—a belief which was responsible for the destruction of the remnant of the library of Alexandria when the city was sacked by the Arabs in AD 642—learning was pursued throughout the Arab world and centres of culture were established in Baghdad, Cairo, Cordoba in Spain, and Samarkand in Turkestan. The Jewish communities of the Mediterranean region readily assimilated with their Semitic cousins the Arabs, to the benefit of the learning which was to follow. Moreover, the influence of the Indian philosophies and mathematics, brought about by contact between Arab and Indian in South-west Asia, was profound.

Notable amongst the Arabs for the encouragement he gave to the progress of science, and in particular astronomy, was the Caliph Al Mamun. Al Mamun flourished during the 8th century AD, and was the successor to the famous Harun al Raschid who was largely instrumental in having many of the works of the Greeks—Aristotle and others—translated into Arabic. Ptolemy's *Syntaxis* was to form, in its Arabic translation, the foundation of Arabian astronomy. In the mid-9th century Al Battani (Albategnus), a Syrian, produced astronomical tables of the motions of the Moon and the planets which were an improvement in accuracy on those of Ptolemy.

The great interest in astronomy during the brilliant period of the Arabs is reflected in many of our present-day star names: Aldebaran, Algol, Mizar and Alphard, to name but a few. In the study of optics the names of Al Kindi and Al Hazen are noteworthy. Al Hazen (965–1038) found that Ptolemy's law of atmospheric refraction holds good for small angles only, and expressed the view that the refraction for small zenith distances varied directly as the zenith distance. In fact the refraction for small zenith distances varies as the sine of the zenith distance; but it is to be noted that for small angles the angle in radians and the sine of the angle are very nearly equal to one another.

Not the least of the important contributions to science made by the Arabs was their introduction to the West of a simple numeration system. The Indian, or Arabic, figures, which are now universally used, have done infinite service in mathematics; and mathematics, of course, is the handmaiden of astronomy. The cumbersome Roman numerals were superseded by the Arabic figures; the zero sign was introduced; and a notation in which the value of a digit depended upon its place in a line of digits made the rules of arithmetic accessible even to a child.

By the 10th century, the great Arab Empire began to crumble. In the eastern part many provinces seceded. The focus of Arab learning was transferred to the Western Mediterranean area, and academies and libraries were set up at Cordoba and Toledo in Spain. It was mainly through, and from, these centres that Arabic learning spread over Western Europe. The Greek works which had been translated into Arabic were now translated from Arabic into Latin. Moreover, the secret of the manufacture of paper had been acquired by the Arabs from the East. This, in due course, was to make possible the printed book. Fortunately the Arab impact on Western Europe was made before the great upsurge of Christian religious fervour, manifested by the Crusades, which was to sweep Islam and the infidels from Europe.

7. EARLY RENAISSANCE SCHOLARS

The legacy of the Arabs paved the way for a great flowering of learning in Western Europe, and many of our own countrymen were to play important roles in this scientific awakening. The Jewish communities in Spain, in particular, were active in the field of mathematics, astronomy and instrument-making. The mediaeval universities of Paris, Oxford and many others were established in the early 13th century; and the so-called schoolmen engaged themselves in philosophical discussions. In addition to the European universities, the monastic orders known as Franciscan and

Dominican respectively, were founded at about the same time; and these religious bodies had a great influence on western science.

The Franciscans and Dominicans devoted a great deal of their energies to the acquisition of knowledge in order to refute the numerous heresies of the time. The giant among many monastic scholars was the Franciscan friar Roger Bacon (1214–1294). His principal contribution to science was his insistence upon experimenting in order to further scientific knowledge. Roger has often been regarded as being the founder of the scientific method, a method by which scientific laws are discovered by experiment. The seeds which Roger had sown were not, however, to bear fruit until about two centuries after his death: but important it is that seeds were sown and fruit was to be borne.

The year AD 1252 saw the publication of a set of astronomical tables which were sponsored by the Castilian King Alphonso X (the Learned). The *Alphonsine Tables* were computed by a team of fifty Jews of Toledo from Arabic observations. The Arabic notation was used in the Alphonsine Tables; and the wide use of the tables was, in no small way, responsible for the Arabic notation becoming widely known and generally adopted. At about the time of the publication of the *Alphonsine Tables* an important textbook on spherical trigonometry and astronomy, the *Sphaera Mundi*, was published. This book, by our compatriot John Holywood—known as Sacrobosco—remained the standard textbook on the subject for many centuries.

Purbach (1423–1461) and John Müller, or Regiomontanus as he is familiarly known (1436–1476)—both of Nuremberg—discovered errors in the *Alphonsine Tables*. They set to work to improve observational instruments, so that faithful observations could be made with the aim of improving the tables. This aim was, unfortunately, not brought to fruition, both men dying in the flower of their lives. Bernard Walther, more usually known as Waltherus of Nuremberg (1430–1504), devoted much of his great wealth in furthering the study of astronomy. He was instrumental in having an observatory built at Nuremberg for the use of Regiomontanus. Waltherus also established a printing press which gave birth to numerous calendars and ephemerides. These were to be of great value in the hands of the leaders of the voyages of discovery which were initiated by the Portuguese under the sponsorship of Prince Henry the Navigator.

It is interesting to note that mechanical clocks were used for the first time for astronomical observations at the observatory at Nuremberg. This marked a great stride forward in the science of astronomical observations.

The German geographer Martin of Bohemia (Martin Behaim) was largely responsible for introducing the ephemerides of Regiomontanus to the Portuguese. It was Martin Behaim who is believed to have first suggested the astrolabe for nautical use. We shall have more to say about this, as well as about the Portuguese navigators, in a later chapter. In the meantime, let us return to our discourse on the progress of astronomy as it affected, or was to affect, the art of astronomical navigation.

8. COPERNICUS, TYCHO BRAHE

The years 1473–1543 mark the birth and death respectively of the Polish scholar Copernicus. After a prolonged education at the universities of Cracow, Bologna and Padua, his knowledge had great breadth, embracing, as it did, mathematics, astronomy, medicine and theology. His readings of the classical works of the Greeks made him familiar with the views of the universe presented by Pythagoras, Hicetas and Aristarchus, amongst others, who postulated a revolving and/or rotating Earth.

The apparent diurnal and annual movements of the heavens could be explained, according to Copernicus, by the real rotation of the Earth about her polar axis, and the real revolution of the Earth around a centrally located Sun.

Copernicus' great work is entitled *De Revolutionibus Orbium Coelestium*. In the dedication of his book, which was addressed to the Pope, he pointed out that any observed change of position of a heavenly body is due to the motion of the observed body or of the observer, or both; and that if the Earth possesses motion it should be noticeable in a body outside the Earth, the apparent motion of which would be equal to magnitude but opposite in direction to the real motion of the Earth.

Copernicus firmly planted the Sun at the centre of a system consisting of circular planetary orbits, the Earth being regarded as a planet. Although the germ of the idea was simple, the grand simplicity of a moving Earth and a fixed Sun was masked by a complexity of details which the author of the system introduced in an attempt to fit the observed movements of the members of the Solar system to his plan. Great difficulty was experienced in endeavouring to achieve compatibility between the plan and the observed motions. The attempt to achieve this was due largely to the deep-rooted belief that circular motion was the natural motion.

The Copernican system, although simple in essentials, was complicated by the elaborations of epicycles and ex-centrics; and it is for this reason that many writers have regarded Copernicus as the supreme exponent of the epi-cycle theory, and that his system was designed, as were the earlier systems of planetary motion, merely to facilitate the computation of astronomical tables. There is every reason to believe that Copernicus' system was not simply a mathematical abstraction: it was evolved along logical lines of argument, and there is no doubt that he believed implicitly in his proposed system. He refuted Ptolemy's argument for a fixed Earth in the clearest and obvious manner; and he proved beyond doubt that man's home in the universe did not, as was generally supposed, occupy the important place which man, in his self-glorification, had believed.

Copernicus is regarded as having been responsible for the first great change in scientific outlook which came after the Renaissance, the great movement of intellectual development in science as well as in the arts, which swept through Western Europe during the period between the 14th and 17th centuries.

The method of experiment, advocated by Roger Bacon, was brought to fruition by William Gilbert of Colchester. Gilbert (1540–1603) is the founder of the science of magnetism and electricity. In his famous book *De Magnete*, Gilbert pointed out the value of the results of his experiments with magnets for the purpose of navigation.

John Werner of Nuremberg is considered to have been the greatest astronomer of his time. He is credited with the introduction of the cross-staff, an instrument adapted specially for seamen for observing altitudes. It is believed that Werner was the first to suggest that longitude could be determined by measuring the angle between the Moon and a fixed star lying in the Moon's monthly path around the celestial sphere. In making this suggestion, in 1514, he argued that the angle between the Moon and a star in its path changes relatively rapidly with time; and that if accurate predictions of the Moon's celestial position could be furnished for a particular meridian, the time at this meridian could be found and the difference between this and the local time would be a measure of the difference between the longitudes of the local meridian and that for which the predictions were given. This method of finding longitude was to become a standard method as soon as accurate tables of the Moon's motion became available.

Gemma Frisius (1510–1555), in a tract entitled *De Principiis Astronomiae et Cosmographiae*, which was printed in Antwerp in 1530, recommended the use of a clock or watch set to the time of a standard meridian, to facilitate finding the difference of longitude between a given meridian and the standard meridian. His suggestion was not to bear fruit until the practical problems related to chronometer-making had been solved; and this was not to be achieved until the time of John Harrison (1693–1776). Gemma is credited with recommending for the use of seamen an improved cross-staff which he had contrived. In his *De Principiis* he delivers several nautical axioms, as he called them, and we shall discuss these in due course.

The year 1545 marked the appearance of an important manual on the subject of navigation. This was the *Arte de Navegar*, a Spanish treatise which was published in Valladolid by its author Pedro de Medina. Six years later, in 1551, another navigation book, which was composed in Cadiz in the year 1545 by Martin Cortes, was published in Seville. This was *Breve Compendio de la Sphera y de la Arte de Navegar con nuevos Instrumentes y Reglas*.

At about the time when these early textbooks of navigation appeared, the most distinguished, diligent and skilful astronomical observer of all time was born. This was the renowned Tycho Brahe (1546–1601), whose birth took place three years after the death of Copernicus. Tycho, after studying mathematics and astronomy at Copenhagen, Leipzig and Basle, received the patronage of Frederick II of Denmark. Tycho was granted a pension and an island in the Danish archipelago on which the famous observatory Uraniborg was built. He was possessed of great mechanical skill and many of the observational instruments with which his observatory was equipped he designed and made. Tycho lived before the days of telescopes and accurate clocks, yet his observations of the celestial bodies were incredibly accurate. He recognized that the best instrument is imperfect, and was the first observer to realize the importance of averaging the results of several observations to arrive at a value in which observational errors were virtually eliminated. His marvellous, ingeniously-contrived instruments, and his method of eliminating errors, resulted in new determinations of the constants of astronomy and stellar positions having an accuracy hitherto unsurpassed.

The appearance of a *nova*, or new star, in the constellation Cassiopeia in AD 1572 gave occasion (as a similar event did for Hipparchus) for Tycho to compose a catalogue of the stars. During the preparation of this catalogue he used accurate values of atmospheric refraction deduced from his own observations. On comparing his observations of the Moon's position with the values tabulated in the lunar tables of Copernicus he discovered errors in Copernicus' tables of as much as 2°.

Tycho maintained that observations should precede theory. He opposed the Copernican theory, being influenced by the Ptolemaic objection that the stars did not change their positions, which would be the case if the Earth moved. The great distances of the stars from the Earth are such that stellar parallaxes, being so minutely small, could not be detected by the relatively crude instruments of the early astronomers. Many philosophers, including Ptolemy, believed that if parallax could not be detected it did not exist: therefore the Earth is fixed! False and illogical reasoning to be sure.

Tycho's great service to astronomy was due to his skill as an observer rather than to his mathematical ability and powers of reasoning. The most important and most fruitful work of this marvellous astronomical observer was his record of his observations of the planets—especially those of the planet Mars. This record laid the foundation of the important work of the famous Kepler.

9. KEPLER, GALILEO, NEWTON

Johannes Kepler (1571–1630) of Stuttgart, at the age of twenty-four years published a defence of the doctrines of Copernicus. He was convinced that the plan of the universe was grand but simple; and his work entitled *Prodromus Dissertationum Cosmographicarum seu Mysterium Cosmographicum* was drawn to the attention of Tycho Brahe who, recognizing the author's intellectual powers, invited Kepler to become his assistant at Uraniborg. Kepler, no doubt, was quick to see that, with Tycho's accurate observation records, any planetary theory advanced could be put to the test.

Kepler was entrusted with the compilation of a new set of astronomical tables—the *Rudolphine Tables*—named after the Emperor Rudolf who had authorized their publication. While engaged in this work Kepler, using epi-cycles, deferents and ex-centrics, found that calculations did not agree with the corresponding observed values as determined by Tycho. Having complete confidence in Tycho as an observer, he refused to accept the basis of uniform motion for any theory of planetary motion.

It is recorded that Kepler, in trying to fit the observations of Tycho to the Copernican doctrine, was left in respect of the planet Mars with a discrepancy of 8'. 'Out of this eight minutes,' he is reported to have stated, 'I will devise a new theory that will explain the motions of all the planets.'

Kepler's discovery that the orbits of the planets are elliptical, and that the Sun is located at one of the focal points of each planetary orbit, was of great moment in the progress of astronomy.

Although the significance of the laws of Kepler could not be understood until they had been explained by Newton's dynamics, it is clear that Kepler saw more than the mere geometrical facts of his discovery. He realized that the planets move in their orbits under the action of a force which is directed towards the Sun; and he wondered if this force was similar to the force under which a stone falls to the ground. He postulated universal gravitation which he described as 'a mutual affection between bodies.' He likened this to magnetism and referred to the work of Gilbert.

The first two laws of planetary motion discovered by Kepler, applied to the planet Mars, are:

1. Mars moves in an elliptic orbit which has the Sun at one of the foci.
2. The line joining Mars to the Sun sweeps out equal areas in equal time intervals.

These laws were announced in Kepler's famous work entitled *Astronomia Nova*, published in 1609. In 1618, in another book entitled *Epitome Astronomiae Copernicae*, he announced the extension of the laws to the other planets, to the Moon, and to the four newly-discovered satellites of Jupiter. In the following year —1619—in his *Harmonicis Mundi*, the third law of planetary motion was published. This law is:

3. The square of the orbital period of a planet is proportional to the cube of its distance from the Sun.

Kepler's three laws contain the law of universal gravitation which was propounded by Newton, the mathematical genius who was born in 1642, twelve years after the death of Kepler.

Galileo (1564–1642), who died in the year of Newton's birth, devoted his energies mainly to the subject of mechanics. He discovered the isochronism of the pendulum; and this knowledge, in the hands of Huyghens in the mid-17th century, led to the invention of the pendulum clock—an instrument of great benefit to astronomy. Galileo's close attention to astronomy resulted from his use of the telescope—an invention of a Dutch spectacle-maker named Lippershey in the first decade of the 17th century. Galileo was first to use a telescope for observing the sky. His discoveries were no less amazing than were their consequences far-reaching. Jupiter's satellites proved that the Earth was not alone in having an attendant Moon: Jupiter was seen to have four. The phases of Venus proved conclusively that the Ptolemaic hypothesis was wrong. The Moon's surface was observed to be rugged; and a vast multitude of stars, which were invisible to the unaided eyes, were observed. Sun spots were observed, and these were to prove that the Sun rotated about a diameter.

In response to the offer of a handsome prize by King Philip III of Spain to anyone who invented a method of fixing a ship when out of sight of land, Galileo gave much thought to the problem. He pointed out that if the positions of Jupiters' satellites could be predicted for a standard meridian, a seaman provided with these predictions would, in effect, have a means for determining the time at the standard meridian. This, compared with local time, would give a measure of the difference of longitude between the standard meridian and the meridian of the observer.

The practical application of the process of reasoning in astronomy was greatly facilitated by the invention of logarithms. Logarithms are the incontestable invention, in the year 1614, of Baron Napier of Merchiston.

What has been regarded as having been the most important event in the history of astronomy, was the publication, in 1687, of Newton's *Principia*. Sir Isaac Newton (1642–1727) was born in Lincolnshire. After early school at Grantham, he entered the university at Cambridge in 1661. Newton was the genius of his age, and it was his brilliant investigations that led to the formulation of the laws of motion and the law of universal gravitation. After establishing the law of gravitation Newton proceeded to investigate some of its consequences. He explained: Kepler's laws of planetary motion; the precession of the equinoxes; the ellipsoidal shape of the Earth before the fact was verified; and a theory of tides, all based on the so-called Newtonian principle.

The dynamical period of astronomy, which was initiated by Galileo, Kepler and Newton, was one in which astronomical tables of the motions of the Moon and the planets were brought to a stage of high accuracy, thus making possible the determination of accurate positions at sea, as well as on land.

10. THE DAWN OF MODERN NAUTICAL ASTRONOMY

We have already mentioned the need that existed for a suitable method of finding longitude at sea, and the offer of a reward by Philip III of Spain, in 1598, for the invention of a method. The method of finding longitude by lunar observation had been proposed many times, but imperfections of lunar tables rendered the method unworkable. In 1674 the English King Charles II was pressed by Sir Jonas Moore and Sir Christopher Wren to establish an observatory for the benefit of navigation, and particularly for the making of careful observations of the Moon's motion so that accurate lunar tables could be drawn up for a year in advance. Flamsteed, who was to become the first Astronomer Royal, had pronounced that lunar tables extant were almost useless. Flamsteed also pointed out that the star positions published in the almanacs of the time were erroneous, and that navigators could derive little benefit from them for finding position at sea. The king decided to establish an observatory, mainly for the improvement of navigation, in Greenwich park, and in 1675, Flamsteed was appointed astronomical observer. Signal work was done by Flamsteed and his successors, and we shall deal in some detail with this significant part of our history.

Lunar tables were improved to a degree sufficient for the needs of ocean navigation, largely through the efforts of Tobias Mayer of Göttingen. Mayer's tables were used by Nevil Maskelyne, who was appointed Astronomer Royal in 1765, for the *Nautical Almanac* and *Astronomical Ephemeris*, which was published for the first time in 1765 for 1767 by order of the Commissioners of Longitude. The *Nautical Almanac*, which is published annually, provides the seaman with astronomical data of use for finding position when out of sight of land. The principal table contained in the earlier *Nautical Almanacs* was one in which angular distances between the Moon's centre and certain fixed stars and the Sun were given against Greenwich time. By measuring the angle between the Moon and one of the given stars or the Sun, and comparing it—after first calculating the angle between the Moon's centre and the star or Sun at the Earth's centre (a process known as *clearing the distance*)—the difference between the local time and Greenwich time could be found by inspection. This difference in time corresponded to the difference between the longitude of Greenwich and the longitude of the ship at the time of observation.

The method of finding longitude by timepiece, which had been suggested by Gemma Frisius as far back as 1530, was perfected by John Harrison, the ingenious Yorkshire clockmaker, at about the same time as the method of finding longitude by lunar distance reached a state of perfection. We shall discuss these methods of finding longitude in detail in Chapter VI.

With the introduction of relatively complex mathematical methods of finding longitude at sea grew the need for the better education of seamen ashore. Nautical academies sprang up at many ports at which seamen could receive instruction on how to find the 'place of the ship' when out of sight of land. Numerous astronomical problems were reduced to complex rules which were to tax the memories of seamen of the 18th and 19th centuries. In particular the necessary rules and calculations in the lunar problem of finding longitude were considered to be most tedious and difficult, and many efforts were made to reduce the labour of calculation. Several of the methods which were designed for the attention of the seamen will be discussed in Chapter VI.

The early timekeepers designed for finding longitude at sea were costly and unreliable. These factors were responsible for the long-standing popularity of the lunar method of finding longitude. Towards the close of the 19th century improvements in techniques of chronometer manufacture, resulting in increased reliability and reduction in price, spelt doom for the lunar method, and lunar tables were tabulated for the last time in a British *Nautical Almanac* for the year 1906.

The principal defect in the method of finding longitude by means of a chronometer was related to the difficulty of finding the local time at the place of the ship. This, as we shall see later, depended upon the solution of a spherical triangle, one side of which was the complement of an estimated latitude of the observer. The accuracy of the calculated local time was dependent upon how close to the actual latitude of the ship was the estimated latitude used in the calculation of the astronomical—or PZX—triangle.

Largely as a result of a far-reaching discovery made by an American sea-captain named Thomas Sumner, the longitude-by-chronometer problem was systematized and simplied. It was due to Sumner that the seaman was introduced to the concept of position-line navigation, whereby he is able to fix his position by what are regarded as cross bearings of celestial objects. Sumner's method, which we shall explain in Chapter VII, was discovered in 1837. An improvement in Sumner's method was made by a French naval officer, Marcq St Hilaire, in 1875.

Notable features in the progress of astronomical navigation, which stemmed from the so-called New Navigation of Sumner and Marcq St Hilaire, were the increasingly popular practice of observing stars for the determination of position at sea, and the standardization of the methods of computation of the astronomical —or PZX—triangle.

To relieve the navigator of the tedium of calculation, many navigational tables were invented to facilitate the solution of the longitude problem. Many of these tables are based on original and ingenious ideas. In addition to the so-called *short-method tables*, many mechanical devices have been invented to facilitate astronomical navigation. We shall have occasion to discuss the more important of the short method tables and mechanical navigation machines in Chapter VIII.

During the present century, the application of electronic principles to the requirements of the seaman has resulted in rapid and significant changes in the art of practical navigation. The radio time-signal, first used in 1908, has virtually superseded the mechanical chronometer. Radio direction-finding, which was introduced in 1911, aids the navigator, particularly when he makes his landfall in thick weather. Hyperbolic navigation systems, such as the Decca Navigator, Consol and Loran systems, provide the navigator with the means of fixing his ship's position when out of sight of land with an accuracy hitherto thought impossible. Inertial and doppler systems of navigation are being developed; and these, when they are available, will enable a navigator to pinpoint his ship almost to the nearest yard.

We live in a technological age when the artist of nautical astronomy is fast becoming a part of history. Few would deny that astronomical navigation is a decaying craft. The story of its development through past ages to the present epoch, when the perfected methods of astronav are being cast aside for more accurate electronic methods of navigation, is a story that surely can never fail to excite the student of the history of science.

CHAPTER II

Astronomical methods of time-measuring at sea

1. THE UNITS OF TIME

The passage of time for technical and scientific purposes is perceived by ever-repeating astronomical phenomena which recur at regular, or nearly regular, intervals. The most important recurring astronomical phenomena in respect of time-keeping are events such as: sunrise and sunset; cosmical risings and settings of planets and bright stars; star, Moon and Sun culminations. All of these phenomena are the direct results of the Earth's rotation.

The period of the Earth's rotation is, for all practical purposes, regarded as being constant. The time taken for the Earth to make one rotation of exactly 360° about her polar axis is known as a *sidereal day*, because it is manifested by the apparent diurnal revolutions of the fixed stars. The appellation 'fixed' is used because the stars are imagined to lie on the inside surface of a sphere of infinite radius. This imaginary sphere is the celestial sphere and, because it has infinite radius, the distance between the Earth and Sun, for some purposes, is regarded as being of no consequence. For many purposes the Earth is regarded as being located at the centre of the celestial sphere, but sometimes the Sun is assumed to occupy the central position.

The Sun, because of his light and heat, governs, to a large extent, the workaday lives of men; and for ordinary purposes a period of time known as a solar day is the fundamental unit of time. A *solar day* is the rotation period of the Earth relative to the Sun. Because the Earth moves along her orbit in the same direction as that of her axial rotation, the interval between successive culminations* of the Sun at any position on the Earth is slightly longer—about four minutes—than the interval between successive meridian passages of a fixed star. That is to say, the day by the Sun, or solar day, is slightly longer than the sidereal day.

* A celestial body culminates when it is at meridian passage, at which time the body bears due north or south and attains its greatest daily altitude.

A great circle on the celestial sphere which lies in the plane of the Earth's rotation is the *equinoctial*. The plane of the Earth's spin and, therefore, the plane of the equinoctial, is inclined to the plane of the Earth's orbit. During the course of a year, during which time the Earth makes one circuit of her orbit, the Sun appears to describe a great circle on the celestial sphere. This great circle is the ecliptic; and the angle between the planes of the equinoctial and the ecliptic—which is about $23\frac{1}{2}°$—is known as the *obliquity of the ecliptic*.

On two occasions each year the Earth occupies positions in her orbit when the Sun appears to be on the equinoctial. As the name equinoctial suggests, on these occasions the lengths of daylight and darkness all over the Earth are each twelve hours. During the half year when the Sun lies north of the equinoctial he is said to have *north declination*. For the other half year the name of the Sun's declination is south. Great circles on the celestial sphere which cut the equinoctial at right angles meet at two points which lie at the projected positions of the extremities of the Earth's spin axis. These points are the *celestial poles*, and the semi-great circles which meet at the celestial poles are *celestial meridians*. The declination of a celestial object is the numerical value of the arc of a celestial meridian intercepted between the object and the equinoctial. The maximum declination of the Sun is numerically the same as the obliquity of the ecliptic, and is $23\frac{1}{2}°$ N. or S. When the Sun's declination ceases to increase and commences to decrease he is said to be at a *solstitial point*. The two solstitial points—one in each of the northern and southern celestial hemispheres—are known as the *summer* and *winter solstices* respectively.

The points of intersection of the ecliptic and the equinoctial are known as the *spring* and *autumnal equinoxes* respectively. The spring equinox marks the position of the Sun when his declination changes from southerly to northerly; and the autumnal equinox marks the position when the Sun's declination changes from northerly to southerly. It is the spring equinox—a fixed point in the celestial sphere more commonly called the First Point of Aries—which serves as a datum point for the measuring of sidereal time. A sidereal day is defined as the interval that elapses between successive transits, or culminations, of the First Point of Aries.

Two factors combine which result in the length of the solar day, as determined by successive transits of the Sun over any given meridian, being a variable period of time. The two factors are: first, the apparent annual path of the Sun is along the ecliptic and not along the equinoctial because the planes of the Earth's rotation and revolution are inclined to one another at an angle of $23\frac{1}{2}°$; and second, the Earth's orbit is an ellipse having the Sun at one focus: this resulting in the Earth's rate of motion around the Sun varying, being greatest when she is at *perihelion* (the point in the orbit nearest to the Sun), and least when she is at *aphelion* (the point in the orbit most remote from the Sun).

Mechanical clocks, for ease of both manufacture and use, are made to keep uniform or regular time. A fictitious body, known as the *Mean Sun*, was invented to facilitate this. The Mean Sun is a celestial point which moves, during the course of a year, along the equinoctial at a uniform rate.

There are four instants each year when time by the Mean Sun coincides with time by the real or *True Sun*. At all other instants time measured by a clock set correctly to Mean time differs from True Sun time. The difference between Mean time and True Sun time at any instant is known as the *equation of time*; and this is essentially the correction to apply to Mean time to give True solar time. Its value lies between $+15$ and -18 minutes.

The units of time so far discussed, namely the sidereal and solar days, form the basis of a highly accurate system of timekeeping which demands astronomical observations of star and Sun transits. It appears from historical evidence that accurate time-measuring, for which star-transit observations were used, began with the Chaldean astronomers of Babylon in the 3rd century BC.

Just as the solar day is the most important unit of time marked by the Earth's rotation, the most important period of time marked by the Earth's orbital motion is the solar year. The *solar year* is defined as the period which elapses between successive instants when the Sun occupies a particular point on the ecliptic. To the nearest day, a solar year comprises 365 days. The cycle of the seasons is considered to commence when the Sun is at the spring equinox; so that the solar year is usually defined as being the interval of time which elapses between successive instants when the Sun is at the spring- or vernal-equinox. This period is 365 days 5 hours 49 minutes, that is, about 6 hours or a quarter of a day in excess of 365 days.

In ancient times, the phases of the Moon must have been regarded as being spectacles of great interest, and it is little wonder that the period of the Moon's recurring phases—a period known as a *lunation*—was used as an important unit of time.

To the mediaeval seaman, as well as to seamen of later ages, a knowledge of the phases of the Moon (which are related to the rising and falling of sea-level and the related flood and ebb of the sea) was of great importance.

The solar day, the solar year and the lunation are often regarded as being the natural units of time; and all other divisions, such as hours, weeks and civil months, are considered to be artificial.

2. THE CALENDAR

The problem of calendar-making, in which attempts are made to fit the days and months into solar years so that with the passage of time the seasons do not fall out of step with the Sun, was one of great complexity on account of the incompatible nature of the periods involved.

The earliest method of describing positions of the Sun, Moon and planets—bodies which have comparatively complex motions relative to the background of the fixed stars—was to relate positions in respect of bright zodiacal stars or constellations which lie on, or near to, the apparent paths of the wandering members of the solar system. This method of describing positions is not conducive to numerical computation. An improvement in the method of describing celestial positions came with the introduction of the ecliptic as a circle of reference. It is unknown, when, and by whom, this improvement was made; but it is believed to have been due to Chaldean astronomers of a period five hundred years before the birth of Christ.

Associated with the civil calendar of the ancient Egyptians were certain bright stars (or star-groups) which collectively formed a star-clock system. These stars are located on the celestial sphere in the vicinity of the ecliptic; and each star or group belonging to the system rose heliacally ten days before or after the adjacent member of the clock-system. The ten-day period was known as a *dekade*, and the star associated with the commencement of each dekade was known as a *dekan*—to use the Greek names.

Now the Earth rotates about $365\frac{1}{4}$ times relative to the Sun during a solar year; but relative to the fixed stars the number of rotations is $366\frac{1}{4}$. In other words $365\frac{1}{4}$ solar days are equivalent to $366\frac{1}{4}$ sidereal days. As the Earth revolves in her orbit the Sun appears to move eastwards across the background of fixed stars at an average rate of $360/365\frac{1}{4}°$ per day. It follows that the fixed stars appear to move westwards across the sky relative to the Sun at the rate of about $1°$ per day. And so it is that stars rise, culminate, and set, about four minutes earlier on successive days.

A particular clock-star or dekan, which rises heliacally to mark the commencement of a dekade, rises at the end of the dekade, about forty minutes earlier; and, to mark the beginning of the succeeding dekade, another dekan would, at this time, rise heliacally.

The consecutive heliacal risings of dekans were used to mark the passage of uniform periods of darkness; and, because the night was marked by the passage across the sky of about twelve dekans between dusk and dawn, the whole day was divided into twenty-four units of time, each unit being an hour. The origin of the twenty-four-hour day is of great antiquity and clearly belongs to the Egyptians. The sexagesimal system, which originated in Babylon, was later combined with the Egyptian twenty-four-hour day. Each hour was divided into sixty minutes, each minute being subdivided into sixty seconds. And so the system of time-keeping used at the present time owes its origin to the combination of important aspects of two very ancient cultures.

3. THE GNOMON

The simplest method of marking the passage of time during the daytime, when the Sun is not obscured by cloud, is by means of a shadow cast by a rod which is planted vertically in the ground. The length of the shadow cast by this simple gnomon decreases as the Sun approaches meridian passage (at which instant the shadow is shortest), and increases as the Sun's altitude decreases during the afternoon. Not only does the length of the shadow change during the course of the day, but its direction changes as well; and the changing direction of the shadow may be used to mark the passage of the hours of daylight.

As well as for measuring time during the day the gnomon was used for measuring the march of the seasons. Polewards of the tropics, the midday shadow of a gnomon is shortest and longest on the days of the solstices.

The simple means afforded by the gnomon for time-measuring led to the invention of Sun-dials of numerous designs. In ancient times there was no great demand for a high degree of accuracy in time-measuring; and the fact that the Sun is a very irregular time-keeper did not detract from its value as a great natural time-piece. Of course, Sun-dials were useless unless the Sun was shining; and when the sky was clouded or during night, water- or sand-clocks were used.

4. THE DIVISIONS OF THE DAY

For astronomical purposes the ancients made use of *horae equinoctales* or mean time hours.

We find mention of the term 'watches' as early as the time when, as recorded in the Old Testament *Book of Exodus*, the Israelites left Egypt. In Roman times the nights were divided into four watches. It was during the fourth watch, it may be remembered, when the disciples saw Jesus walking on the Sea of Galilee, as we are told in St Mark's gospel.

On board the Spanish vessels of the 15th century watches were set by 'half-watch' glasses, which ran for two hours. It is clear, therefore, that at this period the seaman's day was divided into six watches of four hours each. Glasses, both half-hour, as well as two-hour glasses, were used for time-measuring in the British Navy, down to the middle of the 19th century. The best 'sand' glasses contained not sand but finely-ground eggshell; and the expressions 'warming the glass' and 'warming the bell' arose from the belief that if the glass was nursed and kept warm, the 'sand' ran more quickly than would otherwise be the case; and if this were done the watch would be shortened. The end of each half-hour after the commencement of the watch was (and still is) marked by ringing the bell—one stroke for each half-hour, so that eight bells marked (as it still does) the end of one watch and the beginning of the next.

Time-keeping at sea in ancient times followed the same general lines as did time-keeping on land. The stars and Sun provided the principal means of checking the running times of sand glasses.

The star clock of the earliest European navigators involved the use not of the zodiacal stars but that of the constellation of the Lesser Bear (*Ursa Minor*) or the Cynosure of the Phoenicians. The most important star of this constellation is *Stella Maris*—the sailor's star, also known as *Polaris* on account of its proximity to the celestial pole, or hub of the celestial sphere. The extent and location of the Lesser Bear are such that it is circumpolar in all but very low latitudes; and in places where it is circumpolar it is always above the horizon, never rising or setting as do constellations or stars which lie near to or on the equinoctial.

During the course of the year, because of the motion of the stars relative to the Sun, the positions of the constellations, relative to the meridian and horizon of an observer, change by swinging in circular paths centred at the celestial pole at the rate of 360° per year: that is, at the rate of about 1° per day. The change of position of any particular star, because of this, may be interpreted as an angular motion of the celestial meridian on which the star is located, the rate of motion being about 1° per day. If, therefore, the position of a particular star relative to the meridian and horizon (or relative to the celestial pole) is known for a particular time of the year, that star, provided that it is visible, may be used for determining the time of day on any day of the year.

The annual retrograde revolution of any fixed star results in the time at which it occupies a given position relative to the horizon and meridian—that is, when it has a given altitude and azimuth—being earlier to the extent of four minutes per day. This is equivalent to two hours per thirty-day month. Thus, for example, if a star crosses the meridian of a place at say 3 a.m. on January 16th, it will cross the same meridian at 2 a.m. a fortnight later; and at 1 a.m. on February 15th.

The manner in which the star clock was used by early seamen—probably as early as the 13th century AD—was to employ the two stars of the Lesser Bear. These are known as the Guards. The brighter of the two—the foremost guard as it is called—is *Kochab*, which has a magnitude of 2·2* and is only slightly less bright than *Polaris*. It was necessary to remember the positions of the Guards relative to the celestial pole for midnight on the days which mark the beginnings of the months. To facilitate time-telling by the Lesser Bear without instrumental aid, a human figure was imagined to stand vertically in the heavens looking down on the

---

* The magnitude of a star is an expression of its apparent brightness. Magnitude numbers increase as apparent brightness diminishes. A star which is just visible to the naked eye has a magnitude of 6·0 and has one-hundredth of the apparent brightness of a star of magnitude 1·0.

observer, having the *Stella Maris* in the region of his midriff. The head of this figure was above, and the feet below, the celestial pole, and his left and right forearms were imagined to extend to the east and west respectively of the celestial pole. By relating the midnight position of the Guards of the Lesser Bear to this figure, the time of night could be ascertained with reasonable accuracy. For example, the Guards are in the head at midnight during mid-April; and they are in the right arm at midnight in mid-July, so that at 6 a.m. in mid-April the Guards are in the left arm, and so on.

5. THE NOCTURNAL

In the 16th century an instrument was invented by the use of which the time by the Lesser Bear could be ascertained with an accuracy greater than that possible by using the eye alone. This instrument is the *nocturnal*, which was first described by Coignet in 1581. The earliest nocturnals consist of two concentric plates of wood or brass. The outermost plate is divided on its circumference into twelve equal parts corresponding to the months, each part being subdivided into sixths representing five-day periods. The circumference of the inner plate is divided into twenty-four equal parts, each part corresponding to an hour of the day. The outer plate carries a handle, the axis of which corresponds with the date on which the Guards of the Lesser Bear have the same Right Ascension as that of the Sun. On this date the Guards cross the observer's meridian at noon, that is at the same time as the Sun crosses the meridian. The inner plate carries a long index, one end of which is pivoted to the centre of the plate. To ascertain the time by means of the nocturnal, the projecting tooth marking 12 o'clock on the inner plate is turned to coincide with the date on the outer plate. The instrument is then held at arm's length and the Pole star is observed through the hole at the centre. The long index bar is then turned until the bevelled edge coincides with the line joining the Guards. The time of night is then read off the scale of hours on the inner plate.

In latitudes where the Great Bear (*Ursa Major*—the *Hellice* of the Greeks) is circumpolar, the two stars known as the Pointers—Dubhe and Merak, may be used in the same way as the Guards of the Lesser Bear for finding the time of night. Nocturnals of the 17th century had two scales—one for use with the Guards of the

Lesser Bear and the other for use with the Pointers of the Great Bear. (See Plate 3.)

Nocturnals, as well as providing the means for ascertaining the time of night, provided the means for computing the time of High Water at any place for which the *establishment* was known. The 'establishment of a port', otherwise known as the H.W.F. & C. constant (High Water Full and Change constant), is the interval of time which elapses between the time of meridian passage of the Full Moon or New Moon (midnight and midday respectively) and the time of the following High Water.

The Full Moon bears south in European latitudes at midnight, and it crosses the meridian of a stationary observer later each day to the extent of about fifty minutes or four-fifths of an hour. It follows, therefore, that the Moon at the end of the Third Quarter, that is at about a week after the time of Full Moon, bears south at about 6 a.m. At the end of the First Quarter, that is about a week after the time of New or Change of the Moon, the Moon souths at about 6 p.m. It is a simple matter to relate the time of the Moon's southing and the *age of the Moon*—that is the number of days since New Moon occurred—by means of circular scales on the nocturnal. By applying the establishment—which was obtained from tide tables—to the time of the Moon's southing, the approximate time of High Water, or Full Sea as it was called, could readily be found.

To find the time of the Moon's meridian passage for any day of the year without instrumental aid demanded knowledge of the *epact*—this being the age of the Moon on January 1st. Because twelve lunations amount to 354 days, which is eleven days short of a year, the epact increases by eleven days each successive year.

A period known as the *cycle of Meton*, named after its discoverer, is one of nineteen years consisting of 235 lunations, after which the phases of the Moon recur on the same day of the solar year. The number of the year in the Metonic cycle is known as the *Golden Number*.

An interesting description of how to find the time of the Moon's meridian passage is contained in *Compendium Artis Nauticae*, written by John Collier and first published in 1729—a book in which each problem of navigation, according to the author, is 'rendered intelligible to the meanest capacity.'

ASTRONOMICAL METHODS OF TIME-MEASURING 41

'Rule: divide the Date of the year by 19, add one to the remainder, and you have the Golden Number; multiply that Golden Number by 11, and divide the product by 30, the Remainder is the epact:... To the epact add, in the month of January 0, February 2, March 1, April 0, May 3, June 4, July 5, August 6, September 8, October 8, November 10, December 10, and the Day of the Month, if this sum is less than 30 it is the Moon's age; if greater than 30, take 30 from it, and the Remainder is the Moon's age. Multiply the Moon's Age by 4 and divide the product by 5, the quotient are the hours, and the remainder are the minutes of the Moon's southing.'

The author adds laconically:

'Note: While the Moon is in the increase she souths before midnight; while she is decreasing she souths before noon. These things are known by every Cabbin-Boy, Collier's Nag, and Waterman's Servant, therefore needs no farther explanation.'

During the daytime, the Sun, when visible, was used to ascertain the time of day. The Sun crosses the meridian of any observer at midday; and when he is on the meridian he bears north or south. North of the tropic of Cancer the Sun crosses the meridian bearing south on every day of the year. Northern seamen, therefore, could readily find the time of noon by means of a magnetic compass. Near the time of noon, the rate of change of the Sun's bearing, or azimuth, is greatest, because his rate of change of altitude is small, being zero at the instant of noon.

6. SUN TIME AND THE RING DIAL

On the days of the equinoxes the Sun rises bearing due east and sets bearing due west; that is to say, his bearing *amplitude* is 0°. The time of sunrise on the days of the equinoxes is 6 a.m.; and the time of sunset is 6 p.m. Sunrise and sunset observations on the days of the equinoxes provided, therefore, the means for finding time. By assuming that the length of daylight changes at a uniform and known rate after the days of the equinoxes, the times of sunrise and sunset could be estimated. It was the practice of some mariners in the 16th century to divide the outer margin of the compass card into twenty-four equal parts which they reckoned as hours, so that time could be estimated by compass bearing of the Sun. This was an extremely rough-and-ready method of time reckoning, but it gave reasonably accurate results in high latitudes.

The common instrument for reckoning time during daytime, before the advent of mechanical clocks and watches, was the *ring dial*. The invention of the ring dial is sometimes attributed to Gemma Frisius, although Gemma himself, in his dedication of his 'astronomical ring' to the secretary of the king of Hungary, admitted that it was not entirely his own invention.

Gemma's ring dial, which was introduced in 1534, consists of three brass rings. One of these represents the meridian; and one quadrant of this ring is graduated with a scale of latitude from 0° to 90°. A second ring, fitted at right angles to the plane of the first, represents the equinoctial, and the upper surface of this is divided into twenty-four equal divisions representing hours. On the inner side of this ring are marked the months of the year. A third ring is fitted within the first, this being free to rotate about a polar axis. The third ring is fitted with a groove which carries a movable sight.

To find the hour of the day using Gemma's ring, the instrument is suspended from a point corresponding to the latitude of the observer on the meridian ring. The sights on the inner ring are set to the angle of the Sun's declination, and this ring is turned freely to point to the Sun. When this has been done, an index opposite the equinoctial circle will indicate the hour of the day. The plane of the meridian ring, at the same time, will indicate the directions of north and south.

A universal ring dial described in Seller's *Practical Navigation*, which was published towards the end of the 17th century, was thought by the author to have been contrived by Edward Wright a century before Seller described it; and certain it is that Wright described the construction and use of this dial in his *Certaine Errors of Navigation* . . . first published in 1599.

Ring dials, despite the ingenuity of their inventors, are not reliable indicators of time, especially when used on a lively ship at sea; but no better instrumental means for measuring time was available until mechanical timepieces came into general use.

7. MECHANICAL CLOCKS

Mechanical clocks were first devised in the 13th century. They were weight-driven and were, therefore, fixed. An invention of the 15th century, in which the driving mechanism of the clock employed the use of a coiled spring instead of a falling weight, made possible the portable clock. The earliest portable watches were called Nuremberg eggs—a name due to mistranslation of *uhrlein* (little clock) for *eierlein* (little egg).

To equalize the force transmitted from a clock spring to the gear train of a watch, a mechanism known as a *fusee* was invented in 1477.

It was not until the early 18th century, when the pendulum was first applied to clock-making (first done by the Dutch scientist Huyghens), that the notice of people generally was brought to the difference between the lengths of the apparent and mean solar days. Tables were, therefore, pasted on the insides of clock-cases by means of which the necesssary correction could be lifted in order to find true solar time from the mean solar time registered by the clock were it set correctly and working perfectly.

Dr John Dee, a figure famous in the history of navigation, produced in 1570 an English translation of *Arte de Navegar* by Martin Cortes. This was the first textbook on navigation printed in the English language. In the preface to his translation Dee mentions, as part of the art of navigation, the subject of 'Horometrie', which he described as

'... an arte mathematical which demonstrateth how, at any times appointed, the precise usual denomination of time, may be knowen, for any place. ...'

Dee also gave a list of nautical instruments and devices, the use and construction of which should be understood by the pilot. Among these, he included '... Clocks with springs, houre, half, and three-houre glasses.'

8. ARITHMETICAL NAVIGATION

Now the mathematical arts—to use Dee's phrase—including horometrie, were arts unknown to seamen by and large. In fact, simple addition and subtraction and the golden rule of three constituted the whole of the mathematical knowledge of the generality of seamen until Elizabethan times when, through necessity, geometry and trigonometry were introduced to—or rather forced upon—the seaman who would navigate his ship across the ocean.

The learned doctor played a part of signal importance in introducing to the English seamen of his day the elements of mathematics beyond the stage of the golden rule of three. The application of mathematics—trigonometry in particular—to navigational problems, both terrestrial and celestial, owed much to Dr Dee, who may rightly be regarded as being the founder of arithmetical navigation. His teaching paved the way for the later Elizabethan scholars, including Edward Wright, Richard Hues, Thomas Hariot and Edmund Gunter, who devised methods for calculating navigational problems of importance hitherto impossible of solution by the non-mathematical sailor.

Although the invention of trigonometry is attributed to Hipparchus, who used a table of chords for calculating the unknown parts of triangles, modern practical trigonometry dates from the time of the introduction of sine tables in the 15th century. The *sine* of an angle is equivalent to half the chord of twice the angle in a circle of unit radius; and it is clearly equal to the ratio between two sides of a right-angled triangle which contains the angle: the sides being respectively the side opposite to the angle and the hypotenuse. The ratio between the adjacent to an angle in a right-angled triangle and the hypotenuse of the triangle is equivalent to the sine of the complement of the angle. This is so because the two non-90° angles of a right-angled triangle together make 90°. This ratio is known as the *cosine* of an angle. Other ratios of sides of right-angled triangles include *tangent, cotangent, secant* and *cosecant*. Tables of these ratios, or *trigonometrical functions* as they are called, simplify practical computations of the unknown parts of triangles; and the introduction of trigonometrical tables to seamen played a most important part in the advancement of the science of navigation.

The first systematic trigonometrical tables of sines was introduced by Purbach (1423–1461) and his pupil Müller—better known as Regiomontanus because he came from Konigsberg. Müller (1436–1476) and Purbach were mathematicians of the University of Vienna. Their sine tables were constructed on the basis of a radius of 10,000 units, and were designed to facilitate the solution of astronomical problems. Regiomontanus, after the early death of Purbach, published tables of tangents and secants on the same basis as those of the sines.

The first set of trigonometrical tables published in England were those of *Blundeville*, and were printed in his well-known *Exercises . . . necessarie to be read and learned of all young Gentlemen . . .*' first published in 1594. Blundeville was responsible, in no small way, for drawing the attention of practical seamen to the relative ease with which certain navigational problems could be solved arithmetically by using the table of sines.

Before the introduction of mathematics to seamen, many nautical astronomical problems were solved by means of globes. Several works on the use of globes were published. The first work on the globes was written by Thomas Hood and was published in 1592. Two years later, in 1594, a learned treatise in Latin on the same subject was published under the authorship of Richard Hues. Hues declared, in effect, that although the astronomical problems of navigation could be solved by mathematical methods, the use of the celestial globe provided the essential practical solution without the need to labour with tedious calculations.

'The use of the Globe,' wrote John Davis in his famous work *The Seaman's Secret*, 'is of so great ease, certainty and pleasure, as that the commendations thereof cannot sufficiently be expressed: for of all instruments it is the most rare and excellent.'

And certainly true it is that the globe enables a navigator to see in his mind's eye the triangles to be solved, and this facilitates their solution. However, with the introduction of tables of sines and other trigonometrical functions, the costly, cumbersome and fragile globe, as a practical instrument of navigation—despite its rarity in the eyes of John Davis—naturally became obsolescent.

One must not be led to believe that by introducing mathematical methods to them, seamen became skilful mathematicians. Nothing of the sort: the mathematical solutions to their problems were reduced to rules, which were learnt by rote and applied with little or no understanding of the problems themselves. For a long period of time this applied; and it is only within recent times that seamen have gained some little understanding of the mathematical problems which, by using rules, they have always, since arithmetical navigation was introduced in the 16th century, been able to solve mechanically.

Mathematical navigation did not gain momentum until the

invention of logarithms by Napier of Merchiston in 1614. Logarithms to base 10—*common logs* as they are called—are due to Henry Briggs, a Gresham Professor of Geometry. Edward Wright, the discoverer of the Mercator principle, and the first to compile tables of meridianal parts, was instrumental in translating Napier's work into English for the benefit of navigation. Napier wrote:

'... It appears that some of our countreymen well affected to [mathematical] studies ... procured a most learned mathematician to translate the same [Napier's great work *Mirifici Logarithmorum Canonis Descriptio*] into our vulgar English tongue.' Wright's translation was found by Napier to be '... most exact and precisely conformable to my mind and the originall.'

Wright's work on logarithms was published posthumously by his son Samuel in 1616.

Edmund Gunter was the first to publish, in 1620, a table of common logs of trigonometrical functions. Gunter, who was a Gresham Professor of Astronomy, was the inventor of the scale which bears his name. Gunter's scale and the 'Plain' scale described and popularized by John Aspley in his *Speculum Nauticum* in 1624, were instruments of great value in the hands of seamen right down to the beginning of the present century for facilitating the solution of trigonometrical and other mathematical problems.

Before the end of the 17th century, collections of nautical tables included logarithms and natural and logarithmic trigonometrical functions, as well as the traditional tables of Sun's declination, amplitudes, Right Ascensions of stars, and tide tables; and these tables were in general use amongst seamen.

The 'Doctrine of the Sphere', together with his navigational tables, enabled the mariner to solve any of the numerous astronomical problems involving spherical triangles, provided that he memorized, or had access to, the appropriate rule. Not the least important of these astronomical problems were those related to finding the hour of the day or night. Problems such as:

1. Given latitude, Sun's declination, and Sun's altitude, find the time of day.

ASTRONOMICAL METHODS OF TIME-MEASURING 47

2. Given latitude, Sun's azimuth, and Sun's altitude, find the hour of the day.
3. Given latitude, Sun's Right Ascension, and star's altitude, find the hour of the night;

are all relatively simple problems of spherical trigonometry, which may be solved readily using the logarithmic trigonometrical functions tables.

Many nautical astronomical problems involved right-angled spherical triangles, and *Napier's mnemonic rules* for solving right-angled spherical triangles were popular devices amongst seamen.

9. THE AZIMUTH COMPASS (See Plate 4)

The *azimuth compass*, in contrast to the *steering compass*, as its name implies, was used for observing the Sun's azimuth. The instrument, of which there were many designs, consisted of an ordinary magnetic compass, the box of which was fitted on its upper surface with a broad brass circle which carried a shadow pin, and which was graduated in degrees.

The principal use of the azimuth compass was for observing the Sun's magnetic azimuth in order to discover the variation, or the north-easting or north-westing, of the needle, as it was sometimes called. This was necessary in order to rectify the course. But the Sun's azimuth at any time, together with the observer's latitude and the Sun's altitude, enabled the mariner to compute the time.

10. NAUTICAL TABLES FOR DETERMINING TIME

An interesting table appears in Wakeley's *Mariner's Compass Rectified*, first published at about the middle of the 17th century, by means of which the exact hour of the day could be determined, the Sun being upon any point of the compass, '... fitting all places upon the earth and sea that lie between 60 degrees of latitude either north or south.'

Wakeley's book, which ran into many editions, and which was published after his death by his apprentice James Atkinson, was evidently a very popular work with seamen of the day. Each page of Wakeley's table covers a particular degree of latitude, and the table on that page is described as 'A sundial for that latitude'.

Tables often included in collections of navigational tables, by means of which time at night could be ascertained, included:

1. Table of Star's Right Ascensions.
2. Table of Sun's Right Ascension for noon.

These two tables, used in conjunction with one another, rendered it possible to find the time of meridian passage of a given fixed star. This is so because the interval between the times of meridian passages of the Sun and the star on any given day is equivalent to the difference between their Right Ascensions expressed in time. The Right Ascension of a heavenly body, it will be remembered, is the arc of the equinoctial between the spring equinox (First Point of Aries), and the celestial meridian of the object, measured eastwards from the spring equinox. Thus, if the Sun's R.A. is less than a given star's R.A. the star will transit later than noon by an amount which is equal to the difference between the R.A.'s of the Sun and the given star. A star whose R.A. is 3 hrs. 20 mins. will cross an observer's meridian on the day of the spring equinox (at which time the Sun's R.A. is 00 hrs. 00 mins.) at 3 hrs. 20 mins. p.m., whereas a star whose R.A. is 20 hrs. 40 mins. will cross an observer's meridian on the same day at 8 hrs. 40 mins. a.m., that is 3 hrs. 20 mins. before noon.

A useful table found in many 17th- and 18th-century navigational manuals was the Table of Southing of Selected Stars at Midnight. This table was constructed from tables of star's R.A.s and Sun's R.A. for each day of the year.

Another table useful for time-measuring at night was one showing the time when a pair of selected stars had the same azimuth. This table, of course, was drawn up for a particular latitude, and was, therefore, not ideally suited for sea use.

II. THE NAUTICAL ALMANAC

In almost every astronomical computation for nautical purposes, the seaman is dependent upon an *ephemeris*, or table of astronomical data such as the daily celestial positions of the Sun, Moon and planets, and the R.A.s of the stars. The first official nautical ephemeris, or almanac, was that published by the French—*Connoissance des Temps*—which dates from 1678. The first British *Nautical Almanac* appeared in 1765 for the year 1767; but ephemerides of the Sun, Moon, planets and stars, for the use of seamen, were published privately long before this time.

Those elements of nautical astronomical computation which

ASTRONOMICAL METHODS OF TIME-MEASURING 49

are in perpetual change, such as the Sun's declination and Right Ascension, are given in a nautical almanac for times corresponding to a particular standard—or prime—meridian; and it is generally necessary to apply corrections to the tabulated elements when the almanac is used in any longitude other than that of the standard meridian.

The standard time used in the early British *Nautical Almanac* is that of the Greenwich meridian, and the times given are described as astronomical time.

The civil day at sea commenced at midnight, and any time (described as the angle at the celestial pole measured westwards from the observer's lower celestial meridian to the meridian of the Sun) was designated a.m. (*ante meridiem*) if the Sun were east of the observer's meridian. If the Sun were west of the observer's upper celestial meridian the time (in this case described as the angle at the celestial pole between the observer's upper celestial meridian measured westwards to the meridian of the Sun) was designated p.m. (*post meridiem*). The astronomical day, in contrast to the civil day, commenced at the instant the Sun crossed the observer's upper celestial meridian, that is, at noon. The civil day commenced, therefore, at the midnight preceding the noon which marked the beginning of the astronomical day.

The *Mean Solar Day* is the interval between successive transits of the Mean Sun across the same celestial meridian. The civil day commenced when the Mean Sun culminated at noon. It is divided into two parts each of twelve hours; and a clock whose dial is divided into twelve divisions, and which is regulated so that the hour hand makes one circuit of the dial in half a mean solar day, and which is set so that the hour hand corresponds to the 12 o'clock position at noon or midnight, is a perfect indicator of civil time.

Since the civil mode of time-reckoning was always twelve hours in advance of astronomical time-reckoning, the seaman found it necessary to be able to reduce civil time to astronomical time at the same instant. If the civil time at ship were p.m., the astronomical time would be the same with the p.m. omitted. Thus, March 3rd at 6 p.m. civil time was the same as March 3rd 06.00 hrs. astronomical time. If the civil time were a.m. the astronomical time was found by adding twelve to the hours and subtracting one from the day of the month. Thus, 4 a.m. on March

15th civil time was the same as 16.00 hours March 14th astronomical time.

The problem of finding the astronomical time and date at Greenwich, in order to extract astronomical elements from the *Nautical Almanac*, involved expressing the ship's time (civil mode) astronomically, and then applying the longitude expressed in time reckoning 1 hour for each 15° of longitude.

For the first time in the 1925 almanac, and in every *Nautical Almanac* published since that year, times styled G.M.T. are reckoned from midnight as in civil usage; and the seamen, since 1925, has no longer been confused by having to use both civil and astronomical times.

In order to ascertain the ship's longitude when at sea, it is necessary to know both the local and the corresponding Greenwich Mean Time. Longitudes have, since 1884, been reckoned almost universally from the meridian of Greenwich, the Greenwich meridian having been adopted as a prime meridian by the members of an international conference held at New York during that year. The French member of the conference dissented, but in 1911 the French adopted the Greenwich meridian.

Longitudes are now reckoned eastwards and westwards from the Greenwich meridian to the 180th meridian, which latter is the antipodal meridian to the prime meridian. Thus, if the Greenwich Mean Time (G.M.T.) at any instant is greater than an observer's Local Mean Time (L.M.T.) at the same instant, the observer is located in the western hemisphere. If the G.M.T. is less than an observer's L.M.T. at the same instant, the observer is in the eastern hemisphere. This is a direct consequence of the Earth's spin towards the east: a motion which is manifested by the apparent diurnal revolution of the celestial concave towards the west. And so it is that diurnally recurring astronomical events, such as sunrise, sunset and Sun's culmination occur on the Greenwich meridian later than they do at places to the east of the Greenwich meridian, and earlier than in places which have west longitude. Hence the seaman's rule:

> Longitude west, Greenwich time best.
> Longitude east, Greenwich time least.

If the L.M.T. at a certain instant is 9 hrs. 20 mins. and the

ASTRONOMICAL METHODS OF TIME-MEASURING 51

G.M.T. at the same instant is 10 hrs. 20 mins., the longitude of the local meridian, reckoning 360° per 24 hours, is 15° W. If, on the other hand, the L.M.T. is 10 hrs. 20 mins. and the corresponding G.M.T. is 9 hrs. 20 mins., the longitude of the local meridian is 15° E.

12. COMPUTING LOCAL TIME

It should be clear, from the foregoing remarks, that the problem of ascertaining longitude by astronomical observations is essentially a problem of finding Local Time and time at some prime meridian for the same instant. We shall discuss, in a later chapter, the methods available to the mariner since the earliest navigations, by means of which he could find the time corresponding to a particular astronomical event at a particular reference, or prime meridian. The problem of finding Local Time of a particular astronomical event, such as the Local Time at which a star or the Sun has a particular altitude, is primarily a problem of spherical trigonometry involving knowledge of the values of arcs and angles of a spherical triangle known as the *astronomical-* or *PZX-triangle*.

Fig. 1 illustrates a typical astronomical triangle. The pairs of adjacent sides of the triangle meet at: the celestial pole (P), the observer's zenith (Z), and the observed heavenly body (X). It should be evident from the figure that a similar triangle may be projected on to the Earth's surface. If the Earth-triangle is denoted by opx, then:

arc PZ = arc op = (90°—latitude of o, the observer)
arc PX = arc px = (90°—declination of X)
arc ZX = ox   = (90°—altitude of X)

The three angles of the PZX triangle are:

P which is known as the *Local Hour Angle* of X
Z which is known as the *Azimuth* of X
X which is known as the *Angle of position* or *Parallactic Angle*

If three of the six parts of the PZX triangle are known, it is possible to calculate the unknown parts by spherical trigonometry. For calculating Local Time, the angle P is required. If the observed body is east of the observer's meridian, the angle P is a measure of the time that must elapse before the body culminates. If, on the other hand, the body is west of the observer's meridian, the angle P is a measure of the time that has elapsed since the body culminated. In the case of Sun observations, the angle P expressed in hours, minutes and seconds, is the interval of solar time before or since noon. Thus, if angle P is 30°, and the Sun is east of the meridian, the Local Time is 10 a.m., that is two hours before

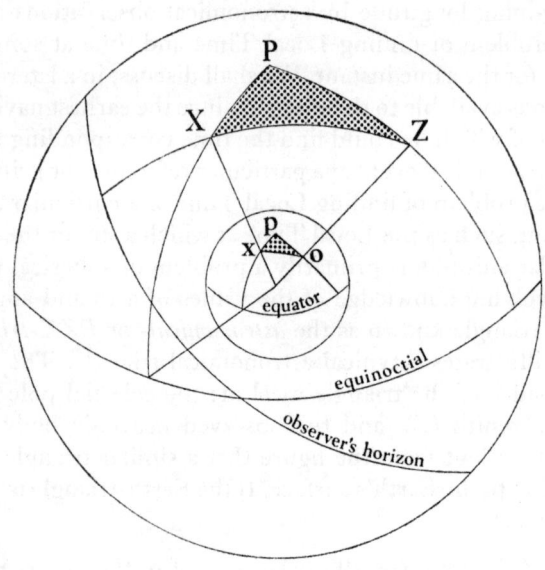

FIGURE I

noon. If the angle P is 30°, and the Sun is west of the observer's meridian, the local time will be 2 p.m., that is to say two hours will have elapsed since noon.

13. THE MARINE CHRONOMETER

In the 18th century, when they were first used on ships for the purpose of finding longitude at sea, chronometers were instruments of great rarity. The high cost of manufacture and the relatively small number of reliable pieces produced were factors that brought chronometers within reach of only the wealthy. It was not until the middle of the 19th century that the mechanical construc-

ASTRONOMICAL METHODS OF TIME-MEASURING

tion of these timekeepers had attained an unexampled and high standard of efficiency, the improvement in manufacture having, at the same time, been accompanied by a reduction in the cost of production. After this time it was not unusual for ocean-going ships to have three, or even more, chronometers on board. The British Admiralty provided all sea-going men-of-war with three chronometers, although a little more than a century ago ships of the Royal Navy were furnished with only one. If, however, a captain supplied a private chronometer the government provided a third. It was argued that if a ship had only one, it would be unwise to trust in it implicitly, and, therefore, great caution would have been necessary. If a ship had two, and they differed in their rates, it would be impossible to determine which one (if either) was correct. If, however, three were carried, then the coincidence of any two would suggest the truth of their results. Moreover, by comparing the three, an irregular one could, in some cases, be detected.

The essential feature of a marine chronometer is the ingenious device known as the *compensated balance*: compensation for temperature changes being achieved by means of a bimetallic balance wheel.

The rate of a chronometer is closely related to temperature. An increase in temperature causes the rate of an uncompensated chronometer to be retarded, whereas a decrease in temperature results in an accelerated rate. The object of the compensated balance is to correct this defect.

The effects of a change in temperature are, first, a change in the tension in the balance spring and, second, a change in the moment of inertia of the balance wheel due to change in the distribution of the mass of the wheel. The tension in the balance spring varies directly as the temperature, whereas the moment of inertia of the balance wheel varies as the square of the temperature. Accordingly there are two, and only two, temperatures at which the temperature compensation is correct. The aim of a chronometer-maker is to construct instruments which are correctly compensated at two standard temperatures—which are usually 45° F. and 75° F. At temperatures between the standards, a compensated chronometer should gain, and at temperatures above the higher, or below the lower, standard temperature, it should lose. A century ago, Captain Charles Shadwell, R.N., in his classic study on the *Management of Marine Chronometers*, observed that the time and trouble expended on compensating a chronometer for temperature was a big factor in the cost of its production. No doubt the same applies at the present time. The author of the 1928 edition of the *Admiralty Manual of Navigation* stated that,

> 'The marine chronometer is simply an enlarged watch ... and its mechanism is by no means complicated, although its construction demands the most accurate workmanship, and its adjustment requires a high degree of skill.'

Cases have been recorded of elaborate attempts having been made to keep the temperature of the air in the chronometer box as nearly uniform as possible. In a French vessel, for example, during a voyage of survey in the year 1816, the air in the chronometer box was maintained at a uniform temperature of 30° C. by means of an oil lamp, the admission of air into the box being regulated by an aperture, the size of which could be adjusted by a sliding shutter. Such precautions were rewarded, wrote Shadwell, 'by the watches performing their functions with extreme regularity.'

It has also been noted that artificially keeping the chronometers at a uniform temperature ensures not only their being kept in dry air, but also the maintenance of their lubricating oils in a state of uniform fluidity.

The rigorous routine relating to the management of a chronometer on board a ship, which was formulated during the early history of the instrument, has persisted, at least in part, to the present time. The winding of the chronometers on board is still regarded as something of a ritual, and to forget to wind them at the proper time is a crime which all self-respecting navigating officers live in fear of committing.

The adoption of a systematic routine for winding the chronometer favoured the uniformity of its rate, as well as reducing the chance of it being accidentally allowed to run down. If more than one chronometer were carried they were wound in the same order at the same time each day. The habit of so doing provided a safeguard against the caprice of memory. In winding, the turns or half-turns were counted, and the last turn or part of a turn made gently but deliberately until the key butted. An oft-told story is related to the case of the over-careful winder who, for fear of injury to the chronometer, never wound up to the butt. This resulted in a little being lost each day, until after a time the chronometer was found to be stopped at the time it was due to have been wound. There is, of course, an indication on the face of a modern chronometer, by means of which the state of winding may readily be seen, so that it is unlikely for a chronometer to run down if a daily check is made on the indicator.

To ensure that the daily duty of winding the chronometer was always carried out, it was the practice in men-of-war for a sentry to report to the captain and the officer-of-the-watch when the time had come for the duty to be performed. The sentry subsequently was not allowed to be relieved until the corporal of the guard had ascertained from the officer in charge of the chronometers that the operation had been performed, and had duly reported the same to the officer-of-the-watch and the captain. The corresponding arrangement in merchant ships normally involved a simple inscription afforded by the magic letters CHRON scribbled —usually with a soap tablet on the mirror fitted in the cabin of the officer in charge of clocks, the inscription serving to remind him of an important daily duty.

When at sea, the regular routine of the ship rendered it relatively difficult to overlook the duty of attending to the chronometers; but when in port, distractions due to a variety of causes, often resulted in the chronometers not being wound, with the consequent possibility of their running down and stopping.

A chronometer which has an irregular rate is not suitable for the purposes of astronomical navigation, unless its rate can be checked by radio time-signal soon before or just after an observation has been made. Before the days of radio time-signals, if the daily rate of a chronometer, even if it were regular, exceeded six or seven seconds a day, it was considered to be unfit for navigational use.

After radio communication had become possible the first electronic aid to navigation was introduced to the seaman in the form of the radio time-signal. At the present time radio time-signals are available, on request, at any time of the day.

Since ships have been equipped with radio gear, the need for a chronometer hardly exists. In fact, a reliable wrist watch, having a sweep seconds hand, may be used to measure the G.M.T. of an astronomical event, provided that its error on G.M.T. may be ascertained at a time not much different from the time of the observation of the event. In other words, the importance of the chronometer as an instrument of navigation has diminished since the advent of radio telegraphy.

Now that the need is almost non-existent a new type of marine chronometer has been produced. This new chronometer is reputed to have an accuracy of one part in a million. This means that its rate is within a little more than a second per month. This type of chronometer, the functioning of which is no way depends upon the memory of the clock officer, employs electronic techniques and a quartz crystal. It marks a significant advance in the construction and degree of accuracy of the marine chronometer.

CHAPTER III

The altitude-measuring instruments of navigation

1. INTRODUCTORY

In this chapter we shall be concerned with the instruments used by the seaman down the ages for taking sights as a preliminary to calculating the latitude or longitude of his ship when out of sight of land. The earliest of these altitude-measuring instruments were adapted from those used by astronomers and surveyors ashore.

The most important astronomical observation made at sea—in ancient as well as in modern times—is the altitude observation, in which the arc of a vertical circle contained between an observed celestial body and the sea horizon vertically below it is measured. The problem of measuring vertical angles ashore is comparatively simple; but at sea, with an unsteady deck from which to observe, the difficulties of making an accurate altitude observation were not entirely overcome until the advent of Hadley's reflecting quadrant in the 18th century.

The earliest instruments used by navigators for observing altitudes were the seaman's *quadrant* and the mariner's *astrolabe*.

2. THE SEAMAN'S QUADRANT (See Plate 5)

It appears that, chronologically, the seaman's quadrant was the first altitude-measuring instrument used by navigators. The instrument is simply a quadrant of wood or metal provided with a plumb-line suspended, when the instrument is in use, from the centre of the arc of the quadrant. One radial edge of the instrument is fitted with two pinnules or sights.

The portable seaman's quadrant was, in all likelihood, adapted from the surveyor's quadrant; this, in turn, was adapted from the fixed astronomer's mural quadrant used for measuring altitudes of celestial bodies, and from the astronomer's hand quadrant which, by means of the Sun's altitude and curved lines engraved on the instrument, enabled the observer to find the time of day.

The mural quadrant of the early astronomers was supported,

against a wall of masonry, in the place of the meridian. One radial edge of the quadrant was plumbed vertically. The shadow of a pin at the centre of the quadrant was cast on a plate held close to the graduated arc, so enabling the observer to measure the altitude of the Sun or Moon.

One disadvantage of the mural quadrant was the difficulty of graduating the arc. The ease with which a straight line could be divided, compared with the difficulty of dividing an arc of a circle, resulted in the *triquet um* being an instrument of greater popularity than the mural quadrant.

The *triquet um*, otherwise known as Ptolemy's Rules, consists of a vertical post at the top end of which, on a horizontal bearing pin, is pivoted an *alidade* or sighting rule fitted with upper and lower pinnules. The lower pinnule is provided with a tiny hole, and the upper pinnacle, or *backsight*, is provided with a large hole. A thin lath, which is pivoted at the lower end of the graduated post, provides the means of measuring the length of the chord of the angle equal to the altitude of the celestial body observed in the sights.

To make an observation with a *triquet um*, the celestial body is sighted through the holes in the pinnules fitted to the sighting rule; and a pin on the alidade, fixed at a distance from the upper pivot equal to the distance on the vertical post between the alidade pivot at the top and the lath pivot at the bottom, is used to make a mark on the lath. After the mark has been made the lath is swung up to the graduated post, and the chord, corresponding to the measured altitude, is read off. A table of chords is then used to ascertain the required angle.

Al Battani, the celebrated Arab astronomer of the 9th century AD, is credited with being the first to suggest graduating the lath to avoid the necessity of transferring the reading of the lath to the scale on the post, so eliminating one source of possible error.

Fig. 1 illustrates a *triquet um*.

When a celestial body is observed through the sights of a *triquet um*, the zenith distance of the body is equivalent to the angle between the vertical post and the alidade. Both Copernicus and the famous Tycho Brahe observed with a *triquet um*.

The arcs of the earliest quadrants used by the Portuguese seamen, during the early part of the Golden Age of Discovery, were not graduated in degrees of altitude (or zenith distance). Angular

ALTITUDE-MEASURING INSTRUMENTS OF NAVIGATION

measure was not to play a part in practical navigation until the mathematical ability of seamen had advanced to a stage beyond that possessed by the first ocean navigators. The earliest practice appears to have been to mark the arc of the seaman's quadrant with the names of important coastal—or island—positions corres-

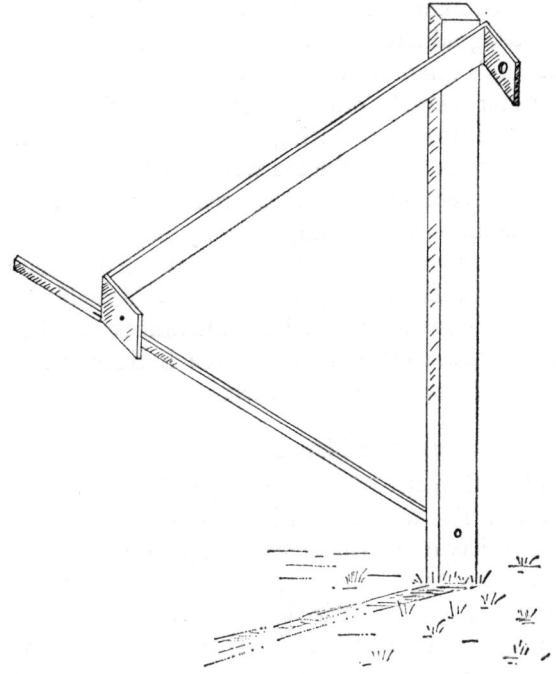

FIGURE I

ponding to the positions of the plumb-line when the Pole Star was observed. In their voyages along the West African coast, during the 14th and 15th centuries, Portuguese navigators knew when they were due west of each of several coastal stations by noting when the plumb-line on the quadrant corresponded with the name of the station engraved on the arc of the quadrant, when observing the Pole Star through the sights.

The first use of the quadrant appears to have been for measuring altitudes of the Pole Star as a means of finding distance south of Lisbon or other port of departure. With the introduction of the table of Sun's declination for navigational purposes, the seaman was taught to find his latitude in degrees north or south of the equator by meridian altitude observation of the Sun. It is not unlikely that at the time when this method became available to the seaman, angular units were introduced to him, the Sun's declination in the table being given in degrees and minutes of arc north or south of the equinoctial. The seaman's quadrant was, from this time onwards, graduated in angular measure.

The seaman's quadrant demanded two observers; one to sight the Sun or star, and the other to note the position of the plumb line. It was an altitude-measuring instrument quite unsuitable for observations at sea unless the sea were smooth and the air calm. The degree of accuracy of the measured altitude was coarse—although this could have been improved by employing instruments of larger radius. It was a simple instrument and one that could be made, without difficulty, by the mariner himself. Using the plumb-line to define the vertical, the quadrant could be used for measuring altitudes even when the sea horizon was obscured because of darkness or thick weather.

3. THE ASTROLABE (See Plate 2)

In about 1480, the astronomer's planispheric astrolabe was adapted for sea use. The word *astrolabe* (ἀστρολαβον) has been used to designate a variety of instruments used by astronomers. Included amongst these are the armillary spheres. An *armillary sphere* consists of a number of concentric rings, each representing one of the principal great circles of the celestial concave. Armillaries were designed to ascertain celestial positions and celestial angles without having to resort to tedious calculations. Equatorial armillaries were used to determine declinations and Right Ascensions of celestial positions; and zodiacal armillaries were used to determine celestial latitudes and celestial longitudes—the coordinates of the ecliptic system. The armillary sphere is said to have been the invention of Eratosthenes in about 250 BC and they appear to have been used by Hipparchus, Ptolemy and Tycho Brahe, amongst others, for mapping the heavens.

The term *planispheric astrolabe* applies to a 'compendium of instruments' as R. T. Gunther describes it in his *Early Science in Oxford*. The planispheric astrolabe consists of an evenly balanced metal disc fitted with a ring or shackle at a point in its circumference from which the instrument may be suspended. Centrally pivoted to the metal disc is an alidade or diametrical sighting rule fitted with a pair of pinnules. By means of the alidade a celestial body may be sighted and its altitude measured, the metal disc being provided with a scale of angles extending from 0° to 90°, one radial edge of the alidade providing the fiducial line or index.

The metal disc of the astrolabe is recessed to accommodate one of a series of thin metal plates, each engraved with a stereographic projection of the celestial sphere appropriate to a particular latitude. Covering the plate is a metal disc in the form of a star map designed in fret work. Below the *rete*, as this openwork star map is called, the interchangeable plate is clearly visible.

The planispheric astrolabe is essentially an astronomer's instrument employed for measuring time, using the Sun's altitude during the day and that of one of a small number of bright stars and planets by night.

The great Hipparchus of Bithynia is usually credited with the invention of the planispheric astrolabe. It is almost certain that the planispheric astrolabe could not have been used before the time of Hipparchus, since scientific astronomy had not advanced to the stage when such an instrument could have been put to profitable use. If, in fact, Hipparchus did invent the planispheric astrolabe, the instrument of his invention could have been but a primitive form of the complex astrolabes of a later age. It is to the astronomers of India, Persia and Arabia that honour is due for the perfecting of the planispheric astrolabe.

The oldest surviving treatise on the astrolabe was written in the 7th century AD by Severus (Sebokht). It was not until the 13th century, when Ancient Greek learning was revived, that the astrolabe was re-introduced into Europe by Arab scholars. The first treatise on the astrolabe written in Britain was that of Geoffrey Chaucer (c. 1340–1400) which he made 'to his sone that callid was Lowis.' This treatise was written in 1390 and appears to be a re-statement of an Arabic work of the 8th century.

Unlike the astronomer's astrolabe, which is essentially a time piece, the mariner's instrument (which has no real right to the appellation astrolabe) is a device used simply for measuring the altitude of Sun or star. The varied uses of the astronomer's astrolabe—or mathematical jewel as it was called—were responsible for its popularity. And with increased popularity we find that astrolabes increasingly became objects of rare beauty reflecting the highest degree of art and skill of the instrument-maker. The exquisite and elaborate astrolabes of the period between the 16th and 19th centuries rank with the finest examples of the art of metal-working. The mariner's astrolabe, of which no more than a dozen or so examples are known to exist, is an instrument having but little beauty, ornamentation or precision. It is simply an astronomer's astrolabe shorn of its astronomical appendages, leaving only the graduated metal ring and alidade.

According to Ramond Lull, the famous alchemist and astronomer of Majorca, the astrolabe was in use among the Majorcan pilots as early as 1295; but Purchas, in his *Pilgrims*, states that Martin Behaim was the first to apply the astrolabe to the art of navigation in the year 1484.

Martin of Bohemia was commissioned by John II, king of Portugal—who was active in advancing scientific navigation—to teach the pilots of Portugal the rudiments of nautical astronomy. There is no doubt that Martin introduced the astrolabe to the Portuguese seamen who initiated the Golden Age of Discovery.

The earliest record of a description of how an astrolabe is made and used for sea purposes appears in *Arte de Navegar* by Martin Cortes. The earliest seaman's astrolabe was a massive open ring, usually of metal, so that it would hang vertically and steady. It was of relatively small diameter so that, when in use, it offered but little resistance to wind. It was held vertically by means of a metal ring or shackle, or by a thread fitted to the top of the instrument. An alidade, having two sights, completed the instrument. One of the quadrants was graduated in degrees of altitude (or zenith distance).

Master Thomas Blundeville, in his *Exercises* ... informs us that the astrolabes of the Spanish were: '... not much above 5 inches broad and yet doe weigh at least 4 pounds....'

On the other hand, Blundeville mentions that:

'English pilots ... that be skillful, do make their sea astrolabes 6 or 7 inches broad and therewith verie massive and heavie, not easie to be moved with winde, in which the spaces be the larger and thereby the truer.'

The astrolabe was used for measuring altitudes of the noon-day

ALTITUDE-MEASURING INSTRUMENTS OF NAVIGATION 63

Sun by day, and the Pole Star by night. Each of the sighting vanes described by Cortes carried two holes: one, a relatively large hole for use when observing the Pole Star; and the other, a tiny hole, for use when observing the Sun. For observing the altitude of the Sun the ring was held lined up with the plane of the vertical circle through the Sun. The alidade was then turned to a position so

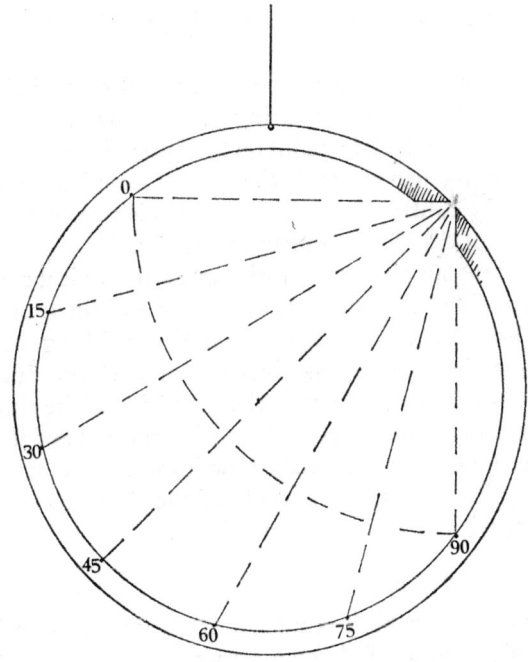

FIGURE 2

that a beam of sunlight passed through the hole in the upper sighting vane and fell near the corresponding hole in the lower sighting vane. At times when the Sun was partially obscured by thin clouds, it was necessary to observe him direct through the larger holes in the sighting vanes, as was done when sighting the Pole Star.

An improvement on the earliest type of astrolabe was the engraving of a second quadrant, thus providing a means of checking and eliminating errors due to faulty graduation and centering of the alidade. The degree of accuracy of altitudes obtained from astrolabe observations made at sea was coarse, and was unlikely to have been better than to the nearest degree of arc.

The seaman's astrolabe was often called an *astronomical ring*, although the latter name is often used to describe a modified form of mariner's astrolabe. The astronomical ring was used solely for measuring the meridian altitude of the Sun, and was preferred to the simple astrolabe because the divisions on the ring were larger, and, therefore, more accurately cut than those on the astrolabe.

The astronomical ring consisted of a metal ring of about nine inches in diameter fitted with a ring, shackle, or thread, so that it may be held vertically. At a point on the outer side of the ring, 45° from the point of suspension, is the apex of a conical shaped hole. This lies at the centre of a quadrant which is projected on to the inner surface of the ring as illustrated in Fig. 2.

When making an observation with an astronomical ring the instrument is held at the point of suspension and lined up with the Sun when he is on the meridian. The sunbeam falling through the conical shaped hole appears as a bright spot on the graduated scale of altitude (or zenith distance) on the inner side of the ring.

The seaman's astrolabe was a clumsy instrument and ill-adapted for sea use. One may appreciate how impossible it would be to measure, by its means, the altitude of the Sun or a star from a rolling ship, with any degree of precision.

4. THE CROSS-STAFF

An improvement on both seaman's quadrant and astrolabe was the seaman's *cross-staff*, known variously as *baculus Jacobi* or Jacob's staff, *virga visoria, radius astronomicus*, and by the Portuguese and Spanish as *balhestila*, meaning cross-bow.

The invention of the cross-staff is often attributed to Levi ben Gerson (1288–1344), a Jew of Provence. It is doubtless true that Levi was first to describe the instrument in writing; but the invention appears to belong to Jacob ben Makir, who flourished during the 13th century.

In its simplest form the *baculus Jacobi* described by Levi ben Gerson consists of a square-sectioned graduated staff having a cross-piece, or transom or transversary, set at right angles to the staff, along which it could be slid. One end of the staff was held at the eye and the two ends of the transom, when correctly set, pro-

ALTITUDE-MEASURING INSTRUMENTS OF NAVIGATION 65

vided lines of sight terminating respectively at the two objects between which the angle is required.

In 1470, Regiomontanus described the cross-staff under the name *radius astronomicus*. Pedro Nuñez also described the instrument in an essay published in 1537. Our own countryman William Bourne described the use of the cross-staff in his *Regiment for the Sea*, first published in 1574. Bourne described the cross-staff under the name *balla stella*, and his remarks on its use are interesting:

> '... that it is beste to take the height of the Sunne with the crosse staffe, when the Sun is under 50 degrees in heigthe above the Horizon, for two causes. The one is this: till the Sunne be 50 degrees in heigthe the degrees be largely marked uppon the crosse staffe, but after (the Sunne being above 50 degrees high) they are lesser marked. The other is, for that the Sunne being under 50 degrees in heigthe, you may easily take the height, because you may easilie see or viewe the upper end and the nether end of the crosse staffe bothe at one time: but if it doe exceed 50 degrees, then by the meanes of casting your eye upwardes and downwardes so muche, you may soone commit error, and then in like manner the degrees be so small marked, that if the Sunne dothe passe 50 or 60 degrees in heigth, you must leave the cross staffe and use the Mariner's Ring, called by them the Astrolaby which they ought to call the Astrolabe.'

Bourne goes on to say:

> 'The Astrolabe is best to take the height of the Sun, if the Sunne be very high at 60, 70, or 80 degrees, and the cause is this: the Sunne coming so neere unto your zenith, hathe great power of light, for to pearce the two sights of the Alhidada of the Astrolabe, and then it is not good to use the crosse-staffe, for that the Sunne hurteth the eyes of a man, and besides that it is to high to occupy the crosse staffe (as before is declared) so that this way you may very well preserve your eyes. If you have not glasses upon your staffe (to save your eyes when taking the heigth of the Sunne) but be unprovided of them, do thus: take and cover the Sunne with the end of the transitorie of the crosse staffe, unto the very upper edge or brinke of the Sunne (so shall you not need to beholde the brightnesse of it), and with the other end

of the transitorie to take the horizon truely, and that being done, and that the Sunne is 30 or 31 minutes in diameter or breadth, therefore you shall rebate 15 minutes from the altitude or heighte of the Sunne.'

The cross-staff does not appear to have been used by seamen until the early 16th century. It is interesting to note that neither Columbus nor Vasco da Gama used it, both having navigated by of astrolabe and quadrant.

Martin Cortes, in his *Arte de Navegar*, first published in 1551, and translated into English by Richard Eden in 1561, appears to have been the first to describe the cross-staff specifically for the seaman's use, and to give instructions for making, graduating and using it.

The principal defect of the cross-staff rested in the fact that unless the eye were placed in the exact position for observation, an error known as *eye-* or *ocular-parallax* affected the observation. To avoid or reduce this error many a navigator had his staff specially shaped so that the observing end fitted snugly over his cheek-bone when the instrument was held in the correct attitude for observing. Bourne, in his *Regiment for the Sea*, had given this advice; but later writers drew attention to the possibility of this recommended cure making matters worse instead of better.

Richard Hariot, the famous Elizabethan mathematician, wrote on ocular parallax as it applied to the cross-staff, and found that an error of as much as $1\frac{1}{2}°$ may result because of it.

A difficulty in measuring an altitude with a cross-staff is due to the need for seeing in two directions simultaneously; and the greater is the angle to be measured the greater is this difficulty. When used for measuring the altitude of the Sun, the glare of this luminary, unless he happened to be partially obscured by a veil of cloud, resulted in the temporary blindness of the observer. This could be overcome by the use of smoked glass, as had been advised by Bourne; but this remedy often resulted in error due to the composition of the glass not being uniform, or through the surfaces not being ground parallel. Hariot, like Bourne, suggested covering the Sun with the top part of the transom so that, by measuring the altitude of the Sun's upper limb, the temporary blindness, which would otherwise result, is prevented.

One advantage of the cross-staff over both quadrant and astrolabe lies in the relative ease with which the straight staff may be graduated. The distances of the graduations from the zero position on the staff are equivalent to the natural cotangents of the corresponding half altitudes, the half cross being equivalent to the radius or unity. Fig. 3 illustrates the method of graduating the cross-staff.

The distance d from the eye end of the staff, and the graduated mark corresponding to the altitude of a heavenly body (α, β, γ, etc.) is given by the formula:

$$d = r \cot (\alpha/2, \beta/2, \gamma/2, \text{etc.})$$

where r = half the length of the transom.

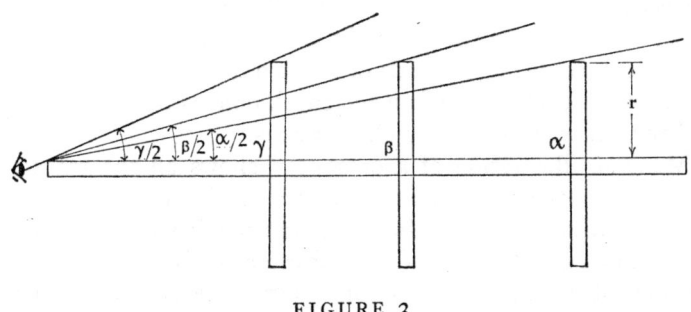

FIGURE 3

The cross-staff was most suitable for measuring altitudes of more than about 20°, and of less than about 60°. For altitudes greater than about 60°, not only was it difficult to set the transom properly on account of lining up the ends with the horizon and Sun's limb simultaneously, but the distances between the successive graduations become increasingly smaller as the altitude increases. For altitudes of less than about 20°, the length of the transom would have to be abnormally small, or that of the staff abnormally great. The cross-staff by itself, therefore, was not sufficient for the navigator's needs.

Michel Coignet is credited with being the first to describe and illustrate, in 1581, a cross-staff having more than one transversary. From the beginning of the 17th century it became common to provide the cross-staff with three transoms designated the 15°, 30° and 60° transoms respectively. Each transom belonged to one of three scales engraved on each of three sides of the square-sectioned staff. The 15° transom belonged to the side graduated up to 15°, and this, the longest transom, was used for measuring small altitudes. The 30° transom was used with the scale that extended from about 10° up to about 35°; and the smallest, or 60° transom, was used with the scale that extended from about 55° upwards to about 80°.

It appears that John Davis, the famous English navigator, was the first to explain how to deal with ocular parallax as it applied to a cross-staff provided with more than one transversary. The problem of avoiding this error is one of finding the exact spot on the cheek-bone at which to place the eye-end of the staff when observing. To find this position, Davis explained that the navigator had merely to take two transoms and to set them on the staff at the correct positions for a common angle that could be measured by either transom; and then to sight along a line to a star or other suitable object using the ends of the two transoms. The position of the eye-end of the staff on the cheek bone is then to be noted and remembered for future observations.

Many an illustration of a cross-staff shows all the transoms belonging to the instrument set on the staff, thus leading to the mistaken idea that more than one transom is used to make an observation. It is to be noted that the cross-staff was used with one transom at a time, the selection being made according to the altitude of the object to be measured.

Robert Fludd of Christchurch, Oxford, designed a cross-staff of novel design in 1617. The staff was three feet long; and the transom, which was one foot in length, had a three-inch slot or groove at each end, so that the central part of the transom—of length six inches—could be used for measuring small altitudes.

In 1624, Edmund Gunter of Oxford published a work entitled *De Sectore et Radio*, in which he described a cross-staff. His cross-staff—a yard long so that it could be used as a convenient linear measuring device—provided a convenient instrument on which to engrave his famous scales for facilitating the solutions of navigational problems.

Although, in many respects, the cross-staff is a better instrument than either the seaman's quadrant or astrolabe, the problem of measuring altitudes accurately was recognized as a difficult one; and the attention of many astronomers and navigators was directed towards improving the seaman's altitude-measuring instruments.

A noteworthy step forward in the technique of measuring the

ALTITUDE-MEASURING INSTRUMENTS OF NAVIGATION

altitude of the Sun came with a novel cross-staff designed by Thomas Hood, a mathematician who was engaged to lecture on navigation in London in 1588. In 1590 Hood published a small work in which he described his measuring staff.

Hood's staff consisted of a staff and transom, square sectioned and of equal length. A specially designed socket with two thumbscrews provided the means for setting staff and transom at right angles to one another. The staff was graduated from 90° to 45°, and the transom from 0° to 45°. To measure the Sun's altitude when less than 45°, the staff is held horizontally in line with the direction of the Sun, with the transom standing vertically above the staff. The transom is then lowered (one of the thumb-screws in the socket allowing this to be done) until the edge of the shadow cast by a vane fitted at the top of the transom strikes the end of the staff. In this position the reading on the transom is the Sun's altitude. For measuring the Sun's altitude when greater than 45°, the transom is set to 45° and, with the staff horizontal and pointing in the direction of the Sun, the transom is slid along the staff to a position at which the shadow cast by the vane at the top of the transom strikes the end of the staff. The reading on the staff is then the Sun's altitude. Two observers are needed to make an observation using Hood's staff: one to hold the staff horizontally, and the other to manipulate the transom.

Hood appears to have been the first to have employed an instrument for measuring the Sun's altitude using a shadow cast by a vane. This idea was used by many other inventors during the decades following the publication of Hood's description of his staff.

5. THE KAMAL

A navigational instrument of great antiquity, the principle of which is the same as that of the cross-staff, is the *kamal* (= guide). In its simplest form the kamal consists of a rectangular board to the mid-point of which is fastened a thin cord. The cord is knotted at points corresponding to the positions of trading stations lying on a navigator's route. If, when holding a particular knot at the eye, and holding the tablet at arm's length, the upper and lower edges are coincident with the directions of the Pole Star and the horizon respectively, the navigator knows that his latitude is the same as that of the trading station which lies due east or west of his ship, and which corresponds to the particular knot. The kamal provided the ancient Arab navigators of the Red Sea and Indian Ocean with the means for navigating by the Pole Star. The instrument became known to European navigators through Vasco da Gama, after he had rounded the African Cape in 1497.

The ancient kamal, in a modified form, is used by Arab navigators of the present time for navigating the dhows often to be seen in the Red Sea and off the East African coast.

6. THE BACK-STAFF

The most fruitful attempt made during the 16th century to overcome the difficulties of taking sights at sea was that of John Davis, the inventor of the *back-staff*. Two variations of a back-staff were described by Davis in his famous *Seaman's Secrets*, first published in 1595.

The more simple of Davis's back-staves consisted of a graduated staff to which is fitted a sliding half-transom in the form of an arc of a circle. At the fore end of the staff is fitted a horizon vane with a slit through which the horizon may be observed. The back-staff could be used for measuring the altitude of the Sun only when the horizon was visible. It was an instrument essentially for daytime use for measuring the meridian altitude of the Sun. To take a sight of the Sun the staff is held horizontally, the observer having his back to the Sun (hence the name back-staff). The half-cross, which is held vertically above the staff, is then moved along the staff to a position at which the shadow of the top end of the staff cast by the Sun struck the horizon vane and made coincidence with the horizon viewed through the slit in the vane.

Not only did Davis overcome the difficulty associated with the temporary blindness of the observer when observing the Sun directly; he also achieved the means of measuring the Sun's altitude without having to look in two directions simultaneously, as is the case when using the cross-staff.

The simple back-staff described above was graduated up to 45°, and was useful for observing the Sun at small altitudes only. For the northern Atlantic voyages of the English explorers of Elizabethan times, this was sufficient; but for southern voyaging, in which the Sun's altitude may reach 90°, Davis suggested the use of a different type of back-staff. The 90° back-staff—as this type was called to distinguish it from the 45° staff described above—

comprised two half-crosses: one straight and the other arcuate. The straight half-cross is fitted perpendicularly to the staff and is designed to slide along the staff in the same manner as the half-cross fitted to the 45° staff. The arcuate half-cross is fitted to the lower side of the staff, and is provided with a sighting vane. The fore end of the staff is fitted with a horizon vane through which the horizon is viewed when making an observation.

The principle of the 90° back-staff is demonstrated in Fig. 4.

In Fig. 4, AH represents the graduated staff fitted with the horizon vane at H. VB represents the transom or half-cross, designed to slide along the staff, and which is fitted with the shadow

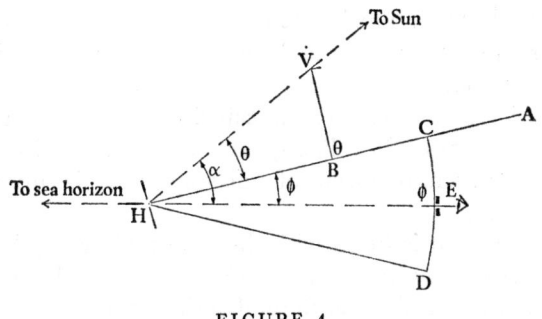

FIGURE 4

vane V. CD represents the fixed arcuate half-cross fitted with a sliding eye vane at E.

To take a sight with the 90° back-staff the half-cross VB is set at a graduation on the staff corresponding to a few degrees less than the Sun's altitude. The observer then holds the instrument vertically, and, with his back to the Sun and eye at the sighting vane, he slides the sighting vane to a position on the arc so that he may view the horizon through the slit in the horizon vane at the same time as he sees the shadow of the edge of the shadow vane coincident with the slit.

From Fig. 4:

$$\text{Sun's altitude} = \text{VHE}$$
$$= \alpha$$
$$= \theta + \phi$$

θ is read off the graduated staff and ϕ is read off the graduated arc.

As time passed Davis's 90° back-staff was modified, and before the end of the 17th century it had all but replaced the cross-staff (or fore-staff as it was often called) and other primitive measuring instruments. Davis's back-staff in the modified form became known as the *Davis quadrant*, and by the French and other European seamen as the English quadrant. It was not superseded for nautical astronomical purposes until the reflecting instruments had made their appearance in the middle of the 18th century.

The Davis quadrant that was in common use during the early part of the 18th century consists of two arches, together making 90° (hence the name quadrant). The two arches are fixed in a common plane, one above and the other below a straight bar corresponding to the staff. The length of the straight bar is a little more than the radius of the lower arch and about three times the radius of the upper arch. The upper arch is called the greater arch and it contains 65°. The lower arch is called the smaller—or lesser—arch, and it contains 25°. The greater arch is divided to degrees, and the lesser arch, by means of a diagonal scale, is subdivided to minutes of arc.

The instrument is fitted with three portable vanes known respectively as the *horizon vane*, the *sight vane*, and the *shade vane*. The horizon vane is fitted at the end of the straight bar close to the centre of the two arches. In the horizon vane is a long slit through which the sea horizon may be observed. The shade vane is fitted to slide on the greater arch. The upper edge of the shadow cast by this vane is made to fall coincident with the slit through the horizon vane when making an observation. Some observers used a glass vane instead of the shade vane for measuring altitudes of the Sun. The glass vane is simply a lens which focuses the Sun's rays to a bright spot on the horizon vane when taking a sight.

The sight vane is fitted to slide along the lesser arch. It has a sharp edge to cut the graduated scale of the lesser arch, to facilitate reading off the measured altitude. The sight vane is provided with a sighting hole through which the horizon and shadow line (or bright spot when using the glass vane) are observed.

To take a sight with the Davis quadrant, the shade vane is set to an exact number of degrees on the greater arch, about 10° or 15° less than the Sun's altitude, and the sight vane is placed near the middle of the lesser arch. The observer holds the instrument with the arches in the vertical plane, and with his back to the Sun.

ALTITUDE-MEASURING INSTRUMENTS OF NAVIGATION 73

Then, with his eye at the sight vane, he raises or lowers the quadrant, keeping the eye at the sight vane, until the shadow line is coincident with the slit on the horizon vane, adjusting the position of the sight vane if necessary, until the horizon is also sighted through the slit. The altitude of the Sun is obtained by adding the readings on the two arches. The principle of the Davis quadrant is illustrated in Fig. 5. (See also Plate 6.)

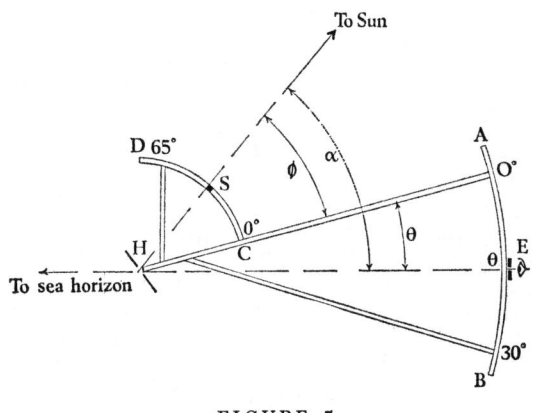

FIGURE 5

In Fig. 5: H represents the horizon vane; E the sight vane; and S the shadow vane. AB represents the lesser arch and CD the greater arch.

$$\text{Sun's altitude} = \text{SHE}$$
$$= \alpha$$
$$= \theta + \phi$$

ϕ is measured on the greater arch and θ on the lesser arch.

The Davis quadrant is not capable of adjustment, so that it was necessary for the observer to ascertain the instrumental error of his quadrant. This was usually done by making meridian altitude observations at places of known latitude, or by comparison with angles measured with an instrument the error of which was known. Having found the error of his quadrant, the observer would apply it to all altitudes measured with it. According as the error tended to increase or decrease the ship's northern latitude the quadrant was, therefore, said to be 'northerly' or 'southerly' respectively by its corresponding error.

7. THE REFLECTING QUADRANT

The Davis quadrant and all other forms of altitude measuring instruments were superseded by the reflecting instrument that became known as Hadley's quadrant.

Although John Hadley is usually credited with the invention of the reflecting quadrant, others before him had designed instruments for measuring altitudes using mirrors.

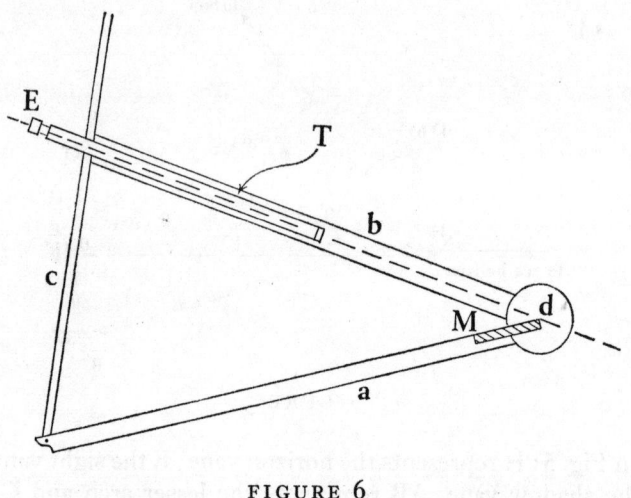

FIGURE 6

It appears that as early as 1666 Robert Hooke described to the Royal Society of London such an instrument. At the request of the society Hooke constructed the instrument and exhibited it before his fellow members later in the year. Hooke's reflecting instrument is illustrated in Fig. 6.

Hooke's instrument consists of three straight arms labelled a, b and c in Fig. 6. Arms a and b are pivoted at joint d. A mirror M is fitted to arm a, one edge of the mirror being coincident with the centre of the pivot d. Arm b is fitted with a telescope T, the axis of which lies in line with the inner edge of the arm. The eyepiece of the telescope is at E in Fig. 6. The third arm c is divided with equidistant graduations so that the angle between arms a and b may be found using a table of chords. This angle is equivalent to half the altitude of an observed object, as demonstrated in Fig. 7.

To measure the altitude of a celestial body using Hooke's

instrument, the arms of the instrument are held in the vertical plane and the telescope is used to sight the reflected image of the celestial body in coincidence with the horizon vertically below it as illustrated in Fig. 7, in which the altitude of the body is shown to be 2θ, the angle θ being equal to the angle between the two arms jointed at M.

The principal disadvantage of Hooke's instrument rested in the fact that the part of the horizon vertically below the observed object is hidden by the mirror unless the image is at the limit of the reflecting surface. There is no evidence that Hooke's instrument

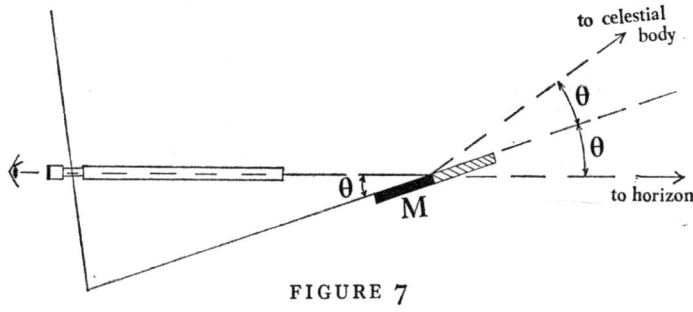

FIGURE 7

was tried at sea; and it appears that his idea of the use of a mirror fitted to an instrument for measuring altitudes at sea was forgotten.

Sir Isaac Newton gave some thought to the question of improving on the nautical quadrant; and he seems to have been the first to suggest the use of two mirrors as Hadley was later to employ for his reflecting quadrant. Little attention was given to Newton's suggestion; and it was not until some fifteen years after his death that Newton's design for a reflecting instrument for measuring altitudes at sea received some publicity. It was Edmund Halley who had remembered Newton's suggestion at the time John Hadley had made public his own invention. Newton's design for a sea quadrant is illustrated in Fig. 8.

Newton's design called for a plate of brass in the form of a sector, the 45° arc of which is graduated from 0° to 90°, each division representing 1° of arc. Pivoted at the centre of the sector or octant is an arm denoted by a in Fig. 8. The fiducial edge of this arm is used to read off the altitude of an observed heavenly body in degrees and minutes of arc. Fitted to one radial edge of the octant is a telescope, T, three or four feet long. Two specula A and B—A fitted to the plate and B fitted to the arm—are parallel to one another when the fiducial edge of the arm indicates zero on the graduated scale. Both specula are fitted perpendicularly to the plane of the graduated arc. The speculum fitted to the plate is set at an angle of 45° to the axis of the telescope.

To take an observation with Newton's octant the instrument is held vertically with the arc held towards the observer. The horizon is then viewed below the edge of the fixed speculum through the

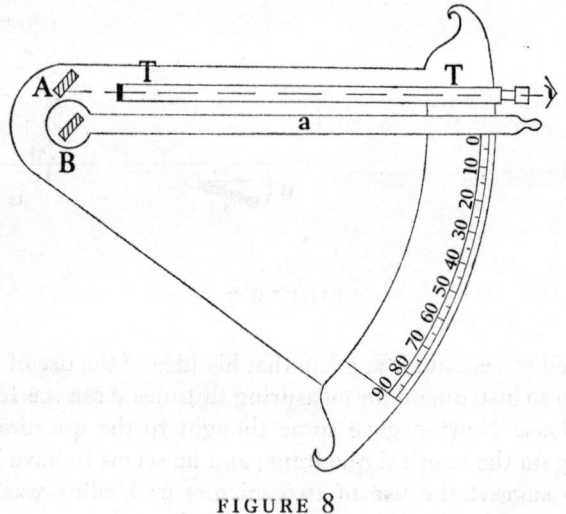

FIGURE 8

telescope. The arm is then swung downwards to a position such that a ray of light from the observed object reaches the observer's eye after having been doubly reflected from the index speculum B and the fixed speculum A.

The geometrical principle of Newton's octant is precisely that of the present-day sextant. Fig. 9 illustrates this principle, viz. the angle denoted by the fiducial edge of the index bar, which is equivalent to the angle between the two reflecting surfaces of the two specula, is equal to half the measured altitude of a celestial body. It follows, therefore, that the 45° arc is divided into 90 divisions each representing 1° of altitude or measured angle.

In Fig. 9 A and B represent the fixed and index specula respectively. A ray of light from a celestial body X enters the observer's eye at E after having been doubly reflected, the zig-zag ray denoted by XYZE.

YN and ZL are normals to the reflecting surfaces at points Y and Z respectively.

By the first law of optics:

$$XYN = NYZ \quad \text{(let this be } \theta\text{)}$$
$$YZL = LZE \quad \text{(let this be } \phi\text{)}$$

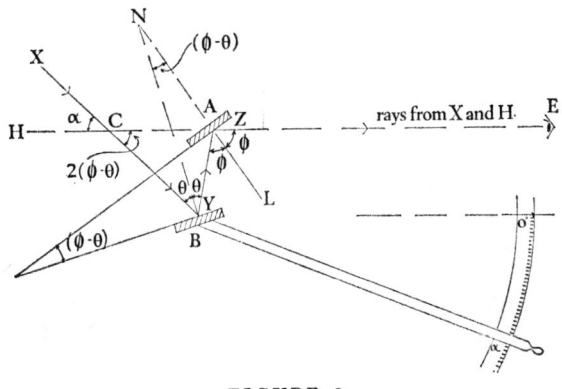

FIGURE 9

The angle between the reflecting surfaces is equal to the angle between the normals NY and ZL; and this is equal to $(\phi - \theta)$ (exterior angle LZY of triangle ZNY is equal to the sum of the interior and opposite angles).

The altitude of the body is denoted by α, and this is clearly equal to $2(\phi - \theta)$ (exterior angle EZY in triangle CZY is equal to the sum of the interior and opposite angles). Therefore, the angle between the reflecting surfaces of the specula is equal to half the altitude of the body X.

8. THE HADLEY QUADRANT

The first account of the reflecting instrument invented by John Hadley (1682–1744) was read before the Royal Society of London on May 13th 1731. The paper was published in Volume 37 of the *Philosophical Transactions of the Royal Society*. Hadley described two reflecting instruments, the design of one of which is very similar to Newton's design described above. Hadley's first instrument is illustrated in Fig. 10.

The instrument illustrated in Fig. 10 consists of a frame in the form of an octant of a circle. The 45° arc is divided into 90 divisions, each representing 1° of arc. An index bar, denoted by I in the figure, which provides for measuring observed angles, is pivoted at the centre of the graduated arc. Fixed to the index bar, perpendicularly to the plane of the arc, is a speculum A. This, the index speculum, is set so that when the pointer on the index bar coincides with the zero graduation on the arc scale, it is parallel to

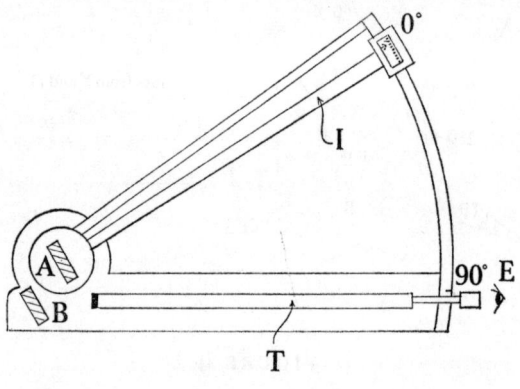

FIGURE 10

a second speculum B, which is fixed to the frame of the octant. A telescope T is fitted to one edge of the frame of the octant. This is set so that, when taking a sight, light from a celestial body is received at the eye after double reflection from the specula A and B, simultaneously with light from the horizon.

To take an observation with this type of Hadley's octant, the instrument is held with the plane of the arc in the vertical plane, with the arc towards the observer. The index bar is then set to a position so that rays of light from the observed object and from the horizon vertically below the object, are received at the eye simultaneously.

The geometrical principle of the octant is the same as that of Newton's instrument. Fig. 11 illustrates the manner in which the octant is used.

ALTITUDE-MEASURING INSTRUMENTS OF NAVIGATION 79

The second type of octant described by Hadley was that adopted by seamen generally. (See Plate 7.)

The telescope in Hadley's second octant is fitted across, instead of parallel to, the radius of the instrument, as is the case with the first type. This type of octant is illustrated in Fig. 12.

The speculum fitted to the index bar is called the index speculum; and that fitted to the frame, the horizon speculum. To use the octant the instrument is held with the arc in the vertical plane and directed downwards. The index bar is moved away from the observer along the arc to a position where the rays of light from

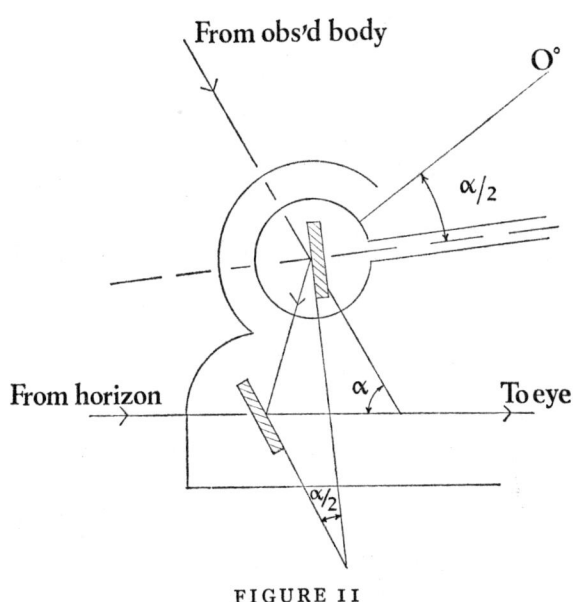

FIGURE 11

the observed celestial body and from the horizon vertically below the body are received at the eye simultaneously. The reading on the arc is then the observed altitude of the body.

The reflecting surfaces of metallic specula, fitted to the original Hadley octants, suffered the serious disadvantage of tarnishing quickly under the influence of sea-air and salt water. In due course, therefore, they were substituted by silvered plates of glass. Good glass mirrors were scarce at the time Hadley invented his octants. Caleb Smith designed an instrument on Hadley's pattern using glass prisms instead of specula, but Smith's octants did not become nearly so popular as those of Hadley's design.

Nevil Maskelyne is credited with being the first to suggest a novel use of glass reflecting-surfaces instead of specula. Maskelyne suggested that the under-surface of the glass block should be ground and painted black. In this event light would be reflected only from the polished surface of the block; and the possibility of double reflection (as is the case with a silvered mirror) with the

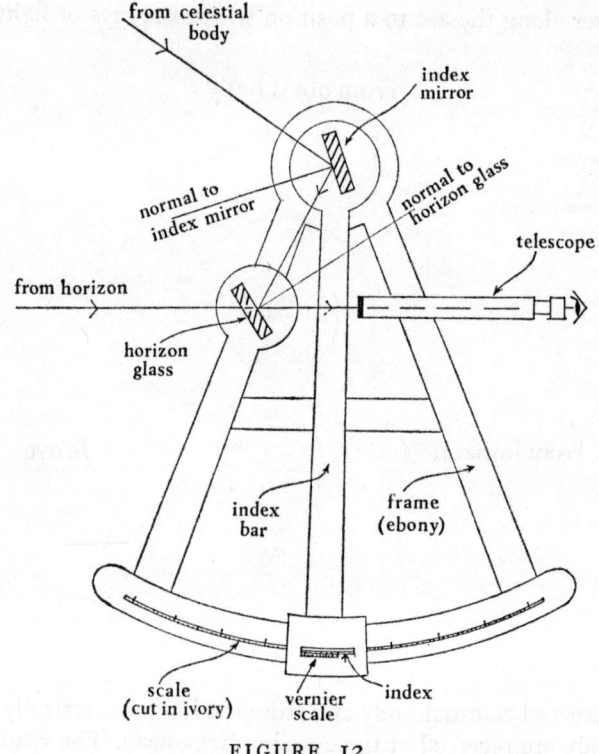

FIGURE 12

possibility of prismatic error due to the two surfaces of the mirror not being perfectly parallel to one another is, thereby, avoided.

The great advantage of Hadley's quadrant over those of Hooke and Newton rests in the fact that the direct image of the horizon and the reflected image of the observed celestial object may be brought into coincidence—both the body and the horizon verti-

ALTITUDE-MEASURING INSTRUMENTS OF NAVIGATION

cally below it being in full view during the time the observation is being made.

Hadley's quadrant was tried at sea in 1732 by James Bradley, the Astronomer Royal, in the presence of John Hadley and his brother. The tests were successful and Hadley's sea octant rapidly became popular with navigators. The instrument is especially suited for use at sea. And even when the ship is unsteady, an observer, with but little practice, is able to make an accurate observation. For its purpose it was ideal; and it was not without reason that the instrument has been described as 'the most perfect appliance that has ever been invented.'

The two mirrors of glass fitted to later Hadley octants are known respectively as the *index mirror* and the *horizon glass*. The horizon glass is half-silvered, the half farther removed from the plane of the instrument being clear glass. When observing, therefore, the reflected image of the observed celestial body is brought into coincidence with the direct image of the horizon at the line separating the silvered from the unsilvered part of the horizon glass.

Hadley's original instrument was suitable for measuring angles up to 90° and, for this reason, was sometimes described as a quadrant. Hadley did, however, provide the means for measuring angles greater than 90°, a third mirror being fitted to the instrument for this purpose.

The great need at the time Hadley's instrument appeared was for an instrument suitable for measuring lunar distances which often exceeded 90°. Captain Campbell of the Royal Navy, who often observed with James Bradley, was prompted to suggest, in 1757, enlarging the arc of the octant to 60°, so that angles up to 120° could be measured. It is to Campbell, therefore, that we owe credit for the introduction of the nautical sextant.

At about the time when John Hadley described his 'new astronomical instrument for making observations of distance (lunar) by reflection,' an American optician Thomas Godfrey was engaged in the problem of measuring altitudes and lunar distances at sea. Godfrey is credited with the invention of a reflecting instrument similar in all respects to the first of the two octants described by Hadley to the Royal Society in 1731.

Hadley's quadrant, unlike Davis's quadrant, was capable of adjustment, so that instrumental errors could be removed. When the instrument is in correct adjustment, the two mirrors are perpendicular to the plane of the instrument and are parallel to one another when the index is set to zero on the arc.

Following the introduction of Hadley's quadrant the seaman relied entirely on having his altitude-measuring instrument made for him by an instrument-maker ashore. No longer was it necessary for him to be provided with instructions in his manuals for making them himself. The mariner, in his role as a prospective purchaser of a quadrant, was advised to pay particular attention to the construction and accuracy of the instrument before parting with his money. He was advised to examine carefully the joints in the several parts of the wooden (usually mahogany, teakwood, ebony or fruitwood) frame; to see that the graduations on the ivory arc and vernier scale were accurately cut; to ensure that the surfaces of the mirrors were plane; and to see that the coloured shades were free from veins.

The index bar of the quadrant is provided with a rectangular aperture through which the graduated scale may be seen. One edge of the aperture carries a dividing scale by means of which angles may be measured to a relatively high degree of accuracy—usually to the nearest minute of arc.

The dividing scale on the earlier Hadley instruments was a diagonal scale, but this was soon superseded by a vernier scale.

The principle of the *vernier scale* was described by Pierre Vernier in a small tract entitled *La Construction, l'Usage et les Propriétés du Quadrant Nouveau de Mathématique*, printed at Brussels in 1638. Vernier is usually credited with the invention of the scale which bears his name; but it appears that Clavius, in a treatise on astrolabes, explained the method in 1611.

The vernier scale on the earlier Hadley quadrants was divided into 20 equal parts, this being equivalent to the space occupied by 21 or 19 equal divisions on the arc. It follows, therefore, that the difference between an arc division and a vernier division is $\frac{1}{20}°$. Each arc division represents $\frac{1}{3}°$; so that angles, therefore, could be measured to the nearest minute of arc.

The Portuguese mathematician Pedro Nuñez (Petrus Nonius) described a method for measuring angles accurately which appeared in print in his *De Arte atque Ratione Navigandi*, in 1522. The name Nonius is sometimes given to Vernier's scale, although the principle of the dividing scale described by Nuñez is different from that of Vernier's.

Nuñez' method consists of 45 concentric equidistant arcs described within the same quadrant. The outermost arc is divided into 90 equal parts; the next into 89 equal parts; the next into 88, and so on. When observing, the radial index (or plumb-line in the case of a mural quadrant provided with a nonius scale) would cross one or other of the graduated arcs at or near a point of division, thus enabling the observer to obtain a very accurate measurement of the observed angle. Nuñez's method of dividing an arc was superseded by the *diagonal scale*, which was described by Thomas Digges in a treatise entitled *Alae sui Scalae Mathematicae* published in London in 1573. Digges gives credit for the invention of the diagonal scale to Richard Chancellor. The diagonal scale was used for dividing the arc of the Davis quadrant.

9. THE REFLECTING CIRCLE

An instrument designed for measuring lunar distances accurately was invented by Tobias Mayer, a figure famous in the history of the method for finding longitude at sea known as the lunar method. Mayer's instrument, called the *simple reflecting circle*, was improved by the French naval officer Borda and by others.

The principle of the simple reflecting circle—familiarly called the circle—is the same as that of the reflecting quadrant. The instrument consists of a circular limb with the index bar pivoted at the centre of the graduated circle and fitted with a vernier scale at each extremity. The simple reflecting circle is illustrated in Plate 8.

The manipulation of the simple reflecting circle is similar to that of the sextant. The mean of the readings of the two diametrically opposite verniers, taken at each observation, will be completely free from error of eccentricity. This error, which results when the pivotal point of the arm of the instrument is not coincident with the centre of the circle of which the arc forms part, was particularly troublesome in the early octants and sextants. One great advantage of the reflecting circle over the sextant, even for measuring angles of less than $90°$, is therefore the elimination of error of eccentricity. At the same time, effects of errors in reading and accidental errors of graduation were diminished, since every result is derived from the mean of two readings at two different divisions of the arc.

Some simple reflecting circles, such as those made by the

instrument-maker Troughton, have three verniers at distances of 120° apart: but, as the eccentricity is fully eliminated by having two verniers, the third increases the accuracy of a result only by diminishing the effect of errors of reading and graduation. Chauvenet, in his *Astronomy*, points out that if ϵ_2 is the probable error of the mean of two readings, and ϵ_3 is that of the mean of three; then:

$$\epsilon_3 = \epsilon_2 \sqrt{\tfrac{2}{3}} = 0 \cdot 8 \epsilon_2$$

So that if two verniers reduce the error to say 5″, the third will only further reduce the error to 4″, an increase of accuracy which for a single observation is not, according to Chauvenet, worth the additional complication and weight, and the extra trouble of reading.

Some simple reflecting circles employed glass prisms instead of specula or glass mirrors. The prismatic reflecting circle constructed by the Berlin firm of Pistor and Martins is illustrated in Fig. 13.

In Fig. 13, ABC represents the arc of the instrument. M is a central mirror on the index arm. m is a glass prism two faces of which are at right angles to one another. The third face of the prism acts as a reflector. The height of the prism above the plane of the arc is half that of the object glass of the telescope T: therefore the direct ray from one observed object passing over the prism can be brought to the same focus as that of the reflected ray from the second observed object. When the central mirror is parallel to the longest side of the prism, the two images are in coincidence and the index error is found as with a sextant, except that every reading is here the mean of the readings of the two verniers.

The *repeating reflecting circle* is an improvement on the simple reflecting circle. In the repeating circle, the horizon glass is not attached to the frame of the circle as it is in the simple instrument. It is attached to a separate arm which may be rotated independently about the centre of the instrument. The telescope, which must always be directed through the horizon glass, is also fitted to this arm. In addition to the arm which carries the telescope and horizon glass, a second arm, on which is mounted the central- or index-mirror, may be turned about the centre of the instrument.

To use the repeating circle, the instrument is held with the plane of the arc in line with the two objects whose angular distance is required. The index on the index mirror arm is clamped to the arc. The right-hand object is then observed direct through the unsilvered part of the horizon glass. The instrument is then rotated, keeping the right-hand object in sight, until the reflected image of the second object is observed through the silvered part of the horizon glass. A fine adjustment is then made to bring the true and reflected images into coincidence. This completes the first part of the observation. The index arm is then unclamped and,

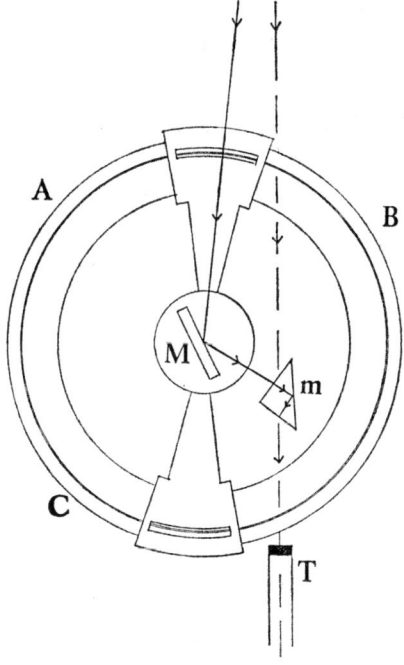

FIGURE 13

leaving the horizon-glass arm in its clamped position, the telescope is directed to the left-hand object. The index-mirror arm is then rotated to a position in which the reflected image of the right-hand object is brought into coincidence with the direct image of the left-hand object. This completes the second part of the observation. The difference between the readings of the index on the index-mirror arm in the two positions is twice the angle between the observed objects. For let R and R_1 be the readings of the index of the index-mirror arm before the first, and after the second, contact. At each contact the angle between the index- and horizon-mirrors is equal to one half the measured angle; and it is evident that the points R and R_1 are at equal distances on each side of that point on the arc at which the index of the index-mirror arm would have stood had its motion been stopped at the instant the mirrors were parallel. It follows that the angle between R and R_1, in the direction of the graduations from R, is equal to twice the angle between the mirrors at either contact. If the measured angle is denoted by γ, we have:

$$2\gamma = R_1 - R$$

If the observations are now recommenced, starting from the last position of the index on the index-mirror arm, this index will be found, after the fourth contact, at a reading R_2, which differs from R_1 by twice the angle γ, so that we have:

$$2\gamma = R_2 - R_1$$

But,

$$2\gamma = R_1 - R$$

It follows, therefore, that:

$$4\gamma = R_2 - R$$

Continuing the process, we shall have, after any even number n of contacts, a reading R_n. Thus:

$$n\gamma = R_n - R$$

and

$$\gamma = \frac{R_n - R}{n}$$

For any number of contacts, it is necessary to read off only before the first, and after the last, observed contact. This led to the great advantage of this instrument, for use on board ship, for measuring lunar distances.

When using the repeating reflecting circle for lunar distance observations, the difference between the first and last readings is the sum of all the individual measures, and the value of the observed distance is found by dividing this sum by the number of observations. This distance corresponds to the mean of the times of the first and last observations, provided that the angular distance is changing uniformly.

The errors of reading and graduation, as well as error of eccentricity, are all nearly eliminated by taking a sufficient number of observations.

In theory, the repeating circle is very nearly a perfect instrument, capable of eliminating its own errors. This theoretical perfection is, however, impossible, owing to the mechanical imperfections arising from the centering of the axes of the two rotating arms one within the other.

The most important improvements in the reflecting circle are due to Chevalier de Borda whose work, *Description et Usage du Cercle de Réflexion*, was first published in Paris in 1787.

The reflecting circle was ideally suited to the lunar distance observations. It provided the highest degree of refinement for this purpose; and many navigators provided themselves with both circle and sextant: the latter for altitude observations, and the former for observing lunar distances.

10. PERFECTED ALTITUDE-MEASURING INSTRUMENTS

Improvements in sextant design and manufacture, coupled with the redundancy of the lunar problem, spelt doom for the reflecting circle during the early part of the 20th century.

The modern sextant (see Plate 9) is an instrument of precision. Some of the numerous improvements made to the early sextants are of great interest and importance.

The manner of making a fine adjustment of the index bar when taking a sight was facilitated by fitting a *tangent screw* to the index bar. The index bar was clamped to the arc when the true and reflected images were observed through the sights or the telescope, after which the tangent screw was used to make a fine adjustment. The clamping of the index bar to the limb, in sextants of the last century, was usually effected by a screw and block piece which travelled over the smooth surface at the back of the limb. This clamp block was attached to the tangent screw in such a way that movement of the vernier is possible when the index bar is clamped. A disadvantage of the early arrangement is due to the limited travel of the tangent screw, rendering it necessary, before observing, to ensure that the tangent screw is not at or near the end of its travel. An improved form of tangent screw was made with a spring on each side of the clamp block, so that when the clamp is released the vernier automatically takes up a central position, and the likelihood of the tangent screw-thread being used up at a critical moment is obviated. This form of tangent screw was ousted during the early part of the present century by the endless tangent screw. The endless tangent screw led to the invention of the micrometer tangent screw which is common to almost all present-day sextants. Amongst the features of this type of tangent screw, not the least important is that it can be read at arm's length.

Many early reflecting instruments were fitted with sight vanes instead of telescopes. Improvements in sextant telescopes have been of great significance. It is not uncommon nowadays for a sextant to be provided with a single all-purpose telescope or monocular. From the middle of the 19th century until the eve of the Second World War most sextants were provided with two telescopes. One, of small magnification and big object glass, was designed for star work. The other, of large magnification and small object glass, was for daylight observations. The star telescope was an erecting telescope, whereas the high-power telescope was an inverting telescope. The slight difficulty of using the inverting telescope is doubtless the reason why it fell into disuse.

Hadley, the inventor of the reflecting quadrant, directed that the line of sight should be parallel to the plane of the instrument; and, for ensuring that this was so, he proposed that two parallel wires should be fixed in the telescope parallel to the plane of the quadrant, and that the contact of the observed objects should be made in the middle of the field of view of the telescope between the two parallel wires. These circumstances have never sufficiently been attended to, although in the inverting telescopes of days gone by, cross-wires were fitted for this purpose and for checking collimation error—error resulting from the line of sight not being parallel to the plane of the sextant arc.

As soon as it became easy to furnish good glass mirrors, these replaced the metal specula of the older reflecting instruments. As every glass mirror gives two reflections, one from the face and the other from the silvered back surface, double reflections may cause confusion with the reflected rays and error may result in the observation. Moreover, if the front and back surfaces of the mirror are not perfectly parallel to one another, the observation may suffer prismatic error. It was to overcome these difficulties that the Rev. Dr Nevil Maskelyne suggested a reflecting surface of plane glass, the back face of which is rough ground and blackened. By this means only rays falling on the polished surface of the glass are reflected.

The glass shades, used when observing the bright Sun, owing to their want of uniformity of colour density, often caused error. In some sextants of the last century it was not uncommon to provide a coloured eyepiece for use when observing the Sun. In others, the coloured screens were designed so that they could be instantaneously reversed, so that, by taking half a set of observations with the shades in one position, and the other half with them reversed, error due to non-parallelism of the surfaces of the shades was eliminated.

The graduations on the earliest instruments were cut in an arc of ivory. Later arcs were of silver, gold or platinum. The most important operation in sextant manufacture is, undoubtedly, the cutting of the arc. The difficulty of graduating an arc of a sextant or other similar astronomical or surveying instrument was not overcome until after the middle of the 18th century. The famous English engineer Jesse Ramsden (1735–1800) invented a remarkable machine, based on the worm and wheel principle, for dividing circular arcs as well as linear scales with precision. *Ramsden's dividing engine* employed principles that had been published by the French academician Duc de Chaulnes (1714–1769), who was the first to use a tangent screw drive for this type of machine.

Names famous in the history of the development of the manufacture of quadrant and sextant, in addition to Hadley and Ramsden, are George Adams (d. 1773), John Bird (1709–1776), and the opticians Troughton and the Dollonds.

George Adams specialized in making quadrants for seamen and, soon after Hadley's invention, Adams was producing instruments at a price well below those of other instrument-makers. John Bird, who had been trained by the famous George Graham (1673–1751), devoted considerable attention to the determination of the most suitable shapes of the several parts and fittings of the quadrant and the best methods of assembling them. Up to about 1760 the frames of quadrants were made from a combination of iron, brass and wood. It was recognized at about this time that accuracy of measurement was impaired largely on account of differences in expansion coefficients of the materials used. Reflecting instruments having brass frames were manufactured long before the dawn of the 19th century, but metal frames were not common until the middle of that century. To facilitate the dividing of the arc of the early instruments the radius of the limb was kept as large as conveniently practicable, and this was often as much as 20 inches. With the introduction of the metal frame and the dividing engine invented by Ramsden, the length of the radius was reduced to a mere 8 inches or even less: that of most modern sextants being no more than about 6 inches.

John Dollond (1706–1761) was the optician who is credited with the invention of the *achromatic lens*. The firm of Dollond's still exists; and this firm, as well as that of Troughton's, produced many fine sextants, circles and other scientific instruments, especially during the 19th century.

As far back as 1894, Lecky described an electric lighting system using a small dry cell and incandescent lamp, for use with a sextant for star observations. This appears to have first been fitted to a sextant by the instrument-maker Cary of the Strand in London. Cary was also instrumental in first using an interrupted thread in order to facilitate the fitting of sextant telescopes.

The fitting of a *Nicol prism* for eliminating horizon glare appears to have first been suggested by a merchant seaman named Mackenzie who communicated his idea to the Royal Astronomical Society. This device is simply a polarizing prism used like a telescope eyepiece, and so placed that when the telescope is screwed home in its collar the polarizing plane of the prism is parallel to the plane of the sextant and, consequently, perpendicular to the horizontal when the sextant is being used for measuring the altitude of a heavenly body. It is used when the glare of the Sun (or Moon) renders it difficult to define the horizon. The prism allows only the 'extraordinary' ray to be transmitted to the eye, the intense glare being refracted upwards out of the prism.

Another device of great value when observing star altitudes is the *Wollaston prism*. This is a pair of prisms in the form of two wedges of different thicknesses or different refractive indices so that two distinct images of an observed star are formed. The Wollaston prism is fitted between the index mirror and the horizon glass. When using the prism the observer brings the reflected images of the observed star to a position where the true image of the horizon lies centrally between them.

A similar, but cheaper, device designed to facilitate star observations is the *lenticular* or *elongating lens*. This, like the Wollaston prism, is fitted between the index mirror and the horizon glass and, being a cylindrical lens, the reflected image of an observed star appears as a line, instead of a point, of light.

To eliminate the uncertainty of the effect of refraction on the dip of the sea horizon, Commander Blish of the United States' Navy invented, in the early part of the 20th century, an attachment for a sextant known as the *Blish prism*. This device has the top and bottom faces bevelled at 45°. It is fitted to the sextant so that the longer of the front and back surfaces faces the observer. This face is provided with two polished surfaces, the lower of which is directly opposite the top of the index mirror, and the higher of which faces over the observer's head towards that part of the horizon 180° away from the part the observer is facing. With the index of the sextant set to zero on the arc the observer looks directly at the sea horizon in front of him and sees, at the same time, the back horizon reflected from the prism. When the fore and back horizons are brought into line, the sextant reading is twice the angle of dip, assuming that the sextant is free from index error.

11. THE ARTIFICIAL HORIZON

To take a sight with a sextant, the sea horizon vertically below the object whose altitude is required must be clear and distinct. Without the horizon, or a horizontal (or vertical) reference, a sight cannot be taken. 'The frequent want of a horizon,' wrote Robertson in his famous 18th-century *Elements of Navigation*, 'is one great inconvenience that mariners have to struggle with at sea.'

Many attempts have been made to provide a means whereby the visible horizon may be dispensed with when taking sights. Hadley himself was the first to provide a remedy for an indistinct or invisible horizon, in the form of a simple spirit level attached to his quadrant.

In 1732, a description of an improved *artificial horizon* appeared in the *Philosophical Transactions of the Royal Society* of London under the name of John Elton. Elton's device consisted of two spirit levels at right angles to each other fitted to the frame of a quadrant. The principle of Hadley's and Elton's bubble horizon is simple, but the difficulty of holding the instrument steady and perfectly vertical during observation rendered it, in its earliest form, impracticable.

The problem of taking a sight with a sextant on dry land without a visible horizon is solved by using an artificial horizon in the form of a calm liquid surface such as a puddle of water or a container of tar, treacle or oil. The more sophisticated artificial horizon of this type, used extensively during the last century for the purpose of finding longitude ashore with the object of checking chronometers, consisted of a trough of mercury, the surface of which provided the reflector. The *mercury artificial horizon* appears to have been invented by the London instrument-maker, George Adams, in about 1738. The equipment consists, in addition to the trough and mercury, of a glass roof designed to prevent the troublesome tremulous motion of the mercury due to wind. The principle of the mercury artificial horizon is based on the first law of optics, which states that the angle of reflection from a mirror is equal to the angle of incidence. When using the artificial horizon the sextant is employed to measure the angle between the Sun and his image on the mercury surface, this angle being equal to twice the Sun's apparent altitude. The apparent altitude is the arc of a vertical circle between the apparent direction of the observed object and the plane of the sensible horizon or horizontal plane on which the observer's eye stands.

The mercury horizon is useless on board ship unless the ship is perfectly steady. The slightest movement of the ship would cause the mercury surface to tremor and become useless for observational purposes.

Robertson, in his *Elements*, gives the following description of a mercury horizon for use on board ship:

'Into a wooden, or iron, circular box, of about $2\frac{1}{2}$ or 3 inches diameter, and about $\frac{1}{2}$ inch deep, pour about a pound or more of quicksilver; and on this lay a metal speculum, or a piece of plain glass, the diameter of which is about a third of an inch less than that of the box; this will float in the quicksilver, and shew the image of the Sun very steady. This apparatus being slung in jimbals will preserve a tolerable good horizon.

'The speculum, or glass, should be homogeneous, and have parallel sides. There are some workmen who can work the two planes of a piece of glass, so that they shall be demonstrably parallel.

'Or the fine surface of the quicksilver will do of itself, when the motion is not great.'

1. Rev. Nevil Maskelyne D.D.: the Father of Nautical Astronomy. From a painting by T. Downman at the National Maritime Museum.

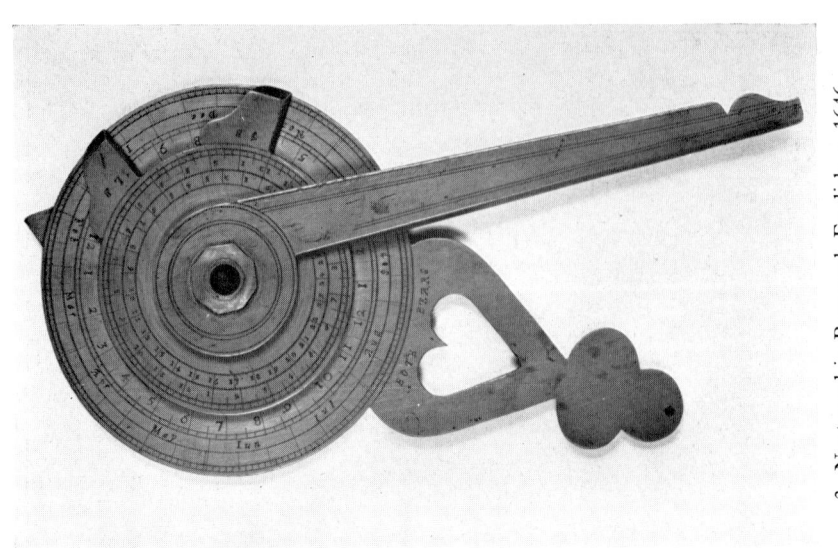

2. Mariner's Astrolabe. Probably Spanish, c. 1585. (Found off the Irish coast in 1845.)

3. Nocturnal in Boxwood. English, c. 1646.

4. Azimuth Compass. English, c. 1720.

5. Mariner's Quadrant. c. 1600.

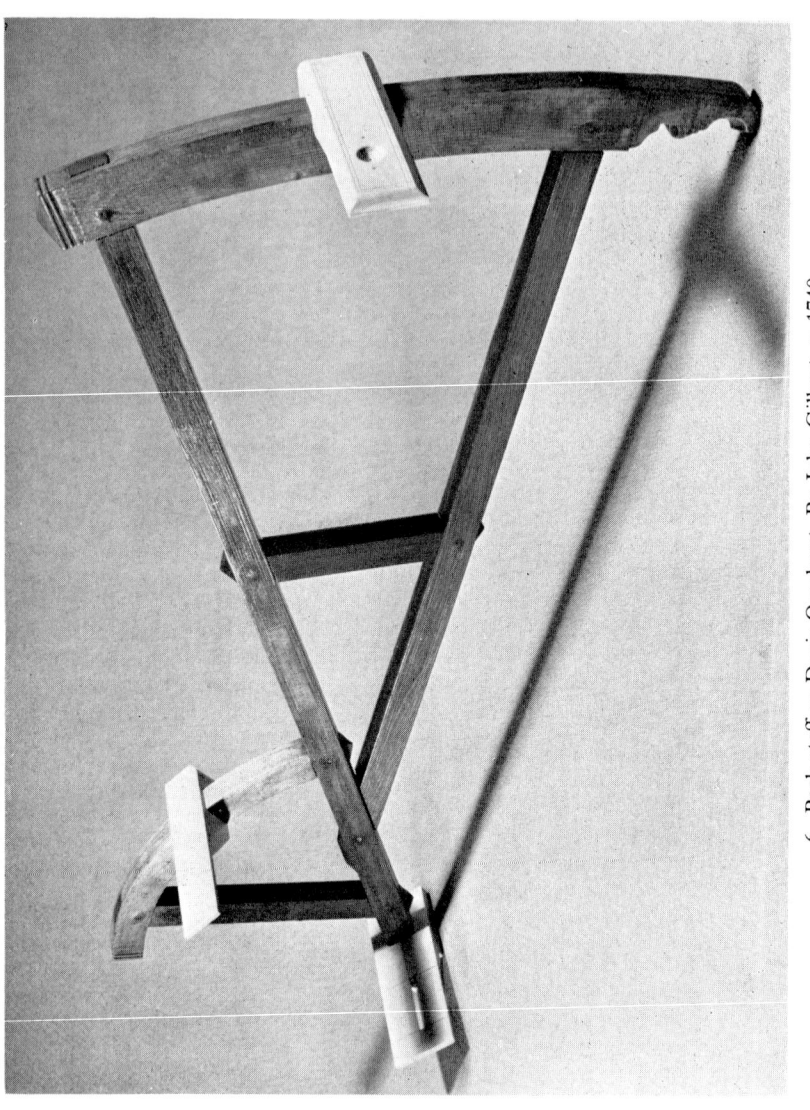

6. Back-staff or Davis Quadrant. By John Gilbert, c. 1740.

A century or so after Robertson had written the above description, John Merrifield, the headmaster of the navigation school at Plymouth, described his attempt at using a mercury artificial horizon on the roof of the school building:

'We found it quite impossible to take observations with the artificial horizon . . . owing to the shaking of the walls of the building by wind or passing vehicles. At the suggestion of the late Commander Walker, R.N., we had a horizon constructed, so that a piece of glass, whose surfaces were perfectly plane and parallel to one another, should float on the mercury, and fit so close to the sides of the trough as to prevent any great motion, yet not so close as to prevent its free action. At first we found very great discrepancies, owing to the glass not being homogeneous, and thus floating slightly deeper at one end than the other: but the idea occurred to us of taking two sights with the instrument reversed, using the means of the altitudes and of the times of the observations. Thus the error due to want of homogeneity was eliminated, and we have since found this to be a very efficient instrument.'

A compact mercury horizon was patented by Captain George, a merchant service officer, during the seventies of the last century. This consists of a circular iron trough containing mercury on which a disc of glass having parallel faces is floated. Before floating the glass, a piece of thin paper was placed on the mercury surface; then the glass was pressed lightly on top of this, the piece of paper being removed at the same time to ensure a perfectly clean mercury surface. The chief advantage of Captain George's instrument is that the whole surface of the artificial horizon is available for observation.

A form of artificial horizon was introduced into the Royal Navy during the early part of the 20th century. This consisted of a shallow trough of metal gilt which was amalgamated, after a first cleaning of the surface with a drop or two of dilute sulphuric acid, by the rubbing into it of a small quantity of mercury until the whole surface was bright. The trough was mounted on three adjustable screws and was provided with a bubble for levelling.

In a very interesting and valuable dissertation on the history of the art of navigation, written by Dr James Wilson, and which appeared in the first edition of Robertson's *Elements of Navigation* of 1772, mention is made of a horizontal top, invented by Serson who was, Wilson informs us . . . 'unfortunately lost at sea aboard the Victory man of war.'

Serson's horizontal top employed the properties of a spinning body. The upper surface of the top was polished metal which formed, when the top was spun, and in obedience to its gyroscopic inertia, a horizontal reflecting surface which could be used as an artificial horizon.

John Smeaton, the renowned English engineer, improved Serson's top, and described his spinning artificial horizon, and how to use it, in the *Philosophical Transactions of the Royal Society* for the year 1752. Smeaton's top had a polished speculum surface of about $3\frac{1}{2}$ inches across. The top was fitted with a brass ring placed at right angles to the axis of the top. The sharp spinning point of the top rested on a cup of a hard substance such as flint or agate. Friction was kept to a minimum, and the top could be made to spin for upwards of about fifteen minutes. There is no evidence to suggest that *Smeaton's gyroscopic horizon* met with any success in practical use.

Raper, in his famous *Practice of Navigation*, refers to the use of a mirror attached to a pendulum which, hanging vertically, provides an 'artificial vertical' which serves the same purpose as an artificial horizon. Raper pointed out the difficulties of using such a device on a rolling ship.

In about 1838 Lieutenant A. B. Becher R.N. invented an artificial horizon attachment for a sextant, which was subsequently made and sold by Cary of the Strand. *Becher's horizon* met with some success. The inventor pointed out that ships are not always in violent motion, and that there are many circumstances of weather and sea in which an instrument such as his artificial horizon has its value. In particular, he referred to the mouth of the English Channel in a southerly wind with a sea horizon obscured by fog and the importance, in these circumstances, of getting a sight for latitude.

Becher's artificial marine horizon was fitted outside the horizon glass in line with the telescope axis. It consisted of a small pendulum the bob of which was suspended in a small cistern of oil, so that the observer could control its movement; which, from the extreme delicacy of its suspension, would otherwise be impossible.

ALTITUDE-MEASURING INSTRUMENTS OF NAVIGATION 95

Fitted to the pendulum and at right angles to it (and to the plane of the instrument when a sight is being taken) is a small arm. Beyond the pendulum, a line for the horizon is formed by the upper edge of a slip of metal at right angles to the plane of the instrument. The pendulum has free motion in any direction; and the observer was required to bring the upper edge of the arm attached to the pendulum, in exact contact with the horizon line formed by the slip of metal. At the same time he was to make his observation by bringing down the image of the reflected object which he observed upon the line of contact.

As the observer has thus to form his horizon at the instant of observation, he was advised, when observing on board, to get into that part of the ship where there is the least motion, and especially into a place screened from the wind.

Becher's marine horizon was fitted with an oil lamp so that observations of stars could be made during hours of darkness.

Admiral F. W. Beechey invented an artificial horizon attachment based on similar principles to that of Becher's. Beechey's device consisted of a balance carrying a glass vane which was fitted in the sextant telescope. The lower half of the glass vane was coloured blue, the horizontal line of demarcation between the coloured and uncoloured parts representing the sea horizon. The reflection of the observed object was brought into coincidence with the artificial horizon. The amount of oscillation above and below the artificial horizon was indicated by divisions on the glass vane, the values of which were determined by the makers. When taking a sight and using Beechey's horizon the observer brings down the reflection of the Sun's limb to the artificial horizon and leaves it there; and then, as the ship rolls, he catches, with his eye, the upper and lower divisions reached by the Sun's limb, and calls them out to an assistant who notes them and records the corresponding times by chronometer. After two or more readings have been taken, the altitude is read off and a correction is made according to the mean of the readings of the vane. *Beechey's artificial horizon* attachment, like that of Becher's, was fitted with a lamp which could be used to illuminate the telescope tube for star observations.

In the early part of the present century, Paget patented an artificial horizon for attachment to a sextant. The *Paget horizon* consists of a short, curved spirit level mounted in a tube with a prism above it designed to throw the image of the bubble into the field of view of the telescope. This type of horizon had the advantage of not being affected by vibration or wind.

Another ingeniously contrived artificial horizon using a small gyroscope was invented by Admiral Fleurais. This was manufactured by the well-known firm of Henry Hughes and Son.

All the artificial horizons described above, and many others besides, were not entirely satisfactory; and they were, accordingly, short-lived.

It was not until comparatively recent times that efficient artificial horizon attachments became available for marine sextants. Amongst these, the *Booth bubble horizon* is noteworthy. The Booth horizon is commonly fitted to air sextants, and good results are obtained by its use.

CHAPTER IV

The altitude corrections

1. INTRODUCTORY

The fundamental process in position-fixing at sea by astronomical methods is the measuring of the altitude of a celestial body. To find the latitude from a meridian altitude observation, for example, the complement of the altitude of a heavenly body at *meridian passage*, that is to say, the body's meridian zenith distance, is combined with the declination of the body to give the required latitude. Moreover, in the general nautical astronomical problem in which the astronomical- or PZX-triangle is to be solved, one of the sides of this spherical triangle is the true zenith distance of the observed object at the time of the observation. The true zenith distance of a heavenly body is the complement of the true altitude of the body, and this is obtained from the measured—or observed —altitude by applying certain altitude corrections. It is the historical account of these several corrections with which we shall be concerned in this chapter.

2. REFRACTION

In astronomical navigation, as in many other scientific activities, the nature and behaviour of light—the phenomenon by which nautical astronomical observations are made possible—are of great importance.

To the query 'What is light?' there is no simple explanation. The physicist regards light as being electromagnetic radiation—a form of energy which travels at the prodigious speed of 300,000,000 metres per second.

Many of the Ancient Greeks and other early peoples considered light to be a fundamental property or accident of nature and usually associated it with the Sun, 'the giver of light and life.' Among many speculations relating to the nature of light made by Ancient Greek scholars, that of Empedocles, who flourished during the 5th century BC, is interesting. The theory of light propounded by Empedocles was described by Aristotle, who flourished during the following century. 'Light is a streaming substance,' stated Empedocles, 'of the movement of which, because of its high speed, we are not conscious.' That light generally travels in straight lines is a fact which has been known since earliest times. The famous Euclid (c. 330–c. 275 BC), in a work on light, laid down the foundations of geometrical optics. Hero of Alexandria (fl. c. 100 BC), amongst many of his scientific activities, produced a work on mirrors. He considered the physical conditions of reflecting surfaces and mentioned the desirability of a polished surface to obtain optimum conditions for light reflection. Hero is often credited with being the author (albeit unwittingly) of the important scientific law known as the 'principle of least action.' This was expressed in relation to the equality of the angles of incidence and reflection of a ray of light striking a point on a reflecting surface. Hero expressed this observed fact by stating that a ray of light makes the shortest route between object and eye.

When light travels through a transparent medium of uniform density, it travels in a straight line. If, however, light travels obliquely from one transparent medium to another, its path is bent at the surface of contact. This phenomenon is known as *refraction*. The remarkable effects of refraction, such as the apparent bend in a straight stick partly immersed in water, have excited the curiosity of men of all times. There is no doubt that Hero, and scholars before him, understood something of this optical phenomenon. Sarton, in his *History of Science*, describes the ancient study of refraction as having been the most remarkable experimental research of antiquity.

Ptolemy, who flourished during the 2nd century AD, in his *Optics*, a work known through a 12th-century translation in Latin which had been translated from Arabic, elucidated certain optical phenomena, amongst which was an approximate law of refraction. According to Ptolemy, the angles which the incident and refracted rays of light make with the perpendicular to the plane separating the two transparent media are directly proportional to one another. This relationship holds good only for small angles of incidence and refraction.

The refraction of light from a celestial body in its passage through the Earth's atmosphere is known as *atmospheric-* or *astronomical-refraction*. Ptolemy explained atmospheric refraction as being due to changes in air density. He also concluded that the apparent position of a star did not always correspond to its true position on account of atmospheric refraction.

After the fall of Alexandria the works on light and optics produced by the Ancient Greek philosophers were developed by the Arabs. Notable amongst the Arab scholars was Al Kindi (c. 800–873), who made a special study of refraction. But the most famous of the Arab physicists who pursued the study of light and optics during the Dark Ages was Ibn al Haithan, better known as Al Hazen (945–1039). Al Hazen demonstrated that the angle of refraction was not proportional to the angle of incidence as had been stated by Ptolemy. Although Al Hazen disagreed with Ptolemy in this respect, he did not give a better law of refraction.

It is believed that Roger Bacon (1214–1294), a disciple of the famous Franciscan Robert Grosseteste (1175–1253), whose discovery of double refraction of light through a lens doubtless paved the way for the invention of spectacles and the telescope, was led to a study of optics through the work of Al Hazen.

The Polish philosopher Vitello (born c. 1230) experimented with the refraction of light passing through air and water, and through air and glass, and determined new values for angles of refraction. Vitello showed that the scintillation of the stars is due to atmospheric effects.

The Arabian physicist Al Farisi (d. c. 1320) gave an interesting explanation of refraction, attributing the phenomenon to a change of speed of light when passing from one medium to another of different optical density.

The renowned Tycho Brahe (1546–1601) is often held to have been the first to have employed atmospheric refraction for correcting astronomical observations. He found the value of atmospheric refraction of light from a celestial body on the horizon to be 33′, and set out to draw up a table of refractions for all altitudes. Tycho was not too clear as to the cause of atmospheric refraction, attributing it to 'the gross vapours that float in the atmosphere.' According to Tycho, the refraction of light from the Sun was different from that of starlight. The former he supposed to extend to an altitude of 45°, and the latter to 20°.

Johannes Kepler (1571–1630), disagreeing with Tycho, stated that atmospheric refraction is the same for all celestial bodies at the same altitude. He also disagreed with the view that refraction is zero for altitude 45°.

The discovery, in 1621, of the true law of refraction is due to the Dutch physicist Willebrord Snell (1591–1626). Snell's law asserts that when light passes from one medium to another the planes of the angles of incidence and refraction and the perpendicular or 'normal' to the common surface of the two media are coincident, and that the sines of the angles of incidence and refraction are in a constant ratio for any two media, this being the *refractive index* for the two media.

The law of refraction was published in 1637, after Snell's death in 1626, by René Descartes, the French philosopher, who may have made an independent discovery of the same law. The French mathematician Fermat (1601–1665) argued that the law of refraction conformed with the idea that the path of light refracted at the common surface of two media was described in the least time. Snell's law implies, therefore, that the velocity of light in a medium is inversely proportional to the refractive index of the medium.

The famous Italian philosopher Dominic Cassini (1625–1712), like Kepler, showed the fallacy of Tycho's refraction doctrine. He proved that atmospheric refraction diminishes from a maximum at the horizon to zero at the zenith. The 18th-century French academician Abraham de la Hire produced a table of atmospheric refraction, and also confounded the idea that refraction of light from the Sun differed from that of light from a star.

The law of refraction propounded by Cassini was based on the hypothesis that the atmosphere is spherical and homogeneous. In the simplest investigation of atmospheric refraction, the Earth is regarded as being flat and the atmosphere is considered to be composed of an infinite number of horizontal parallel layers of air the density of which decreases uniformly with height above the Earth's surface. On this assumption it is readily proved that the effect of atmospheric refraction is the same as if light entering the atmosphere were refracted directly into the lowest layer of the air without traversing the intervening layers.

From Snell's law, a ray of light passes through the atmosphere such that $\mu \sin Z$ is constant for every point in its path; μ being the refractive index at any point, and Z the angle the path makes with the vertical. If Z_0 be the value of Z when the ray enters the atmosphere then, since *in vacuo* the refractive index of light is unity:

THE ALTITUDE CORRECTIONS

$$\mu \sin Z = \sin Z_0$$

If μ and Z are now taken as referring to the position of an observer's eye; and if r is the atmospheric refraction, then:

$$Z_0 = Z + r$$

Hence:

$$\mu \sin Z = \sin (Z + r)$$

i.e.

$$\mu \sin Z = \sin Z \cos r + \cos Z \sin r$$

Since r is a small angle (never more than about $\frac{1}{2}°$)

$$\cos r \simeq 1$$

and

$$\sin r \simeq r \text{ radians}$$

We may, therefore, write:

$$\mu \sin Z = \sin Z + r \cos Z$$

From which:

$$r = (\mu - 1) \tan Z$$

i.e.

$$r = U \cdot \tan Z$$

where $U = (\mu - 1)$ or coefficient of refraction.

This result holds good for small zenith distances; but for small altitudes, by treating sin r and cos r as r radians and 1 respectively, significant error results. Moreover, light from celestial objects at small altitudes has to travel through a considerable length of atmosphere, and we are not justified, therefore, in regarding the layers of air as being bounded by horizontal parallel planes. Cassini recognized this and, accordingly, took into account the Earth's spherical form.

Cassini's formula for atmospheric refraction is explained with reference to Fig. 1.

Fig. 1 represents part of a vertical section through the Earth's centre C and an observer O. XO_1O represents a ray of light from a celestial object X entering an observer's eye at O.

Cassini's hypothesis is that the light undergoes a single refraction on entering the atmosphere at O_1.

Let the apparent zenith distance of the celestial body be θ; and the true zenith distance θ_1. Let the refraction be r radians.

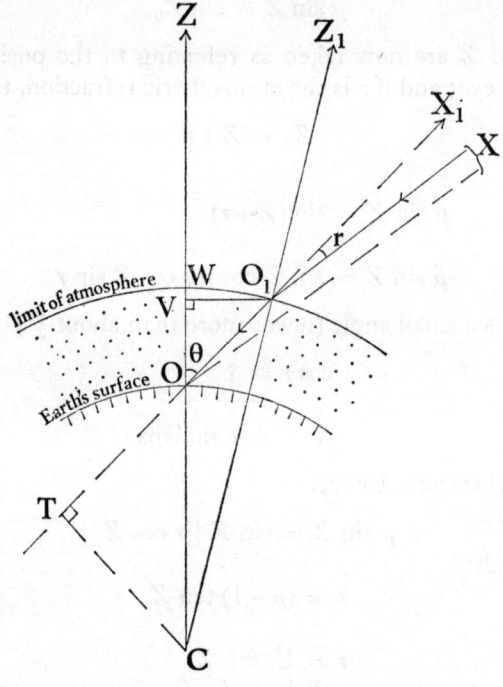

FIGURE 1

If r is small,
$$r = (\mu - 1) \tan \theta_1$$

Cassini expressed $\tan \theta_1$ in terms of $\tan \theta$. This he did by first drawing CT perpendicular to O_1O produced; and O_1V perpendicular to COZ. Then:

$$O_1T \tan \theta_1 = OT \tan \theta$$

i.e.
$$\frac{\tan \theta}{\tan \theta_1} = \frac{O_1T}{OT}$$
$$= 1 + \frac{OO_1}{OT}$$
$$= 1 + \frac{OV \sec \theta}{OC \cos \theta}$$
$$= 1 + \frac{OV}{OC} \sec^2 \theta$$

THE ALTITUDE CORRECTIONS

Now OV is approximately the vertical height of the atmosphere OW, and is, therefore, $x \cdot \text{OC}$, where x is the ratio between the height of the homogeneous atmosphere and the Earth's radius. Therefore:

$$\frac{\tan \theta}{\tan \theta_1} = 1 + x \cdot \sec^2 \theta$$

or

$$\tan \theta_1 = \frac{\tan \theta}{1 + x \cdot \sec^2 \theta}$$

Expanding the denominator $(1 + x \cdot \sec^2 \theta)^{-1}$ by the Binomial Theorem, we have:

$$\tan \theta_1 = \tan \theta (1 - x \cdot \sec^2 \theta + x^2 \cdot \sec^4 \theta \ldots)$$

Since x is a small quantity, powers of x greater than 1 may be ignored without introducing material error.
Thus:

$$\tan \theta_1 = \tan \theta (1 - x \cdot \sec^2 \theta)$$

and

$$r = (\mu - 1) \tan \theta (1 - x \cdot \sec^2 \theta)$$

which is Cassini's formula.

If the value of x is accurately chosen, Cassini's formula gives good results for altitudes not less than about 10°.

Newton, Hooke, Grimaldi, Roemer and Huyghens were the more notable of the 17th-century philosophers who devoted considerable attention to the nature and behaviour of light. To Newton we owe the theory of the spectrum, and to Hooke and Grimaldi the discovery of the phenomenon known as diffraction. The Danish physicist Roemer is credited with being the first to measure the speed of light by comparing the computed and observed times of the eclipses of Jupiter's satellites, and the Dutch physicist Huyghens is credited with being the founder of the wave theory of light.

The table of refractions given in the early editions of the *Connoissance des Temps*, which was first published under royal patents in 1678, was compiled by Jean Picard. The values given in Picard's table agree closely with the tangent law of refraction $r = U \tan$ zenith distance.

One of the best 18th-century tables of refraction is that of de la Caille (1713–1762). De la Caille's table is based on observations of circumpolar stars made at Paris and the Cape of Good Hope. He recognized that atmospheric refraction varies with air pressure and temperature, both these properties affecting the air density and, therefore, the refractive index. De la Caille's table gave mean refractions computed for a standard atmosphere having a specified pressure and temperature at sea level.

Tables of astronomical refraction were compiled by many astronomers during the 18th century, but perhaps the table that

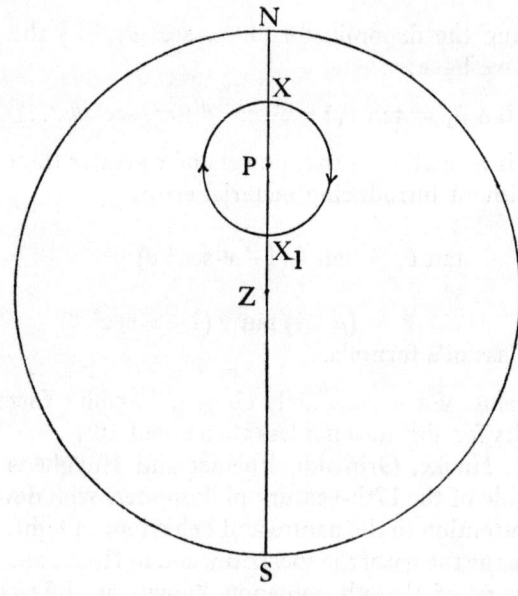

FIGURE 2

was esteemed the best was that of James Bradley (1693–1762). Doctor Bradley's mean refraction table applied to a standard atmosphere having a sea-level pressure and temperature of 29·6 inches of mercury and 50° F. respectively. Bradley furnished an auxiliary table for correcting the mean refraction for use when the atmospheric conditions differed from those for which the mean refractions were tabulated.

It will be of interest to discuss the methods by which atmospheric refraction may be ascertained. The usual method involved the observations of circumpolar stars, and is explained with reference to Fig. 2.

THE ALTITUDE CORRECTIONS

Fig. 2 represents the projection of the celestial sphere on to the plane of the horizon of an observer whose zenith is projected at Z. N and S are the projections of the north and south points of the observer's horizon respectively, and P is that of the celestial pole. X and X_1 are the projections of a circumpolar star at lower and upper meridian passage respectively.

Let z and z_1 be the apparent zenith distances of the star when at lower and upper transit respectively. Let p be the polar distance of the star and the coefficient of refraction $(\mu - 1) = U$. Then:

$$PZ = ZX - PX$$

i.e.
$$PZ = z + U \cdot \tan z - p \quad \ldots \quad (I)$$

Similarly:
$$PZ = ZX_1 + PX_1$$

i.e.
$$PZ = z_1 + U \cdot \tan z_1 + p \quad \ldots \quad (II)$$

Adding (I) and (II):
$$2 \cdot PZ = z + z_1 + U(\tan z + \tan z_1) \quad \ldots \quad (III)$$

In a like manner, if $\bar{z}$ and $\bar{z}_1$ are the apparent zenith distances of another circumpolar star whose declination differs materially from that of the first, we have:

$$2 \cdot PZ = \bar{z} + \bar{z}_1 + U(\tan \bar{z} + \tan \bar{z}_1) \quad \ldots \quad (IV)$$

From (III) and (IV) we have:
$$z + z_1 + U(\tan z + \tan z_1) = \bar{z} + \bar{z}_1 + U(\tan \bar{z} + \tan \bar{z}_1)$$

From which:
$$U = \frac{(z + z_1) \sim (\bar{z} + \bar{z}_1)}{(\tan \bar{z} + \tan \bar{z}_1) \sim (\tan z + \tan z_1)}$$

By repeated observations of circumpolar stars, James Bradley found the value of U to be 57·54 seconds of arc.

Bradley, who was elected a Fellow of the Royal Society in 1718, was appointed Savilian Professor of Astronomy at Oxford in 1721. He made many important contributions to the science of astronomy. He devoted considerable attention to the investigation of stellar parallax.

The philosopher Robert Hooke, who had invented the zenith sector for the purpose of measuring or detecting stellar parallax, had attempted to detect the *annual parallax*.* of the star λ *Draconis*. This star transits in London near the zenith, so that its meridian altitude is not affected by atmospheric refraction. The effect of annual parallax on the apparent position of λ *Draconis* when at meridian passage should be an annual fluctuation in its meridian altitude about a mean value, the maximum departure from the mean value occurring in December and June. Hooke, in his investigations, discovered a fluctuation; but it was regarded as being due to instrumental error and/or faulty observation, rather than as proof of the star's annual parallax. Half a century after Hooke had made his attempt Bradley, and his collaborator Samuel Molyneux, set about tackling the same problem. Their observations showed incontestably that the meridian altitude of λ *Draconis* did fluctuate with an annual period, but the maximum departure from the mean value occurred, not in December and June, but in September and March. In 1728 Bradley, who was then Astronomer Royal, demonstrated that the phenomenon that he had discovered was due to the fact that the speed of light from the star bears a finite ratio to the speed of the Earth in her orbit. When approaching the star its meridian altitude, therefore, is greater than the mean value for the year; and when receding from it, it is less than the mean. This phenomenon, often regarded as being Bradley's greatest discovery, is known as *aberration of light*. Another of Doctor Bradley's great discoveries is the *nutation* of the Earth's axis—a movement similar to the *precession*† of the Earth's axis, but due to the effect of the Moon instead of the Sun.

Bradley's work on atmospheric refraction is often linked with the later work, in the same field, of Friedrich Bessel.

Bessel (1784–1846) started work at an early age in the old Hansa city of Bremen, where he developed a desire to sail abroad as a supercargo. The fulfilment of his desire led him to a study of navigation, mathematics and astronomy, branches of science in which he became so renowned that, in 1818, he was appointed Director of the new Königsberg observatory. Bessel was instrumental in reducing the valuable, but neglected, astronomical observations of Bradley; and it is to Bessel that the discovery of stellar parallax is due.

* That is, the angle at a star contained between diametrically opposite points of the Earth's orbit.
† See Chapter v, p. 123, for a brief discussion on precession.

THE ALTITUDE CORRECTIONS

A formula for astronomical refraction, more exact than the simple tangent law, was formulated by Bradley. The rule, often stated in navigation and astronomy books of the late 18th century, is:

> 'The refraction at any altitude is to 57 seconds, in the direct ratio of the tangent of the apparent zenith distance lessened by three times the estimated refraction, to the radius.'

Expressed in the common mathematical way, this rule is:

$$r_c = 57'' \cdot \tan(z - 3 \cdot r_\theta)$$

In applying this rule, if the calculated refraction r_c differs materially from the estimated refraction r_θ, it is necessary to repeat the calculation using r_c in place of r_θ.

An interesting method of measuring refraction, due to Bradley, involves observing the Sun's apparent zenith distance at noon on the days of the solstices. If z_s and z_w are the summer and winter solstitial zenith distances of the Sun at noon respectively, then the true zenith distances are respectively:

$$z_s + (\mu - 1) \tan z_s$$

and

$$z_w + (\mu - 1) \tan z_w$$

Now the declination of the Sun, when he is at a solstitial point, is equal to the obliquity of the ecliptic ϵ. Therefore, the observations being made north of northern tropic, in latitude λ:

$$z_s + (\mu - 1) \tan z_s = \lambda - \epsilon$$

and

$$z_w + (\mu - 1) \tan z_w = \lambda + \epsilon$$

From which:

$$2\lambda = z_s + z_w + (\mu - 1)(\tan z_s + \tan z_w) \quad \ldots \quad \text{(V)}$$

and

$$(\mu - 1) = \frac{2\lambda - (z_s + z_w)}{\tan z_s + \tan z_w}$$

Alternatively, λ may be eliminated from formula (V) by combining it with formula (III) or (IV), so that $(\mu - 1)$ may be found from combined observations of the Sun at each of the solstices and the zenith distances at upper and lower transit of one circumpolar star. Eliminating λ in this way, using formulae (III) and (IV), we have:

$$(\mu - 1) = \frac{z + z_1 - z_s - z_w}{\tan z_s + \tan z_w - \tan z - \tan z_1}$$

Experiments made in the 18th century—notably by the instrument-maker Francis Hauksbee who flourished during the first decade of the century—showed that atmospheric refraction is proportional to the density of the air. The density of the air varies directly as its pressure and inversely as its heat. Since the pressure and heat of the air are shown by barometer and thermometer respectively, it follows that the mean refraction may be reduced to the actual refraction by allowing for the difference between the actual and mean pressures and temperatures of the air.

The difference of refractions arising from a given difference of temperature may be ascertained by observation. De la Caille made the change of refraction corresponding to a change of 10° on Reamur's thermometer to be 1/27 of the whole. Mayer made the change 1/22 of the whole. According to Bradley:

'The true refraction is to the mean refraction in a direct ratio of the altitude of the barometer to 29·6", and in an inverse ratio of the altitude of the thermometer increased by 350 to the number 400.'

In other words:

$$\text{true refraction} = \text{mean refraction} \cdot \frac{H}{29 \cdot 6} \cdot \frac{400}{(T - 350)}$$

where H is the height of mercury in inches and T is the temperature on Fahrenheit's scale. Bradley's auxiliary table to the table of mean refraction was based on this rule.

Andrew Mackay, in his *Theory of the Longitude* of 1793, explained the physical cause of refraction in a curious, but interesting way:

'It is demonstrable,' he wrote, 'that every body is endowed with an attractive power, which reaches to some distance beyond its surface, as that of cohesion, magnetism, etc. Now a ray of light from a heavenly body will, at its entrance into the terres-

THE ALTITUDE CORRECTIONS

trial atmosphere, be attracted towards the denser parts: and since the density of the atmosphere increases the nearer the Earth's surface, therefore the ray, as it approaches the observer, will be more and more attracted, its velocity accelerated, and of course its rectilineal direction changed. Hence that portion of the ray contained between an observer and the extremity of the atmosphere will be a curve, except in that case when the ray is perpendicular to the refracting medium.'

Let us now turn our attention to atmospheric refraction as a correction to a measured or observed altitude of a celestial body.

It must be appreciated that with the earliest instruments employed by the seaman for measuring altitudes—instruments such as the seaman's quadrant, astrolabe and cross-staff—it was not possible to obtain altitudes to anything but a coarse degree of accuracy. In those circumstances, therefore, the application of altitude corrections amounting to a few minutes of arc to crude observed altitudes measured perhaps to the nearest degree of arc would have been unworthy of consideration.

Edward Wright, in his *Certaine Errors of Navigation*, first published in 1599, brought to the notice of seamen the effect of refraction on altitudes of celestial bodies. He explained the cause of refraction in the same way as did Tycho Brahe; and, in fact, the tables of refraction which he included in his famous work were based on Tycho's observations. Like Tycho, he believed refraction to be different for the Sun than for the stars; and his table of refraction for the Sun extended from 0° to 45° altitude, and that for the stars from 0° to 20°.

Thomas Hariot (1560–1621), the brilliant Elizabethan mathematician who devoted considerable attention to navigational matters, and who was adviser to Sir Walter Raleigh on mathematics and navigation, was of the opinion that atmospheric refraction was trivial, and need not, therefore, be considered by seamen.

An early, and interesting, observation made at sea with the specific object of finding refraction, was made by William Baffin off the west coast of Spitsbergen in 1614. The account of the voyage, during which this observation was made, appears in *Purchas, His Pilgrims*. It was the type of observation in which the famous explorer showed his acute inventive genius. His observations of refraction were based on exact knowledge of the latitude of the place of observation, the declination of the Sun, and the angular diameter of the Sun. Baffin observed the Sun at lower transit at a time at which he estimated that a certain fraction of the Sun's diameter was above the horizon. Knowing his latitude and the declination of the Sun, he calculated the proportion of the Sun's diameter that should be above the horizon. From his results he demonstrated that the angle of atmospheric refraction applicable to his observation was 26'.

The table of refractions given in Maskelyne's *Requisite Tables*, the first edition of which was published in 1781, was based on Bradley's law. This table was reproduced in most of the navigation manuals of the late 18th and 19th centuries—in particular, in those of John Hamilton Moore, J. W. Norie, Mrs Janet Taylor, Andrew Mackay, Edward Riddle and John Riddle.

From about 1825, refraction tables in some navigation manuals were based on a refraction formula due to James Ivory of double altitude fame. An able investigation into the subject of refraction was made by Ivory, and a comprehensive account of the work was printed in the *Philosophical Transactions* for 1823. Ivory's mean refraction table was based on a standard atmosphere having a sea level pressure and temperature of 30 inches of mercury and 50° F. respectively. The celebrated Henry Raper, as well as James Inman, used Ivory's refraction rules in their collections of navigational tables.

It may be remarked that the actual refraction of light from a heavenly body whose altitude exceeds about 10°, is never more than about half a minute of arc different from the mean refraction. It is for this reason that the table of corrections to mean refraction is seldom used by seamen for correcting altitudes. In bygone days however, although the table was of little consequence when finding latitude, it demanded the attention of the seaman who would find longitude by the method of lunar distance.

Refraction of light from heavenly bodies within a few degrees of the horizon can never be known with exactitude. Neither the most refined mathematical investigation nor the most careful observations can remove the uncertainty of refraction at small altitudes. Temperature changes—and therefore density changes—of the air along the line followed by a ray of light from an object near the celestial horizon, are almost always taking place. These changes can never be known with certainty, and no refraction law has yet been formulated which will hold good at all times for altitudes less than about 5°. Testimony to this fact is provided by the frequent investigations made in recent times into the question of refraction at small altitudes.

3. DEPRESSION OR DIP OF THE SEA HORIZON

The *depression*, or *dip*, of the sea horizon is a measure of the angle contained between the plane of the horizontal surface on which rests an observer's eye, and the direction of the visible- or sea-horizon, in the vertical plane through the observer's line of sight. The angle of dip is clearly a function of the elevation of the observer's eye: the greater is the height of eye above sea level, the greater is the angle of dip. Moreover, the greater is the height of eye of an observer the greater is the range of his sea horizon. These facts have been recognized since very early times; and they provided simple and compelling evidence of the Earth's rotundity. It is recorded that the notable Pythagoras, who flourished during the 5th century BC, gave as clear proof of the spherical shape of the Earth the changing appearance of a ship as she heaves into the sight of, or sails away from, an observer standing on the shore.

During the 16th century, when the first of the 'modern' attempts at determining the size of the Earth were made, one method suggested, and in fact employed by the Elizabethan mathematician Edward Wright, was related to the height of an observer's eye above sea level and the corresponding range of his sea horizon. The uncertain effect of terrestrial refraction, that is the bending of light in its passage from a point on the visible horizon to an observer's eye, rendered the method unreliable and inaccurate. The principle of the method is described with reference to Fig. 3.

Fig. 3 illustrates a section of the Earth in the plane of which lies the Earth's centre at C and the observer's eye at O. A is a point vertically below O at sea level; and AB is the Earth's diameter. The observer sighting a point on his theoretical horizon lying in the plane of the section would be sighting along the straight line OH, which is a tangent to the circle ABD at D.

The observer's theoretical horizon, the radius of which is AD, is a small circle which limits the observer's view, assuming that terrestrial refraction is non-existent.

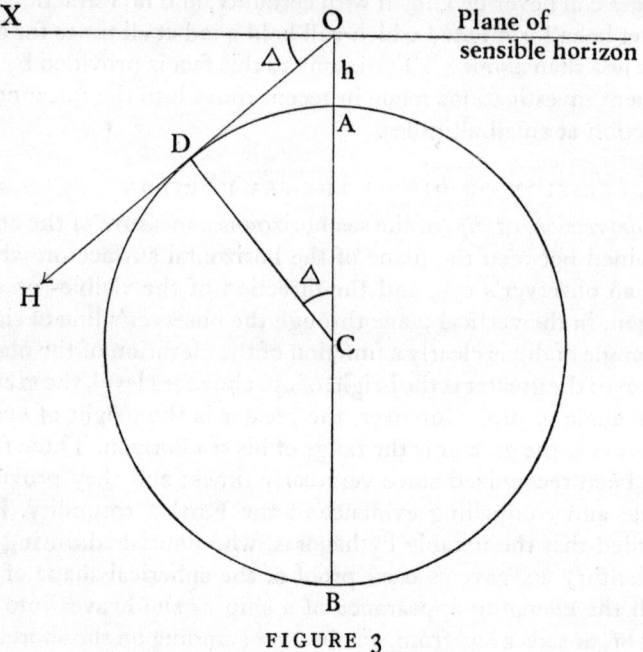

FIGURE 3

Because the height of the observer's eye—h units—above sea level is small compared with the distance of the theoretical horizon AD, it may be assumed that tangent OD is equal to arc AD. From a well-known geometrical theorem:

$$OD^2 = OA \cdot OB$$

i.e.

$$AD^2 = OA \cdot OB$$

and

$$AD = \sqrt{OA \cdot OB}$$

or

distance of theoretical horizon $= \sqrt{2 \cdot R \cdot h}$ (OB = 2R approx.)

It follows, therefore, that if the Earth's diameter and the height of the observer's eye above sea level are known, AD, the distance of the observer's theoretical horizon, may be computed. Conversely, if the distance of the theoretical horizon and the observer's height above sea level are known, the Earth's radius may be computed.

THE ALTITUDE CORRECTIONS

It will be noticed from Fig. 3 that the dip of the theoretical horizon, denoted by Δ, is equal to the angle at the Earth's centre contained between radii terminating at D and A respectively. We have, therefore:

$$\cos \Delta = \frac{R}{R+h}$$

Since the angle of dip is a small angle:

$$1 - \frac{\Delta^2}{2} = 1 - \frac{h}{R}$$

where Δ is expressed in circular measure, and

$$\Delta = \sqrt{\frac{2h}{R}}$$

The effect of atmospheric refraction is for light coming from the actual horizon, the visible- or sea-horizon as it is called, to follow a path concave to the Earth's surface as illustrated in Fig. 4.

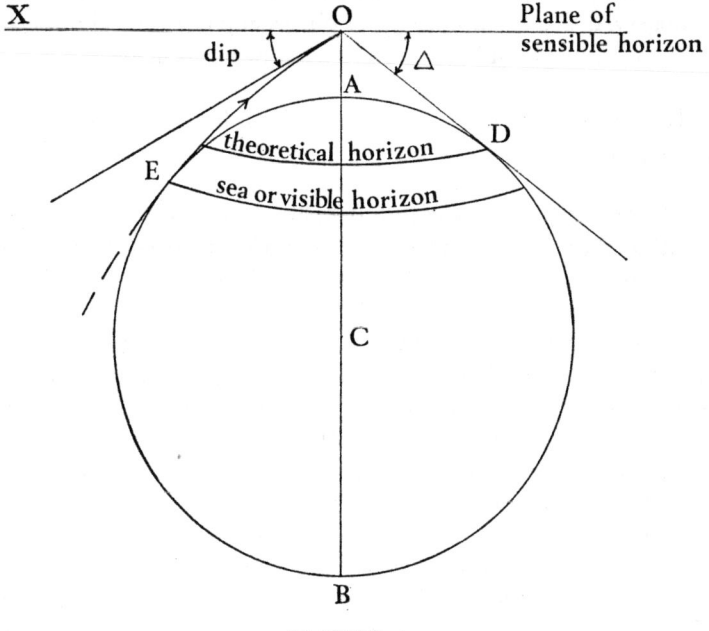

FIGURE 4

Refraction, therefore, causes the sea horizon to have a greater range than that of the theoretical horizon. It also causes the angle of dip to be smaller than that of the theoretical dip.

The effect of terrestrial refraction on dip and distance of the sea horizon received the attention of many 18th-century physicists and astronomers, but there was never general agreement as to the exact effect of refraction. Dr Nevil Maskelyne, under whose direction the first British *Nautical Almanac* was published in 1765, stated that 1/10 of the theoretical dip should be subtracted from the theoretical dip to give the true dip. Other investigators gave fractions generally between 1/9 and 1/15; but according to General Roy (*Philosophical Transactions*, 1790) it varies from 1/3 to 1/24 of the 'comprehended arc'.

Andrew Mackay, in his *Theory of the Longitude*, points out that because the Earth is an oblate spheroid the radius of curvature is variable with the latitude. It follows, therefore, that no single table of dip can answer in all places. Mackay wrote:

'Tables of dip should be calculated for the latitude of the place and the azimuth of the observed object. It however may be observed that the difference of dip arising from the above cause is so inconsiderable as to have been hitherto neglected.'

In the same book, Mackay gave the following rule for computing dip:

'To the constant logarithm 0·4236 add the proportional logarithm of the height of the eye above sea level, in feet; half the sum will be the proportional logarithm of the dip of the horizon.'

When observing the altitude of a heavenly body at sea the measured angle is that contained between the apparent direction of the body and the direction of the point on the visible horizon vertically below the body. This latter point, as we have seen, does not coincide with a point on the true horizon, falling, as it does, below it by an amount which is dependent upon the height of the observer's eye above the sea surface. The angle of depression of the sea horizon must, therefore, be subtracted from the measured altitude to give what is called the *apparent altitude*. The apparent altitude of a heavenly body is defined as the arc of a vertical circle

THE ALTITUDE CORRECTIONS 115

contained between the apparent direction of the observed body and the true or sensible horizon.

The dip varies as the square root of the height of eye of the observer. For a height of eye of 50 feet above sea level, the dip is about 7' of arc: for a 100 feet it is about 10' of arc. For the same reason as the atmospheric refraction correction was not heeded by the early mariners who used relatively crude instruments for their observations, the dip also was ignored. Both refraction and dip corrections are subtractive; but even though they are to be applied in the same direction as each other, the combined correction of dip and refraction for an observation taken from a position on the ship often less than about 15 feet from the sea surface was never more than about 10' of arc for altitudes greater than about 10°.

It appears that the first dip table designed and printed for the use of the seaman was that given by Edward Wright in his remarkable *Certaine Errors in Navigation, Detected and Corrected*, the first edition of which is dated 1599. Wright's table extended from 5 feet to 90 feet height of eye above the sea surface with corresponding dip corrections of 2' and 11' of arc respectively. These corrections compare favourably with those to be found in a modern dip table in which:

dip for 5 feet = 2·57'
dip for 90 feet = 10·91'

Thomas Hariot regarded dip as being sufficiently large to be worth considering for navigational purposes. His table of dip, although ante-dating Wright's table, was never published. Hariot's values for dip are consistently a little too large.

The dip table in Maskelyne's *Requisite Tables* allowed for the effect of terrestrial refraction based on Maskelyne's rule quoted above. Samples of dip taken from this table are:

Height of eye	Dip
10 feet	3·01'
50 feet	6·44'
100 feet	9·33'

Maskelyne's dip table, in which the refraction allowance is 1/10 of the theoretical dip, was given in many of the navigation manuals of the late 18th and early 19th centuries. These included those of Robertson (1788), Norie (1802), John Hamilton Moore (1780), Andrew Mackay (1796) and Janet Taylor (1830).

Edward Riddle's dip table which appears in his *Treatise on Navigation* (1828) is based on an allowance of 1/13 of theoretical dip, the factor 1/13 being that suggested by the French physicist Biot.

Raper, in his *Practice of Navigation* (1840), gave a dip table based on an allowance for refraction amounting to 1/14 of the theoretical dip.

Raper drew attention to the facts that the running of the sea in bad weather causes the sea horizon to be in continual vertical motion; and that the rising and falling of an observer due to rolling, pitching, and heaving of the ship, causes the dip to be in perpetual change. To overcome errors due to these causes the seaman was advised to make a series of observations, instead of a single sight, and to take an average of the results.

Raper also pointed out that the height of eye should be ascertained with some precision, that is to say, within two or three feet; because an error in dip causes a corresponding error of the same amount in altitude. This is of greatest importance when the observer's eye is near the sea surface, because the rate of change of dip with height above sea level is greatest when the height of eye is zero.

In general, the greater the height of the observer's eye the more distinct will be the sea horizon, provided that the air is clear. In misty weather, however, when celestial observations are possible, it is better to observe from a position as near to the sea surface as practicable so as to 'bring' the sea horizon as near as possible to the observer.

It is interesting to note that Raper gave a true dip table as well as a table of apparent dips. The true dip table gives the depression of the theoretical horizon, and it is based on the formula:

$$\text{True dip} = 1 \cdot 063 \sqrt{h}$$

where h is the observer's height of eye in feet.

A footnote to the explanation of the true dip table is interesting:

'As the lower latitudes are more frequented by shipping than the higher, 40° has been assumed as the average latitude. Also, as the curvature of the Earth is different on the prime vertical and on the meridian, the circle of curvature, crossing the meridian at 45° of azimuth has been employed. The depression is accordingly, computed to the radius 20,909,577 feet which gives the length of the average nautical mile 6082 feet nearly.'

When star observations became popular during the closing decades of the 19th century it became common to provide, in collections of nautical tables such as those of Norie, Inman, etc., a combined table of dip and refraction for star altitudes. This table is called the *Stars' Total Correction Table*, no altitude corrections, other than those for dip and refraction, being required for reducing a star's observed altitude to its true altitude.

4. THE SUN'S SEMI-DIAMETER

The correction called *Sun's semi-diameter* is applied to a measured altitude of the Sun, when his limb is observed. It is simply half the angular diameter of the Sun and is to be added to the altitude of the lower limb and subtracted from the altitude of the upper limb, in order to find the altitude of the Sun's centre.

The Sun's semi-diameter varies during the year, being least when the Earth is at aphelion, when its value is 15·8', and greatest when the Earth is at perihelion, when the value is 16·3'. The traditional value for the Sun's semi-diameter is 16', and this is the angle which the seaman was recommended to apply to the measured altitude of the Sun's limb. In fact, some Davis quadrants were graduated on the back edge of the smaller arch in such a manner as to eliminate the need for applying the semi-diameter correction arithmetically.

Before the cross-staff was introduced to the navigator, the only instruments used for measuring altitudes were the seaman's quadrant and astrolabe. When either of these were used for measuring the Sun's altitude, the angle given on the graduated arc—by the plumb-line in the case of the quadrant, and the spot of light in the case of the astrolabe—required no correction for semi-diameter, the measured angle being the altitude of the Sun's centre. Similarly, when using a Davis quadrant fitted with a glass vane instead of a shade vane the same applied.

It was not until the closing decades of the 17th century that observed altitudes of the Sun were corrected for refraction, dip and semi-diameter. Although the errors due to not applying altitude corrections were understood, it was realized that the degree of error associated with the measuring instruments used—especially the seaman's quadrant and astrolabe—was considerably coarser than the small angular values of the corrections. Following the introduction of Hadley's quadrant, the question of altitude corrections sprang to the fore, and altitude correction tables were provided, with corrections often to an unnecessarily high degree of accuracy, in most nautical tables.

The Sun's semi-diameter, and that of the Moon as well, was required when clearing a lunar distance between Sun and Moon; and for this purpose accurate values of semi-diameter were required, because a small error in the cleared distance results in a relatively large error in the calculated longitude.

From the early part of the 19th century, it became customary to provide in nautical table collections a table giving *Sun's total correction*. This gives the combined corrections for refraction, dip and semi-diameter, against observed altitude of Sun's lower limb and height of observer's eye. The table was constructed using 16′ as the semi-diameter correction, although in later tables an auxiliary table giving a monthly correction for the variation in Sun's semi-diameter was provided. In *Norie's Nautical Tables*, the first edition to include a Sun's total correction table was that of 1828. Robertson, in his *Elements of Navigation*, gave a combined table of Sun's semi-diameter, dip and refraction, but left the user to combine the separate corrections.

5. THE MOON'S SEMI-DIAMETER

The Moon's semi-diameter is sensibly affected by her altitude. This follows because the radius of the Earth is a significant proportion of the Moon's distance from the Earth. The Moon's semi-diameter is least when she is at apogee and has an altitude of 0°. It is greatest when she is at perigee and is in the zenith of an observer. The tabulated values of the Moon's semi-diameter are given for altitude 0°, so that a small correction, known as the *augmentation*, is to be applied to the tabulated value. The augmentation of the Moon's semi-diameter is given by the formula:

Augmentation = Moon's semi-diameter × sine apparent altitude

THE ALTITUDE CORRECTIONS

Although the maximum value of the augmentation is no more than about 0·3′ of arc, it formed an important correction when using the method of finding longitude by means of the lunar distance method.

6. PARALLAX

The point in the heavens which a celestial body occupies when viewed from the Earth's surface, is called the *apparent place* of the body. The point which it would occupy were it viewed from the

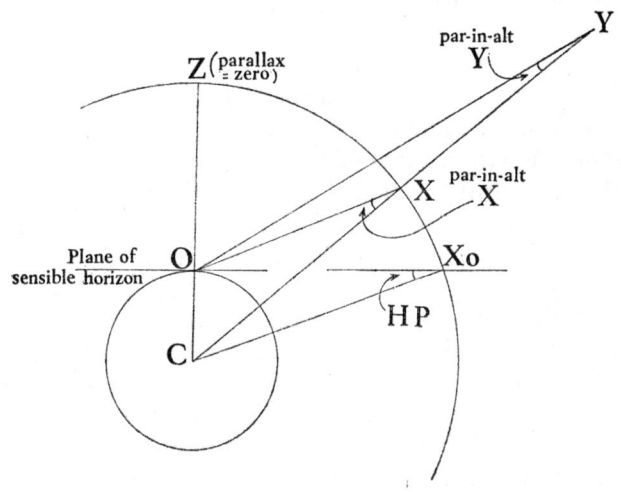

FIGURE 5

Earth's centre is called the *true place* of the body. The angular difference between the apparent and true places of a celestial body at any given instant is equivalent to the angle at the centre of the body contained between lines terminating respectively at the observer's position and the Earth's centre. This angle is called the body's *parallax-in-altitude*. Fig. 5 illustrates how parallax varies with altitude and distance.

It should be clear from Fig. 5 that parallax-in-altitude for any given body such as X, is greatest when the body lies on the horizon of the observer. This maximum value is called *horizontal parallax*. Parallax-in-altitude varies as the cosine of the altitude, being zero when the altitude is 90°. It should also be clear from Fig. 5 that parallax-in-altitude varies inversely as the distance of the body. That for body Y, which has the same apparent place as that of body X, is smaller than it is for X, the nearer body.

For the fixed stars, on account of their great distances compared with the Earth's radius, parallax-in-altitude is infinitesimally small, and is regarded as being zero. For the Sun the value of horizontal parallax varies during the year, being greatest when the distance between the Earth and Sun is least, but it is is never more than about 9" of arc. In practice it is generally ignored when correcting the Sun's altitude, although for clearing a Sun lunar distance Sun's parallax was considered to be a correction of some small importance.

The Moon's parallax-in-altitude is a correction which is of great importance. The horizontal parallax of the Moon is greatest when the Moon is in perigee and least when she is at apogee, the corresponding values being 62' and 53' respectively.

In the early British *Nautical Almanacs*, the Moon's horizontal parallax was given at intervals of twelve hours. The value of horizontal parallax given in the almanac is described as *equatorial horizontal parallax*, for it is the value applicable to an observer located on the equator. For an observer at any latitude other than the equator, the horizontal parallax of the Moon is less than the equatorial horizontal parallax. This is so on account of the spheroidal shape of the Earth. Horizontal parallax is the angle whose tangent is the ratio between the Earth's radius and the distance between the Earth and the centre of the observed object. The Earth's equatorial radius being the maximum results in the equatorial horizontal parallax being maximum. The tabulated value of the Moon's horizontal parallax must be corrected by an amount known as the reduction. The reduction of the Moon's horizontal parallax for the figure of the Earth* is found from a table included in most nautical collections. The reduction is never more than about 0·3' of arc.

Parallax is often regarded as being an error due to observing

---

* The term figure of the Earth is a mathematical expression for the ellipsoidal shape of the Earth. This is given in terms of a quantity called compression (c), where $c=(a-b)/a$ in which a and b are the Earth's equatorial and polar radii respectively. The compression is approximately 1/300, this small fraction indicating that, for most practical purposes, the Earth's shape may be taken to be that of a perfect sphere.

from the wrong position. The term *ocular parallax* applies to the error due to not placing the eye in the exact position when observing with a cross-staff. Ocular parallax is discussed in Chapter III.

As far back as the beginning of the 19th century correction tables were available in which refraction and Moon's parallax were combined. Mendoza del Rios included such a table in his nautical tables which were published in 1802.

Fig. 6 serves to demonstrate the relationship between Moon's horizontal parallax and Moon's semi-diameter.

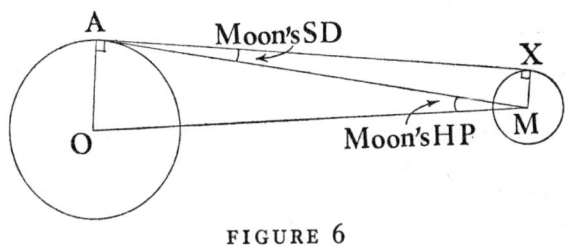

FIGURE 6

In Fig. 6, O and M represent the centres of the Earth and Moon respectively.

The Earth's radius = 4000 miles (approx.)
The Moon's radius = 1000 miles (approx.)

Because MOX and AMO are small angles:

$$\frac{\text{Moon's H.P.}}{\text{Moon's S.D.}} = \frac{AO}{MX} = \frac{1}{4} \text{ (approx.)}$$

Because the Moon's horizontal parallax and her semi-diameter always bear a constant ratio one with the other, it is easy to construct a Moon's altitude correction table in which refraction, semi-diameter and parallax are combined.

7. IRRADIATION

When a bright object, such as the Sun, Moon or star, is viewed against a darker background such as the sky, the bright object appears to be larger than it actually is. This phenomenon is known as *irradiation*. Moreover, the sea horizon, because the sky is generally brighter than the sea, appears to be depressed on account of irradiation. The resulting effect is that altitudes of the

Sun's lower limb are not materially affected by irradiation, because the apparent lowering of the lower limb tends to neutralize the apparent depression of the sea horizon. For upper limb observations, however, the two effects combine to produce an error, the magnitude of which depends upon the relative brightnesses of Sun and sky, and sky and sea. It is only within recent times that an irradiation correction has been made available to seamen.

8. PERSONAL ERROR OR EQUATION

The timing of an event such as the instant when a star, or the Sun's or Moon's limb makes exact contact with the horizon, is affected by the temperament and nervous and physical condition of the observer. Any error due to this cause is called *personal error* or *equation*.

Personal error varies not only between observers, but may vary for different observations made by the same observer.

At the present time little attention is given by practical navigators to the question of personal error. During the last century, however, when great accuracy of observed lunar distances was the aim of all keen navigators, personal equation in nautical astronomy was regarded as a matter of great moment.

CHAPTER V

Methods of finding latitude

1. INTRODUCTORY

Because the Earth spins she possesses, in common with all spinning bodies, the property of gyroscopic inertia. The *gyroscopic inertia* of a spinning body is related to the tendency it has to maintain its plane and axis of spin.

Because of the Earth's gyroscopic inertia, the *celestial poles*—which are the projections of the Earth's poles from the Earth's centre on to the celestial sphere—tend to be fixed points on the celestial sphere. The celestial poles do not, in fact, remain fixed, because the rotating Earth is acted upon by a resultant external couple due to forces between the Sun and Moon and the Earth. This couple results in a phenomenon known as the *precession of the equinoxes*. The effect of precession is for each of the celestial poles to describe an approximately circular path centred at one of the extremities of the axis of the ecliptic.* The radius of each of these circular paths is equal to the maximum declination of the Sun: that is, about $23\frac{1}{2}°$. The movement of the celestial poles around the poles of the ecliptic results in the declination and Right Ascension of every fixed celestial point changing with time.

The precession of the equinoxes, which was discovered by the renowned Hipparchus, is an extremely slow motion amounting to about $50''$ per year. It takes about 26,000 years for the equinoxes to swing through $360°$ of the ecliptic. It is because of the slowness of the precession of the equinoxes that, for most practical purposes, the celestial poles are regarded as being fixed points in space.

The *latitude* of a place is its angular distance north or south of the equator. It is equivalent to an arc of the meridian between the equator and the parallel of latitude through the place. All places which have the same latitude lie on a small circle which is parallel

* The ecliptic is the celestial great circle, co-planar with the Earth's orbit, traced out by the Sun during his apparent annual orbit in the celestial sphere.

124   A HISTORY OF NAUTICAL ASTRONOMY

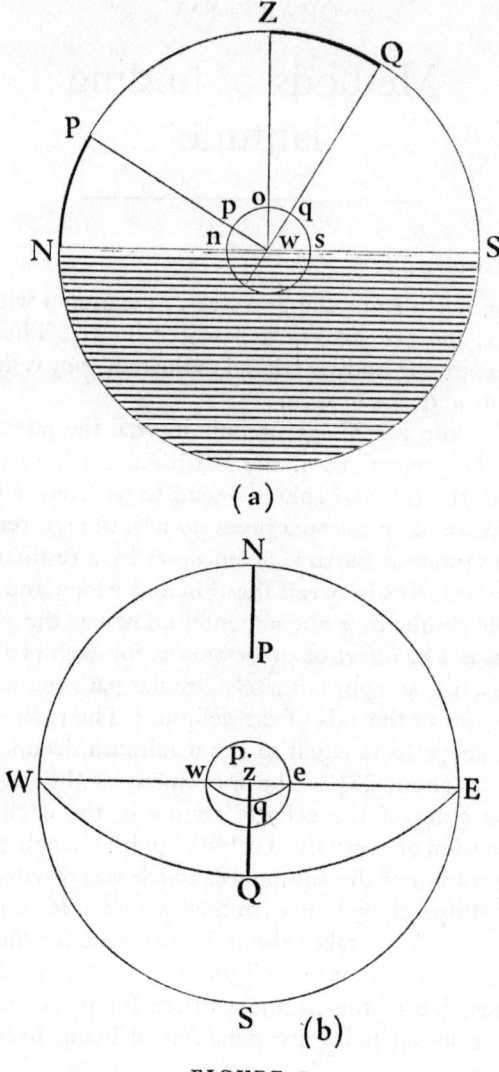

FIGURE 1

to the plane of the equator. These circles are known as *parallels of latitude*.

In sketch (a) of Fig. 1, the celestial sphere is projected on to the plane of the celestial meridian of an observer, the larger circle representing the observer's celestial meridian.

METHODS OF FINDING LATITUDE 125

In sketch (b), the celestial sphere is projected on to the plane of the celestial horizon of the observer. The larger circle, in this sketch, represents the celestial horizon. The smaller circle in both sketches represents the Earth.

o represents the observer and Z his zenith; p represents the Earth's pole and P the celestial pole; N, E, S and W are the projections of the cardinal points of the horizon; wqe represents the equator; and WQE is the projection of the equinoctial.

Observer's latitude = arc qo = arc QZ
Altitude of celestial pole = arc NP

Now

arc PQ = arc NZ = 90°

and

arc QZ = (90 − PZ)

Also,

arc NP = (90—PZ)

Therefore:

arc QZ = arc NP

or

Latitude of observer = altitude of celestial pole

During the course of a day the slow spin of the Earth towards the east results in an apparent revolution of the celestial sphere towards the west. As a consequence of this, the celestial objects tend to describe circular paths which are parallel to the equinoctial, and which are centred on the axis of the equinoctial. These circular paths are known as *diurnal circles*.

The nearer is a star, or other celestial object, to either celestial pole, that is to say, the greater is the declination of a celestial object, the greater is the proportion of its diurnal circle above the celestial horizon of an observer whose zenith lies in the same celestial hemisphere as that in which the object is located. At any place in the northern hemisphere celestial bodies which have north declination are above the horizon for more than twelve hours each day. Celestial objects which have south declination are, correspondingly, above the horizon for less than twelve hours each day. At any place on the equator, all celestial objects, regardless of the name or magnitude of their declinations, are above and below the horizon for exactly twelve hours each day.

A celestial object whose diurnal circle lies wholly above the celestial horizon of an observer is said to be *circumpolar* for the observer's latitude. It should be evident that at the equator no celestial object is circumpolar. Likewise, at the North Pole, all celestial bodies which have north declination, and at the South Pole all celestial bodies which have south declination, are circumpolar.

For a celestial body to be circumpolar at the place of an observer, the *polar distance*, that is the complement of the body's declination, will have to be less than the observer's latitude. This fact should be clear from Fig. 2.

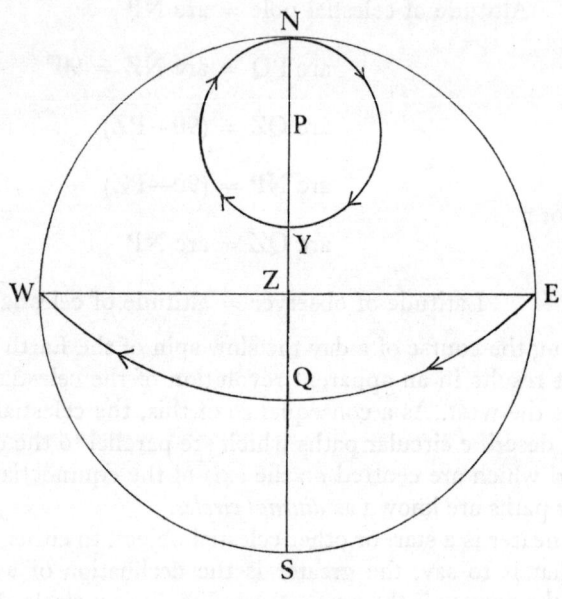

FIGURE 2

Fig. 2 is a projection of the celestial sphere on to the celestial horizon of an observer in the northern hemisphere, whose zenith is projected at Z. The observer's latitude is equal to arc NP. All celestial objects whose north declinations exceed arc QY lie within the parallel of declination (or diurnal circle) centred at P and having a radius equal to arc PY.

Now,
$$\text{arc PY} = 90 - \text{arc QY}$$
and
$$\text{arc PY} = \text{arc NP}$$

METHODS OF FINDING LATITUDE

Therefore, the limiting declination for an object to be circumpolar in the latitude of the observer is arc QY. In other words, for an object to be circumpolar its polar distance (complement of declination) must be less than the latitude of the observer, and the names of the declination and the latitude must be the same.

An important adjunct to the needs of the ocean navigator is a chart on which his ports of departure and destination are plotted, and on which the position of his ship, at any time during the voyage, may be plotted so that her progress may be checked and her course, if necessary, rectified.

The earliest ocean voyagers relied on knowledge of latitude and estimation of courses and distances made good in order to deduce the ship's longitude. This method of navigation by *dead reckoning* (D.R.), or by 'guesstimation' as it has so aptly been described, gave unreliable results because of the difficulty of making accurate estimations of courses and distances made good.

The early navigators, and the ancient geographers as well, were able to find the latitude of a place from astronomical observations; but the practical determination of longitude, apart from that by the method of D.R. navigation, was to remain a mystery until comparatively recent times.

The chart of the early navigator, on which he plotted his D.R. positions and observed latitudes was, like all maps, a representation of part of the spherical Earth's surface on a plane surface, requiring the use of a projection. The earliest forms of map projection, which were geometrical and simple in principle, date from the 4th century BC.

During the 4th century BC the activities of land- and sea-traders led to a great expansion of knowledge of the Earth's surface and of the distribution and positions of the major population centres and trading stations. We have, in Chapter I, referred to Eratosthenes, and noted his attempt to reduce the problem of terrestrial position-fixing to a regular system employing a network of parallels of latitude and meridians. The principal parallel of latitude used by Eratosthenes—a parallel which passes through Gibraltar and Rhodes—was first suggested by Dicearchus, who flourished during the closing decades of the 4th century BC. Eratosthenes established a meridian line on his world map, this passing through Rhodes and Syene.

The terms *latitude* and *longitude*, which mean respectively breadth and length, sprang from the notion that the habitable part of the Earth is broader in the east–west direction than it is in the north–south direction. This is certainly true of the Eastern Mediterranean region in which the ancient Greek and earlier civilized societies flourished.

The latitude of a place may be ascertained without reference to the latitude of any other place. In order to find the latitude of a place in this way, recourse must be made to astronomical observation.

When it became known that the Earth's shape is spherical, and that the Sun's annual path across the celestial concave is inclined at an angle of approximately $23\frac{1}{2}°$ to the plane of the Earth's rotation, it became possible to compare the latitudes of places from the lengths of the midday shadows cast by gnomons set up at the places.

Essentially, a gnomon provides the means whereby the passage of time during sunlight hours may be measured. Because of the changing declination of the Sun, due to the obliquity of the ecliptic, the gnomon may also be used for measuring the march of the seasons.

Pytheas of Marsala, who flourished during the 4th century BC, bestowed considerable attention upon the measurement of latitude by means of the gnomon. He found the latitude of his native city with great accuracy; and during his famous Atlantic voyage the most northerly land reached, according to Pytheas, was located on what we would describe as the parallel of latitude of $66\frac{1}{2}°$ N. or the Arctic Circle.

The term *Arctic Circle* has not always had its present-day meaning. To the ancient astronomers and geographers the term was used to describe a small circle on the celestial sphere which enclosed all the circumpolar stars for any particular latitude.

To the ancient Greeks the word *climata* was used to describe zones bounded by parallels of latitude.

It appears that Hipparchus was the first to suggest that parallels of latitude on world maps should be projected at regular intervals. On the map of Eratosthenes' the parallels of latitude were not plotted systematically and were placed at irregular intervals. Hipparchus introduced the idea of fixing the interval between successive plotted parallels of latitude with reference to the longest —or *solstitial*—day.

METHODS OF FINDING LATITUDE 129

Although Hipparchus referred to the latitudinal zones as *climata*, later writers, including Strabo, referred to the boundaries of the zones as *climata*. On this basis *climata* are parallels of latitude.

The Earth's surface was divided by ancient geographers into twenty-four climatic zones in each of the northern and southern hemispheres. The lengths of the solstitial days on the two boundary parallels of each climatic zone differed by half an hour. The ancient geographers knew no more than nine climatic zones, all of which are located on the northern side of the equator. These nine climatic zones, or climates, were named after the principal cities situated within them. The third climate from the equator, for example, was named after Alexandria, the metropolitan city of Egypt; the fourth was named after Rhodes; and the fifth after Rome; and so on.

We have shown that the latitude of an observer is equivalent to the altitude of the celestial pole; and we have explained how the diurnal circles traced out by the celestial objects, due to the Earth's rotation, are circles centred at the celestial pole. Now the stars which, for purposes of navigation, are regarded as being at an infinite distance from the Earth, maintain their declinations over relatively long periods of time, so that it is an easy matter to find latitude from an observation of a star when it culminates.

The culminating altitude of a star is known as its *meridian altitude*. This is so because the greatest daily altitude of a star at any place is attained when it bears due north or south, that is when it is at *meridian passage*.

We have noted, in Chapter III, that the Arab navigators of the Red Sea and the Indian Ocean have used the *kamal* for finding latitude from very early times. It is not without significance that the Red Sea, in contrast to the Mediterranean Sea, has great latitudinal extent. It must have been evident to the early Semitic seamen that the meridian altitudes of the visible stars changed appreciably during a voyage extending the full length of the Red Sea. Moreover, it occurred to these seamen that the meridian altitudes of selected stars gave a guide to the positions of certain harbours located on the Red Sea coastlands. Star meridian altitudes were used, and still are used, by the Arab traders of the Red Sea, and provided, as they still do provide, valuable aids to their navigations.

There is every reason to believe that the Polynesian seamen of the past navigated their craft for hundreds of miles between the widely-spaced islands of Polynesia by using star observations. These intrepid voyagers used the important fact that the latitude of an observer is equivalent to the declination of a star which culminates at his zenith; or, as D. H. Sadler puts it in a tribute to the late Harold Gatty, 'a star in the zenith is a heavenly beacon lighting up the latitude circle which revolves beneath it.' 'This assuredly,' Mr Sadler goes on to say, 'is the simplest principle of all position-finding methods.' The pattern of stars in the vicinity of the zenith provided the Polynesian navigators with the necessary astronomical information which enabled them to set their courses and make their desired landfalls.

During the period of the discovery of the Atlantic coastlands of Africa by the Portuguese seamen under the sponsorship of Prince Henry the Navigator, it was customary to make astronomical observations ashore for the purpose of finding distance south of Lisbon. Instrumental aid for performing this task on a lively ship at sea was, at the time, not available to the navigator.

2. LATITUDE BY THE POLE STAR

An early method for finding latitude was afforded by the Pole Star—the *Stella Maris* of the early seamen. The Pole Star, on account of its large declination describes, during the course of a day, a tiny circle of angular radius which is equal to the small polar distance of the star and which is centred at the celestial pole. On two occasions each day the altitude of the Pole Star is equal to an observer's latitude; but never does its altitude differ by more than a small angle from the latitude. Henry's navigators were taught to observe the altitude of the Pole Star at the place from which they departed for their voyages of discovery, at a time during the night when the bright star *Kochab*, the foremost of the so-called Guards of the Lesser Bear, was in a particular position relative to the Pole Star. On sailing southwards the altitude of the celestial pole decreases in proportion to the change in latitude. In order to ascertain how far south of the departure point he had sailed, the navigator would observe the altitude of the Pole Star at a time when *Kochab* occupied the same position relative to the Pole Star as it had when the departure-position observation was made. The difference between the two altitudes in degrees, when multiplied by the number of leagues in a degree—16⅔ according to the Portuguese reckoning—gave the distance in leagues between the present and departure positions. The earliest observations were made ashore using the seaman's quadrant and a plumb-line, instruments which have been described in Chapter III.

After the West African coast and the Atlantic islands of the Azores and Madeiras had been discovered, the Portuguese made a practice of marking the arc of the quadrant at points corresponding to the positions of the plumb-line for latitudes of certain islands and important coastal stations. From the time when this practice began, navigators employed the method of 'running down the latitude.' By this is meant sailing southwards or northwards until the parallel of latitude of the destination is reached, and then sailing along the parallel, that is due east or west, until the required landfall is made.

The charts used by the early Portuguese navigators did not have a scale of latitude, and the terms *altura* (altitude or height) and 'running down the altitude' were used in reference to the latitude of a particular place. When seamen became accustomed to using the degree as an angular unit it became convenient to provide a scale of latitude on the maritime chart. When this stage of development had been reached, seamen began to use a *Rule*, or *Regiment, of the North Star*, by means of which they could translate the altitude of the Pole Star into latitude.

To use the North Star for finding latitude the seaman was required to memorize, or have access to, the corrections necessary to apply to the altitude of the Pole Star, according to the position of *Kochab* relative to the Pole Star itself.

When the Pole Star is on the celestial meridian of an observer above the celestial pole, the latitude of the observer is equal to the altitude of the star minus its polar distance. When it is on the meridian below the pole, the latitude is equal to the altitude plus the polar distance. When the local hour angle* of the Pole Star is more than six hours and less than eighteen hours its altitude is less than the latitude; and when the hour angle is more than eighteen hours or less than six hours its altitude is greater than the latitude. The correction to be applied to the altitude of the Pole Star in order to obtain the latitude is clearly dependent upon the position of the Pole Star in its diurnal circle. This, in turn,

* See Appendix 2.

affects the position of *Kochab* relative to the position of the Pole Star.

Martin Cortes, in his *Breve Compendio de la Sphera*, based his Rule of the North Star on the erroneous assumption that the polar distance of the Pole Star was 4° 9'. This value was too great, the

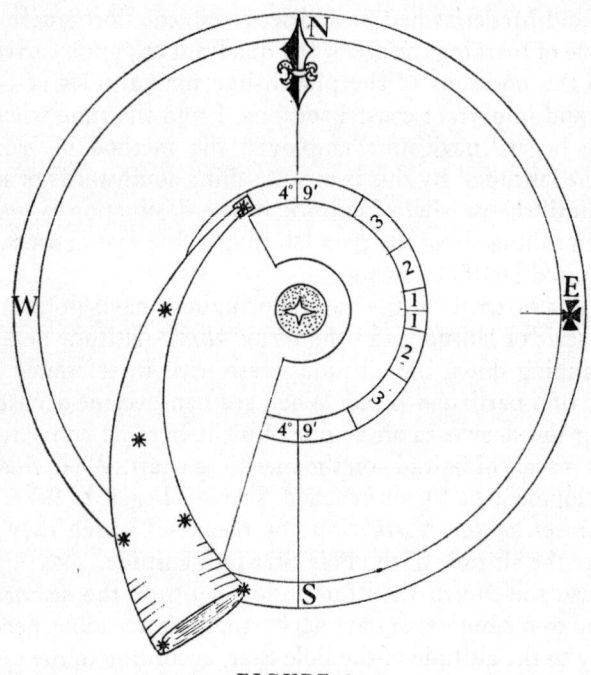

FIGURE 3

correct value at the time (1556) being 3° 30'. Cortes devised a simple instrument by means of which the correction to apply to the altitude of the Pole Star to find the latitude could readily be found. The instrument is similar to a nocturnal, and consists of a circular plate having a hole in the middle through which the Pole Star could be sighted. Pivoted at the centre of the plate is a movable indicator in the shape of a hunting horn—'The Horn' being the equivalent of the Portuguese name for the Lesser Bear. The instrument was to be held at arm's length and the Pole Star sighted. The horn indicator was then turned to a position corresponding to the positions of the stars of the Lesser Bear, where-

METHODS OF FINDING LATITUDE

upon the mouthpiece end of the horn indicated the correction to apply to the altitude of the Pole Star to find the latitude of the observer. Fig. 3 illustrates the instrument.

In the commonly-used Rules of the North Star, the position of the foremost guard was described in terms of a compass direction from the Pole Star. Edward Wright devoted Chapter 12 of his *Certaine Errors in Navigation Detected and Corrected*—first published in 1599—to a discussion on the position of the North Star and the Guards. This chapter runs as follows:

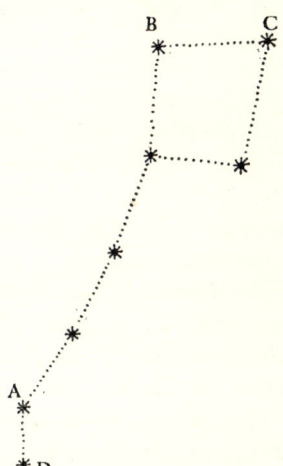

'Among the 48 constellations which the astronomers place in the heavens the neerest unto the Pole of the World is that which they call the lesser Beare, and the Mariner's Bozina or ye Horne, in regard to the shape thereof: which Constellation consisteth of 7 Stars, which are placed after this manner: and of these Stars the three greatest marked A B C doe serve especially for our purpose. And so A is called the North Starre; B the foremost Guard, and C the Guard behinde. And they are so-called because by that force of the motion of the first moveable Heaven the star B goeth alwaies before and the starre C behinde.

'Every of these Starres, as well as al others in the Heavens besides, describe their circles round the Pole with the motion of the first or highest moveable Heaven; in which motion; sometimes the two Starres A B are iust of one height above the Horizon; and when they are said to be E and W one from another. Sometimes they are in a perpendicular line to the Horizon according to our sight; and then they are said to be N and S. And sometimes also the two Guards B C are E and W one from another, and then the foremost Guard beareth from the North Starre NE and SW. And when these two Guards be in a perpendicular line one above another, the foremost Guard beareth from the North Starre SE and NW. In so much that from the foure Positions do arise eight

Rules for the eight Rhumbes, wherein the foremost Guard may stand, being considered in respect of the North Starre. And so presupposing that the North Starre is distant from the Pole three degrees and a halfe (according to the opinion of some mariners, who love numbers that have no fractions) sometimes the North Starre shall bee as high as the Pole it selfe, sometimes three degrees and one halfe lower or higher than the Pole, and sometimes three degrees, and sometimes halfe a degree.'

It will be noticed that Wright used 3° 30′ as the polar distance of the Pole Star. The same value was used by William Bourne in his *Regiment for the Sea*, a work which was designed to supplement the work of Cortes, and which was first published in 1577. Other writers, following Cortes, used the erroneous value of 4° 9′.

The declination and Right Ascension of the Pole Star change comparatively rapidly, on account of the precession of the equinoxes. The use of an out-of-date polar distance in the Rule of the North Star, therefore, resulted in an error in the latitude obtained from an observed altitude of the Pole Star. Edward Wright included as one of the errors of navigation, that caused by using a false Rule of the North Star. Wright advised seamen to use a polar distance of 3° 8′, although in his rule, given above, he used 3° 30′. Seller, whose *Practical Navigation* appeared in the late 17th century, used 2° 9′ in his 'Rule of the North Starre'.

Chapter 15 of Wright's book is entitled 'Other Things to be noted in Observing the Height of the Pole'. It reads thus:

'Next unto the Constellation of the Horne there is a Starre which is called by the Spaniards el Guion signified before [in the accompanying diagram] by the letter D, which standeth east and west from the North Starre giveth you to understand that it and the North Starre, and the very Pole, are east and west. And so taking the heigth of the North Starre when it is thus situate in regard of the Guion, without making any other account, you have the iust heigth of the Pole and the Distance from the Equinoctial.'

Accurate Pole Star tables of the type published in seamen's almanacs, were first published in the *British Nautical Almanac* for

METHODS OF FINDING LATITUDE

the year 1834. These tables were calculated using a formula supplied in 1822 by the astronomer Littrow.

The formula used for finding the latitude from an observation of the altitude of the Pole Star is:

$$l = a - p \cos h + \tfrac{1}{2} \sin 1'' \, (p \sin h)^2 \tan a$$

where l is latitude, a is altitude, h is local hour angle and p is polar distance of *Polaris*.

The local hour angle h, is found by combining the Local Sidereal Time of the observation and the Right Ascension of the star. That is:

$$h = \text{L.S.T.} - \text{R.A.} *$$

Both the polar distance p and the Right Ascension of *Polaris* change rapidly because of precession and nutation, so that average values of p and R.A. are used in compiling Pole Star tables. For the year 1941, for example, the average polar distance of *Polaris* was 61′ with a maximum deviation from this value of 0·3′. The corresponding values for the Right Ascension were 01 hrs. 44 mins. and 2 seconds.

The First Correction, giving values for $-p \cos h$, were tabulated in Table 1, and the Second Correction, giving values for $+\tfrac{1}{2} \sin 1'' \, (p \sin h)^2 \tan a$, were tabulated in Table 2, assuming average values for p and R.A. A Third Correction, given in Table 3, allowed for the differences between the true and assumed values of p and R.A.

The earliest Pole Star tables gave only the First Correction, that is $-p \cos h$. This was sufficient for the early navigators whose instruments were not accurate enough to measure altitudes to a very high degree of accuracy.

The Portuguese mathematician Nuñez pointed out, in the early 15th century, that an error in the Rule of the North Star results from the assumption that the area bounded by the diurnal circle of the Pole Star is a plane circle. The error, however, amounts to no more than a couple of minutes of arc; and it is interesting to note that, as late as 1824, Edward Riddle, in his *Treatise on Navigation*, stated that the distance of the Pole Star from the celestial pole is so small that the circle which it describes may, without causing material error, be considered as a plane. His solution to the Pole Star problem involves the First Correction only.

The famous Elizabethan mathematician Thomas Hariot used a Pole Star table, which was not published for general use, in which he incorporated a correction for the effect of latitude. This correction corresponds to the Second Correction, and it will be noted from the formula given above, that the Second Correction varies as the tangent of the altitude (or latitude).

We are informed by Staff Commander W. R. Martin in his *Navigation and Nautical Astronomy*, which was first published in 1888, that good Pole Star tables were published in 1810 by Mr J. Stevens of the East India Company's Service.

The Regiment of the North Star formed part of the stock-in-trade of ocean navigators of the 15th–17th centuries. It is a noteworthy fact that the navigation manuals of the 18th century and the early 19th century give no instructions for finding latitude from an observation of the Pole Star. Robertson, whose *Elements of Navigation* was probably the most well-used navigation manual of the 18th century, makes no reference to the Pole Star. Neither does Mackay in his well-known navigation manuals; nor does J. W. Norie in the first edition of his *Epitome*. It is not without significance that the period when the Pole Star seems to have been out of favour was one during which the double altitude problem became popular.

It is interesting to note that Carl Zeiss invented an instrument in 1938 designed for finding position at sea by means of an observation of *Polaris*. The *Pol Fernohr*, as the instrument is called is, in some respects, similar to the ancient nocturnal. It employs a small pane of glass on which are engraved two marks corresponding to α and β *Ursa Minoris* (The Guards of the Lesser Bear). The instrument is held pointing to the north celestial pole and then adjusted so that the two marks correspond with the appropriate stars, whereupon the latitude and Local Sidereal Time may be read off the instrument.

The Regiment of the North Star and Pole Star tables are applicable to the northern hemisphere only. So that after the Portuguese navigators, in their voyaging southwards along the West African coast, had crossed the equator, an alternative method for finding latitude became a necessity. As early as 1484, Prince Henry's royal father, King John of Portugal, was instrumental in forming a commission to investigate the problems of position-finding, especially in the southern hemisphere. From the investigations of

METHODS OF FINDING LATITUDE 137

the astronomers and mathematicians who formed this commission, the method for finding latitude by meridian altitude of the Sun was introduced to seamen.

3. LATITUDE BY MERIDIAN ALTITUDE OF THE SUN

Finding latitude from an observation of the Sun on the meridian demands knowledge of the Sun's declination. Solar Tables were, therefore, prepared for sea use. These tables contained the Sun's declination for every day of the year. Henry's navigators were taught how to apply the Sun's declination to the midday height

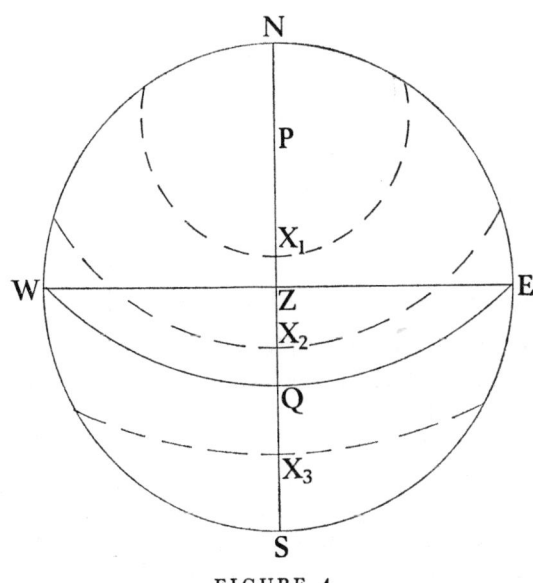

FIGURE 4

of the Sun, in order to obtain the latitude, or height of the pole, as it was generally called.

The latitude of an observer, which is equal to the altitude of the celestial pole at the observer's position, is equal to the angular distance between his zenith and the equinoctial measured in the plane of the observer's celestial meridian. It follows, therefore, that the latitude of an observer is equal to a combination of the declination of a heavenly body and its meridian zenith distance. Fig. 4 illustrates this.

Fig. 4 is a projection of the celestial sphere on to the plane of the celestial horizon of an observer whose zenith is projected at Z. N, E, S and W are the projections of the cardinal points of the horizon. P is the projection of the elevated celestial pole, and WQE is that of the equinoctial. X_1 represents a celestial body which is culminating north of the observer's zenith; X_2 represents a celestial body which is culminating south of the observer's zenith but north of the equinoctial; X_3 represents a celestial body which is culminating south of the equinoctial.

Now,
$$\text{Latitude of observer} = NP = ZQ$$

also,
$$ZQ = QX_1 - ZX_1$$

also,
$$ZQ = QX_2 + ZX_2$$

In general:
$$ZQ = ZX_3 - QX_3$$

or:
$$ZQ = QX \mp ZX$$

Latitude = Meridian Zenith Distance * ∓ Declination *

The *Rules*, or *Regiment*, *of the Sun* were more complex than those of the North Star. The navigator, when using them, had to consider whether the declination of the Sun was north or south; and whether the Sun crossed his celestial meridian to the north or the south of his zenith. In addition to the general cases—three for each hemisphere, making six—there were special cases applying when the Sun's declination was 0°, and when the Sun had a meridian altitude of 90°, to make confusion doubly sure for the non-mathematical navigator.

The Rules of the Sun required knowledge of the meridian altitude, the declination and the 'shadow'. Chapter 3 of Wright's book is entitled 'Of the Shadowes'. He tells us:

'The shadows being compared with the Sun, may be of three sorts: for at high Noon the shadow falleth either towards that part of ye Worlde to which the Sunne declineth, or towards the contra part, or else we make no shadow at all. The first and second sort are, when the heigth of the Sunne is less than 90°, and the third is when it is iust 90°. . . . The rule of the Shadowes

METHODS OF FINDING LATITUDE   139

is that wee looke well to the lower fane of the Astrolabe when we are taking the height of the Sunne at Noone. . . .'

The earliest Rules of the Sun used meridian altitude instead of meridian zenith distance for finding latitude. Typical of the meridian zenith-distance rules were those of John Seller, which appeared in his *Practical Navigation*, first published in the late 17th century.

'Rule 1. If the Sun comes to the meridian in the South, and have South Declination, subtract the Declination from the complement of the Meridian Altitude. The Remainder is the Latitude of the Place of Observation Northerly. But if the Declination exceed the Zenith Distance, then subtract the Zenith Distance from the Declination, the Remainder is the Latitude Southerly.

'Example 1. Admit you are at Sea, and the Sun being on the Meridian in the South is 37 degs 30 mins distance from the Zenith, and at the same time hath 12 degs 00 mins South Declination; I demand the Latitude of the Place.

'The Operation:
'Complement of the Meridian Altitude . . 37° 30'
The Sun's Declination south subtract . . 12° 00'
The Latitude of the Place . . . . . 25° 30' North

'Rule 2. If the Sun be upon the Meridian in the South, and hath North Declination, then add the Declination to the Zenith Distance, the Sum is the Latitude Northerly.

etc. etc.'

The quadrant, with its plumb-line, was not a suitable instrument for measuring altitudes at sea from the deck of a heaving ship. The need for a more suitable altitude-measuring instrument resulted in the adaptation of the cross-staff and astrolabe for the seaman's use. Wright informs us in his *Certaine Errors* . . . that the Portuguese seamen of his time marked their astrolabes 'in reverse,' so that the angle given on the arc was the zenith distance. Hence the need for alternative rules for finding latitude by meridian altitude of the Sun.

4. LATITUDE BY MERIDIAN ALTITUDE OF A STAR

The method for finding latitude by meridian altitude of the Sun applied equally well to the stars, provided that the navigator was provided with a table of stars' declinations.

William Bourne, in his *Regiment for the Sea*, explained how star observations could be used to find latitude. Tables of stars' declinations were drawn up for the seaman's attention and these, in due course, made their appearance in navigation manuals. Provided that a mariner could measure the meridian altitude of one of the tabulated stars during morning or evening twilight, he had the means of finding the latitude of his ship. The quadrant and the astrolabe were quite useless for star observations, relying as they did on a shadow being cast. The cross-staff, therefore, became the normal altitude-measuring instrument for star sights.

In the northern hemisphere, at least to the north of the tenth parallel of latitude, the Pole Star is always available for observations made during twilight so long as the sky and horizon are clear. The North Star, therefore, was regarded by seamen up to the end of the 17th century as being superior to other stars for finding latitude.

5. LATITUDE BY THE SOUTHERN CROSS

The constellation of the Southern Cross was used by seamen in southern waters to find latitude. As early as 1505, the Portuguese navigators were provided with a rule for doing so. This rule was copied, in its essentials, by many writers of navigation manuals. Edward Wright's remarks on the use of the Southern Cross, or Crozier as he called it, are interesting. He regarded the polar distance of the star at the foot of the cross (the Cocke's Foot as he called it) as being 30°. He pointed out that when the stars at the head and the foot of the Cross are perpendicular to the horizon, at which time, Wright informs us, they have their greatest altitudes, the latitude may be found by measuring the height or altitude. The rules are:

> 'For if the said Heigth be thirtie degrees, then wee are in the very equinoctial, and if it be more than thirtie degrees then we bee so much past the equinoctial towards the South. And if it be less than thirtie degrees, so much as it wanteth are we to the north of the equinoctial.'

The Right Ascensions of the stars at the head and foot of the

METHODS OF FINDING LATITUDE 141

Southern Cross are approximately equal to one another. This is the reason why they cross the meridian, or culminate, at about the same time. The values of the declinations of these two stars used by John Seller are 28° 45' and 34° 45' respectively. These values are those determined by Edmund Halley. The latitude was found by subtracting 28° 45' or 34° 45' from the meridian altitude of the Head of the Cross and the Crow's foot, as Seller called the Foot of the Cross, respectively. The seaman was advised, in order to ascertain the exact time for observing, to hold up a thread and plummet: if the thread cuts both stars simultaneously, they are at meridian passage and suitably placed, therefore, for a latitude observation.

Observation of the Southern Cross stars for latitude could be used north of the equator as well as in the southern hemisphere. In fact, between the parallels of about 10° and 20° north, the Pole Star and the Southern Cross could be used for this purpose. Within these parallels of latitude the navigator was, therefore, afforded a means of checking his observations.

6. DECLINATION TABLES

An advantage of the method for finding latitude by meridian altitude of a star over that for finding latitude by a noon-day Sun sight is that a star's declination is constant (or, at least, practically so) for both time and longitude. The Sun's declination, on the other hand, varies with the time of year and also with longitude or local time of an observer's position. The earliest tables of declination of the Sun gave declination for noon on each day of the year for a standard—or reference—meridian that of Lisbon or London for example. Because the Sun crosses any given meridian at local noon (which differs from noon at any other meridian by an amount which is proportional to the difference of longitude between the local and standard meridians) it is necessary to apply a correction to the tabulated value of the Sun's declination in order to find the declination for the time of local noon. Edward Wright referred to this correction as the 'aequation of the Sunne's Declination.' It is interesting to note that he remarked that: 'They which sail in the moneth of Iune and December, need not much to make an aequation.'

John Davis, the famous English navigator, in his *Seaman's Secrets*, explained how to correct the tabulated declination for longitude. Davis also repeated the method described by William Bourne for finding latitude from an observation of the Sun when it crossed the observer's celestial meridian below the pole. This method is applicable only in latitudes where the Sun is circumpolar, and this can be so at no place on the equatorial side of the Arctic (and Antarctic) circle. It was a method of great importance to the polar navigators of Elizabethan times, during their searches for a northern route to the fabulous spice islands of the East.

The table of the Sun's declination, so essential for finding latitude from an observation of the Sun on the meridian, reached a stage of perfection only after Kepler had discovered the true form of the orbit of the Earth, and the manner in which the Earth revolves around the Sun. Following the careful observations of the fixed stars by astronomers—notably those of the renowned Tycho Brahe who is generally esteemed as being the most skilful astronomical observer of all time—tables of declinations and Right Ascensions of the fixed stars were improved and perfected.

The apparent motions of the Moon and the planets relative to the background of fixed stars perplexed astronomers, and it was not until the 18th century, after the illustrious Newton had formulated, and given to the world, his law of universal gravitation, that tables of declinations and Right Ascensions of the Moon and the visible planets became part of the stock-in-trade of the navigator.

Martin Cortes, in his work on navigation, gave a single table of the Sun's declination for every day of a leap year. In addition to this, a secondary table of corrections was provided, by means of which the declination of the Sun for any day in a non-leap year could be found. During the 17th and 18th centuries it was the common practice of writers of navigation textbooks to provide four tables of Sun's declination: one for leap years; and a second, third and fourth for the first, second and third years after leap years respectively. The seaman, in order to know which of the four tables to use, required a rule for finding whether the year in question was a leap year, or first, second, or third year after leap year. It is interesting to note the curious and ponderous instructions given by Edward Wright, pertaining to this simple problem.

> '... I will set down a rule whereby you may know whether the present year be leape year, or whether it bee the first, second or third yeare after the leape year.

METHODS OF FINDING LATITUDE 143

'And the rule is this, that taking from the yeeres of our Lord (which run in our common account) the number of 1600, if the remainder thereof be an even number, and halfe of the remainder an even number, then that yeere is leape yeere: and if the remainder be even, and the halfe thereof odde then that yeere is the second yeere after the leape yeere. But if the remainder of the yeeres number be odde we must trie the yeere next going before, to see whether the remainder thereof, and halfe the remainder be even numbers, for then the present yeere is the first yeere after the leape yeere. And if the remainder of the yeere going before be even, and the halfe thereof odde, then the present yeere is the third yeere after the leape yeere.'

Pedro Nuñez (Nonius), the famous Portuguese mathematician, is credited with being the first to suggest providing tables of the Sun's declination for a four-year period.

7. THE DOUBLE-ALTITUDE PROBLEM

Nuñez, in his book *De Arte et Ratione Navigandi*, proposed a method for finding the latitude from two observations of the Sun's altitude together with the intermediate azimuth. Nuñez's solution to the double-altitude problem (as it later became known) was published in 1537. It involved the use of a globe, as did the solution to the same problem put forward by our countryman Richard Hues in his *Tractatus de Globis*, published in 1594. An English version of Hues's book on the globes was published in 1638 under the hand of John Chilmead M.A. of Christchurch in Oxford.

At the time when Hues published his treatise on the use of the globes mechanical watches, which kept time with a fair accuracy, had become available. Hues's solution to the double-altitude problem—a problem which was to engage the attention of numerous mathematicians and astronomers in later centuries—required two altitudes of the Sun together with the elapsed time between the observations.

John Davis, in his *Seaman's Secrets*, repeated Hues's method. The method consisted in drawing two circles on the celestial globe. The first is centred at the Sun's position on the ecliptic, its radius being equal to the complement of the first observed altitude. The second circle is centred at the intersection of the parallel of the Sun's declination and the hour circle which made an angle with the celestial meridian at the time of the first observation equal to the elapsed time between the two observations. The radius of the second circle is to be equal to the complement of the Sun's altitude at the time of the second observation. The declination of one of the points of intersection of the two circles of zenith distance is equal to the observer's latitude. Which of the two points of intersection is equal to the observer's latitude is obvious from knowledge of the latitude by account.

Blundeville, in his *Exercises* ... which was first published in 1594, described a method for finding latitude from simultaneous altitudes of two stars, one on each side of the meridian. Master Blundeville's method involved adjusting the globe so that the altitudes of the two stars corresponded to the two observed altitudes, whereupon the latitude could be measured on the globe: the altitude of the celestial pole, which is equal to the observer's latitude, indicating this.

The use of globes for navigational purposes was short-lived. Not only were globes cumbersome and expensive, but the degree of accuracy of the problems solved by their use was crude—even for standards of the time. With the improvements in mathematical methods for the seaman's use, a new breed of navigator developed. Methods of computation for the several astronomical problems related to position-finding at sea were devised and these were adapted for the use of seamen, usually in the form of complex rules. The navigator, having access to the rules from his manual or his memory, was able, therefore, to solve the relatively complex astronomical problems without it being necessary for him to understand the principles involved. Seamen, it appears, at no time in the history of navigation have taken kindly to the mathematical arts—the Black Arts, as they were sometimes called. The application of mathematical principles of position-finding at sea were forced upon the non-mathematical mariner; and it is only within comparatively recent times that navigators have had some little understanding of the principles of the processes involved in nautical astronomy. By and large, seamen at all times have been content to 'proceed according to the Rules.' They are concerned more with finding the answer than with knowing the underlying principles of the methods they use for finding it.

In the preface to his well-known *Elements of Navigation*, which was the principal English navigation manual of the 18th century,

John Robertson F.R.S. who wrote his very fine book for the use of 'the children of the Royal Mathematical School,' drew attention to the fact that:

> 'The common treatises of navigation, which on account of their small bulk and easy price, are vended among the British mariners, seem not to be written with an intention to excite in their readers a desire to pursue the sciences, farther than they are handled in those books; so that it is no wonder our seamen in general had so little mathematical knowledge; for the person who could keep a trite journal, formed on the most easy occurrences, has been reckoned a good artist; but whenever those occurrences have not happened, the journalist has been at a loss, and unable to find the ship's place with any tolerable degree of precision; and such accidents have probably contributed to the distress which many ships crews have experienced, and which a little more knowledge among them might have prevented, or at least have lessened.'

The problem of finding latitude from two observations of the Sun or star, using mathematical devices other than a globe, received the attention of many scientific men and writers on navigation during the 18th and 19th centuries. The so-called double-altitude problem is one of great interest, and one for which many ingenious solutions were furnished. The name *double altitude* was given to a method of finding latitude. In modern navigational practice the term applies to the method of fixing the ship, that is to say, finding her latitude *and* longitude from astronomical observations.

The meritorious Raper, the author of the 19th-century classic *The Practice of Navigation*, objected to the name double altitude. '... it is defective,' he wrote, 'inasmuch as the word double means twice the same.' He suggested the use of the term *combined altitudes*.

The calculations involved in the solution to the double-altitude problem are more complex than those for finding latitude from an observation of the meridian altitude of a heavenly body. Practical seamen never favoured the double-altitude problem until relatively recent times, when the method was perfected for finding both latitude and longitude.

Given two altitudes of a heavenly body and the interval of time between the observations, the latitude may be computed by the so-called direct method of double altitude. In using the direct method, the latitude is computed by the rigorous process of spherical trigonometrical calculation. Many writers, including the famous French astronomer Lalande, whose excellent writings on navigation are well known, preferred the rigorous process and advocated its use in preference to any of the many indirect methods that were devised.

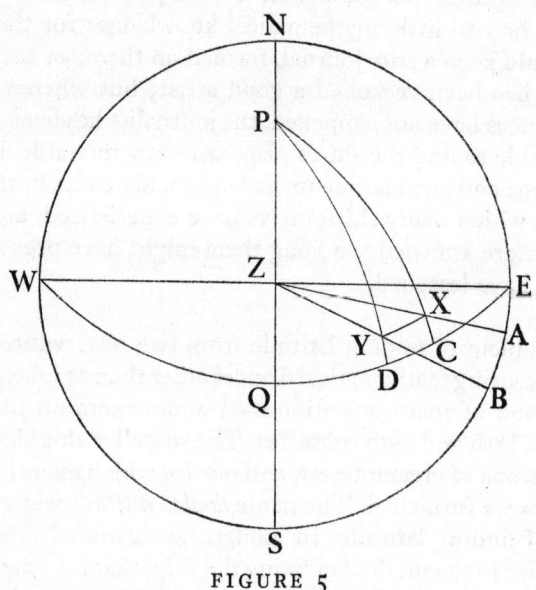

FIGURE 5

The earliest double-altitude problems were related to the Sun. Nicholas Facio Duillier F.R.S. is credited with being the first to devise a mathematical method for solving the problem. In 1728 Duillier published a pamphlet entitled *Navigation Improved*. He discussed the double-altitude problem in some detail and, in his solution, he took into account the movement of the ship during the interval between the times of observation. Duillier's method required a considerable amount of tedious computation and, although it was improved by several writers and teachers of navigation, it was not generally considered as providing a practical solution to the problem. The method is explained with reference to Fig. 5.

METHODS OF FINDING LATITUDE

Fig. 5 is a projection of the celestial sphere on to the plane of the celestial horizon of an observer whose zenith is projected at Z. N, E, S and W, are the projections of the cardinal points of the horizon. WQE is the projection of the equinoctial and P is that of the elevated celestial pole. X and Y are the projections of the object at the times of the first and second observations respectively. The arc XY is a great-circle arc through X and Y. Arcs PC and PD represent the hour circles through the body at the times of the observations.

The following arcs and angles are known:

$\quad$ PX = polar distance of observed object
$\quad$ PY = polar distance of observed object
$\quad$ ZX = first zenith distance
$\quad$ ZY = second zenith distance
$\quad$ XPY = elapsed time

The successive steps in the computation in order to find PZ, which is the complement of the observer's latitude, are as follows:

1. In triangle PXY $\quad$ Using P, PX and PY, find XY.
2. In triangle PXY $\quad$ Using PX, PY and XY, find PXY.
3. In triangle ZXY $\quad$ Using ZX, ZY and XY, find ZXY.
4. From PXY and ZXY find PXZ.
5. In triangle PZX $\quad$ Using PX, ZX and PXZ, find PZ.
6. Latitude = 90 − PZ.

Duillier's method, which is described in Leadbetter's *Astronomy of the Satellites of Jupiter*, published in 1729, although involving tedious calculations, provided a direct and unambiguous solution. Moreover, only two common rules of spherical trigonometry were required, these being the rules for:

1. Finding an angle given the three sides.
2. Finding a side, given the opposite angle and the other two sides.

John Robertson, in his *Elements of Navigation*, describes Duillier's method—which he calls Facio's method—but states that his solution requires too many trigonometrical operations to make it of ready use.

A solution to the double-altitude problem was published by Richard Graham in 1734. Graham's solution is ingenious, despite the remark made by the author of the article on navigation in the ninth edition of the *Encyclopaedia Britannica* to the effect that it was published in the *Philosophical Transactions* 'with much boasting.' Graham's solution was an instrumental operation requiring the use of a beam compass which could be fitted to the meridian ring of the celestial globe. The beam compass, which was to be fitted so that it could be made to slide along the meridian ring, was used to describe arcs of circles of radii equal to the Sun's zenith distances at the times of the observations. Graham claimed that his method was capable of giving the latitude to within a few minutes of arc of the truth and, according to John Robertson, '... with ease and expedition.'

In 1740 Cornelius Douwes, an examiner of sea officers and pilots under the College of Admiralty at Amsterdam, devised a method for finding the latitude by double altitudes. Douwes's method became very popular among British navigators.

W. R. Martin, in his *Navigation and Nautical Astronomy* of 1888, regarded Douwes's method as providing the first practical method for solving the double-altitude problem for the use of seamen.

According to Robertson, manuscript copies of Douwes's method fell into the hands of English officers who, holding the method in high esteem, caused it to be published in 1759 without any demonstration. However, Dr H. Pemberton examined the method and the solar tables devised by Douwes for use with his method. Pemberton communicated his investigations to the Philosophical Society, in whose *Transactions* it was published. Pemberton demonstrated the method and showed its limitations in 1760.

Douwes's method requires the use of an estimated latitude. If the latitude yielded by the calculations differs materially from the estimated latitude used in the calculation, it is necessary to resolve the problem using the calculated latitude in place of the estimated latitude.

As a result of investigations into Douwes's method, improved auxiliary tables were published under the direction of Nevil Maskelyne, Astronomer Royal, in his *Tables Requisite to be used with the Nautical Ephemeris*, second edition, 1781. Maskelyne stated in this work that Douwes transmitted his method to the Lords Commissioners of the English Admiralty, and that he was

METHODS OF FINDING LATITUDE

rewarded with £50 by the Commissioners of Longitude. Douwes's method was again investigated in 1797 by Mendoza Rios whose results, together with improved auxiliary tables, were published in the *Philosophical Transactions* of 1797. Douwes's method is described with reference to Fig 6.

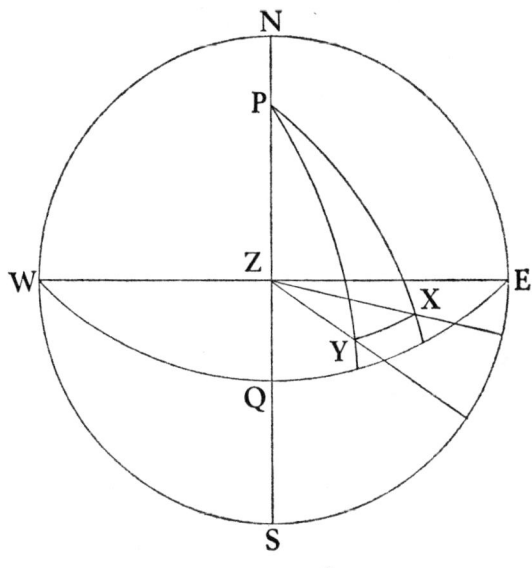

FIGURE 6

Fig. 6 represents the projection of the celestial sphere on to the plane of the celestial horizon of an observer whose zenith is projected at Z. N, E, S and W, are the projections of the cardinal points of the horizon, and P that of the elevated celestial pole. X and Y are the projected positions of the Sun at the times of the first and second observations respectively.

The following arcs and angles are known:

   XPY = elapsed time
   PX = polar distance of Sun
   PY = polar distance of Sun (assumed to be equal to PX)
   ZY = second zenith distance
   ZX = first xenith distance

In triangle PZX:

$\cos ZX = \cos PZ \cos PX + \sin PZ \sin PX \cos ZPX$ . . (1)

In triangle PZY:

$$\cos ZY = \cos PZ \cos PY + \sin PZ \sin PY \cos ZPY \quad . \quad . \quad (2)$$

Subtract (2) from (1):

$$\cos ZX - \cos ZY = \sin PZ \sin PX (\cos ZPX - \cos ZPY) \quad . \quad . \quad (3)$$

that is:

$$2 \sin \tfrac{1}{2}(ZX+ZY) \sin \tfrac{1}{2}(ZX-ZY) = \sin PZ \sin PX \cdot 2 \sin \tfrac{1}{2}(ZPX+ZPY) \sin \tfrac{1}{2}(ZPX-ZPY)$$

From which:

$$\sin \tfrac{1}{2}(ZPX+ZPY) = \frac{\sin \tfrac{1}{2}(ZX+ZY) \sin \tfrac{1}{2}(ZX-ZY)}{\sin PZ \sin PX \sin \tfrac{1}{2}(ZPX-ZPY)} \quad . \quad . \quad (4)$$

The term $(ZPX-ZPY)$ in the denominator of (4), is equal to the elapsed time XPY. Thus, having found $(ZPX+ZPY)$ from equation (4), it is an easy matter to find ZPY.

From (2), we have:

$$\cos ZY = \cos PZ \cos PX + \sin PZ \sin PX \cos ZPY$$

i.e.

$$\cos ZY = \cos PZ \cos PX + \sin PZ \sin PX (1 - 2 \sin^2 \tfrac{1}{2} ZPY)$$

i.e.

$$\cos ZY = \cos (PX - PZ) - 2 \sin PZ \sin PX \sin^2 \tfrac{1}{2} ZPY$$

i.e.

$$\cos (PX - PZ) = \cos ZY + 2 \sin PZ \sin PX \sin^2 \tfrac{1}{2} ZPY$$

Using the latitude by account to give an assumed PZ, $(PX-PZ)$ can be found. From this, PZ—and hence the latitude—can be determined. If the calculated latitude differs materially from the latitude by account, the problem must be re-worked using the latitude found from the first calculation.

The auxiliary tables provided by Maskelyne (and others) gave the logarithms of quantities significant in the computation for 10-second intervals of time. Douwes's original solar tables were given for intervals much coarser than this, thus necessitating troublesome interpolation.

Should the ship's position at the time of the second observation be different from that at the time of the first observation, it is necessary to adjust the first observed altitude to what it would have been had it been observed at the second position. That is to

METHODS OF FINDING LATITUDE 151

say, an allowance for run would have had to have been made. The correction in minutes of arc to apply to the first measured altitude is equal to the distance in miles sailed either towards or away from the Sun during the time which elapsed between the instants of the two observations. Moreover, a change in longitude between the times of observation necessitated a correction to the elapsed time.

A figure famous in the history of astronomical navigation is Samuel Dunn, one-time teacher of mathematics in London. Dunn wrote several works on navigation amongst which were *A New Variation Atlas*, and *A New Epitome of Navigation*, which were published in 1776-1777. In both of these works, which were dedicated to the Honourable East India Company, Dunn introduced a new solution to the double-altitude problem. The problem was entitled 'Of a general method whereby the latitude may be found, having any two altitudes of the Sun and the time elapsed.' The method is as follows:

1. Assume two latitudes differing about a degree or less, and not widely different from the latitude by account.
2. For each co-latitude, together with the co-declination and the co-altitude of the Sun, calculate two hour-angles, making four calculated hour-angles in all.
3. The latitude is then found using the following proportional statement:
   'As the difference of the elapsed times computed from the latitudes is to the difference of those latitudes so is the difference between the true elapsed time and either of the computed elapsed times is to the number of minutes which added to or subtracted from the corresponding assumed latitude (as the case requires) gives the true latitude.'

Dunn's solution was a remarkable discovery, and it formed the basis of the so-called 'Trial and Error' method for finding longitude. It also paved the way for the development of position-line navigation, which we shall deal with in Chapter VII.

James Andrew A.M. described a solution to the double-altitude problem in his *Astronomical and Nautical Tables*, which were published in 1805. Andrew dedicated his work to the Honourable the Court of Directors of the United Company of Merchants of

England trading to the East Indies. The principal feature of his tables was the inclusion of a table of squares of natural semi-chords, that is to say, a table of $\sin^2 \theta/2$ or Haversine θ, for values of θ, at intervals of 10 seconds of arc, from 0° to 120°. In his preface, Andrew declared that the table, which occupied the space of 120 pages, was entirely new. It was designed primarily with the object of facilitating the computation of lunar distances by means of a new method of his invention. He was quick to realize that the table could be employed for solving astronomical prob-

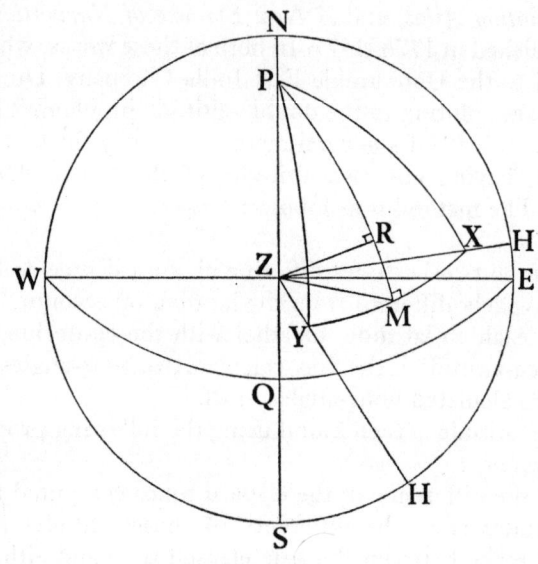

FIGURE 7

lems besides that for which it was specifically designed. In particular, precepts were given for solving the double-altitude problem using the table of squares of natural semi-chords.

An ingenious solution to the double-altitude problem was furnished by James Ivory, Professor of Mathematics at the Royal Military College at Sandhurst. Ivory's method which was well received by seamen, was published in the *Philosophical Transactions* of the Royal Society in 1821.

Ivory's rule applies strictly to bodies whose declinations do not change in the interval between the times of the two observations. In practice, it is available for Sun sights by using the mean of the

METHODS OF FINDING LATITUDE 153

two polar distances proper to the times of observation. The method is described with reference to Fig. 7.

The circle in Fig. 7 represents the projection of the celestial horizon of an observer whose zenith is projected at Z. WQE is the projection of the equinoctial, and P is that of the elevated celestial pole. PX and PY are the projections of the hour circles through the Sun at the times of the first and second observations respectively. PM is perpendicular to the great-circle arc through X and Y. ZR is a great-circle arc at right angles to PM.

In triangle PMX:

$$\sin MX = \sin XPM \sin PX \quad \ldots \quad (1)$$
$$\cos PM = \cos PX \sec MX \quad \ldots \quad (2)$$

In triangle ZMY:

$$\cos ZY = \cos ZM \cos YM + \sin ZM \sin YM \cos ZMY$$

i.e.

$$\cos ZY = \cos ZM \cos YM + \sin ZM \sin YM \sin ZMR \ . \quad (3)$$

In triangle ZMX:

$$\cos ZX = \cos MX \cos ZM + \sin MX \sin ZM \cos ZMX$$

i.e.

$$\cos ZX = \cos MX \cos ZM - \sin MX \sin ZM \cos ZMY$$

i.e.

$$\cos ZX = \cos MX \cos ZM - \sin MX \sin ZM \sin ZMR \quad (4)$$

Add (3) and (4):

$$\cos ZY + \cos ZX = 2 \cos ZM \cos YM \ (YM = MX) \quad . \quad (5)$$

Subtract (4) from (3):

$$\cos ZY - \cos ZX = 2 \sin ZM \sin YM \sin ZMR \ . \ . \ . \quad (6)$$

In triangle ZMR:

$$\cos ZM = \cos ZR \cos MR \quad \ldots \quad (7)$$
$$\sin ZR = \sin ZM \sin ZMR \quad \ldots \quad (8)$$

Substitute $\cos ZR \cos MR$ from (7) for $\cos ZM$ in (5):

$$\cos ZY + \cos ZX = 2 \cos ZR \cos MR \cos YM \quad . \ . \quad (9)$$

Substitute $\sin ZR$ from (8) for $\sin ZM \sin ZMR$ in (6):

$$\cos ZY - \cos ZX = 2 \sin YM \sin ZR \quad . \quad (10)$$

Using complements of ZY and ZX in (9) and (10) we have:
$$\sin HY + \sin HX = 2 \cos ZR \cos MR \cos YM \quad (11)$$
$$\sin HY - \sin HX = 2 \sin YM \sin ZR \quad (12)$$

From (12), by transposition, we have:
$$\sin ZR = \frac{\sin HY - \sin HX}{2 \sin YM}$$

From which:
$$\sin ZR = \frac{\cos \tfrac{1}{2}(HY+HX) \sin \tfrac{1}{2}(HY-HX)}{\sin YM} \quad (13)$$

From (11), by transposition, we have:
$$\cos MR = \frac{\sin HY + \sin HX}{2 \cos ZR \cos YM}$$

From which:
$$\cos MR = \frac{\sin \tfrac{1}{2}(HY+HX) \cos \tfrac{1}{2}(HY-HX)}{\cos ZR \cos YM} \quad (14)$$

From solutions to (2) and (14), PR may be found thus:
$$PR = PM - MR$$

Finally, in triangle PZR:
$$\cos PZ = \cos PR \cos ZR \quad (ZR \text{ found from (13)})$$
or
$$\sin \text{Lat} = \cos PR \cos ZR \quad (15)$$

The above method is described in *The Theory and Practice of Navigation* published in 1900 and written by W. G. Tate, Examiner of Masters and Mates at Tyne and Wear ports.

The following solution is given in *The Extra Master's Guide Book* written by Thomas Ainsley, and published in 1867:

Referring to Fig. 7, in triangle PMX:
$$\sin MX = \sin XPM \sin PX \quad (1)$$
$$\cos PM = \frac{\cos PX}{\cos MX} \quad (2)$$
$$\sin ZR = \frac{\cos \tfrac{1}{2}(ZX+ZY) \sin \tfrac{1}{2}(ZX \sim ZY)}{\sin MX} \quad (3)$$
$$\cos MR = \frac{\sin \tfrac{1}{2}(ZX+ZY) \cos \tfrac{1}{2}(ZX \sim ZY)}{\cos MX \cos ZR} \quad (4)$$

Arc PR is then found by combining PM and MR found from (2) and (4) respectively. Finally, arc PZ which is the co-latitude, is found from PZR. By Napier's rules:

$$\cos PZ = \cos PR \cos ZR$$

or

$$\sin \text{Lat} = \cos PR \cos ZR \quad \ldots \quad (5)$$

It will be noticed that Ainsley's formulae (1), (2), (3), (4) and (5) correspond essentially with those given by Tate and numbered (1), (2), (13), (14) and (15) respectively.

It is interesting to note that Ainsley remarked that the formulae derived above are essentially the same as those given for this case by M. Caillet in his *Manuel de Navigateur*, first published in 1818, that is, three years before Ivory's method made its appearance in the *Philosophical Transactions* of the Royal Society.

The practice of dividing an oblique spherical triangle into two right-angled triangles, as Ivory (and James Andrew) did in his method for solving the double-altitude problem, is one of great antiquity and is, in fact, almost as old as trigonometry itself. It provided the only practical solution for solving oblique spherical triangles up to the time of the invention of the fundamental cosine formula which for any spherical triangle ABC is:

$$\cos A = \frac{\cos BC - \cos AB \cos AC}{\sin AB \sin AC}$$

This formula was known to Bategnius, the celebrated astronomer of Batnae, who died in AD 930.

In the practice of navigation, from the beginning of the second half of the 19th century to the present time, the use of auxiliary right-angled spherical triangles to facilitate the solution of oblique spherical triangles—and especially in its application to the construction of the so-called *short method tables*—has been of prime importance.

Ivory's method was improved in 1822 by Edward Riddle, the master of the mathematical school at Greenwich (formerly the master at the Trinity House school at Newcastle), the year after it was published. Other teachers of navigation, including Mrs Janet Taylor, Lieutenant Henry Raper, and John Riddle, the son of Edward Riddle, who succeeded his father as master at Greenwich, also published modified versions of Ivory's method.

John Riddle is credited with being the first to recognize that an error in the solution to the double-altitude problem may arise through using the polar distance of the Sun for the middle time. He accordingly devised a method for correcting error due to this cause. This was published in his revision for the sixth edition of Edward Riddle's *Treatise on Navigation*, which appeared in 1855.

In 1824 Edward Riddle in his treatise noted that the latitude by double-altitude method could be employed in respect of two altitudes of the same fixed star. Hitherto, Sun altitudes alone had been used for the double-altitude problem. Riddle pointed out that were a star used for this purpose, it was necessary to convert the elapsed interval into sidereal units of time. He supplied a simple rule for doing this. 'Increase the observed interval of solar time,' he wrote, 'by one second for every ten minutes.'

Graphical methods of solving the double-altitude problem were devised, although these did not find favour among practical seamen. As early as 1659 John Collins, in his *Mariner's Scale new Plained*, gave a solution using the stereographic projection. P. Kelly, in 1796, published an ingenious solution by construction, again employing the stereographic projection.

The use of simultaneous observations of two stars for finding latitude was suggested by John Brinkley whose method was first published in the *Nautical Almanac* for 1825.

The great value of the method for finding latitude by simultaneous observations of two stars rested in the fact that the difference between the Right Ascensions of the stars supplied the place of the elapsed time in the case of the normal double-altitude problem in which the Sun is observed. The method is, therefore, independent of the need for measuring time.

If the two star altitudes are measured by the same observer it is necessary to reduce the altitude of the first star observed to the time of the observation of the second star. This is a simple matter, especially if a table giving the rate of change of star's altitude is provided. Ideally two observers should make the observations simultaneously.

A second advantage of the star double altitude applies in cases where the angular distance between the observed stars is known, this arc being required in the solution to the problem.

We find, in Thomas Lynn's *Astronomical Tables* which were first published in 1825, two tables of angular distances between

METHODS OF FINDING LATITUDE

pairs of selected stars for facilitating the problem of finding latitude from simultaneous star observations. One of these tables was designed for use in the northern hemisphere and the other for use in the southern hemisphere. The same tables were published in the 1825 *Nautical Almanac*, in which Brinkley made the following observations:

'Captain Lynn's tables, lately published, in which the Log Rising is given to seconds of time, will be found extremely

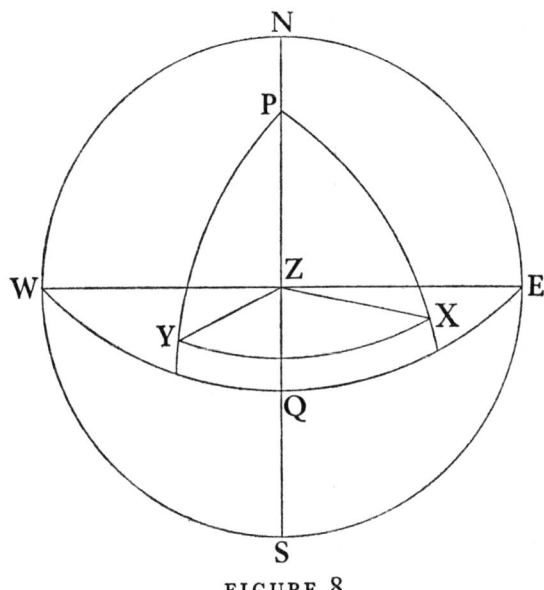

FIGURE 8

convenient for the practice of this method [star's double altitude]. In fact it was his [Lynn's] remarks relative to the opportunities of taking exact altitudes of the brighter stars during the time of twilight, that induced me [Brinkley] to endeavour to investigate an easy process of finding latitude from the altitudes of two stars.'

Three tables, in addition to the two mentioned above, were given by Lynn to facilitate the solution to the double-altitude problem. These tables are similar to those published by Maskelyne (and others) to facilitate the solution to the problem using the modified Douwes's method. Values were given to every second of time so that interpolation, when using the tables, was not necessary.

Commander (later Admiral) Charles Shadwell R.N. favoured the star double-altitude method, and he published tables of angular distances between selected stars in his *Star Tables* which first appeared in 1839, and which were extended in 1849.

The principle of the star double-altitude method is explained with reference to Fig. 8.

The circle in Fig. 8 is a projection of the celestial horizon of an observer whose zenith is projected at Z. N, E, S and W are the projections of the cardinal points of the horizon. WQE is the projection of the equinoctial and P that of the elevated celestial pole. X and Y are the projections of two stars whose zenith distances at the time of the observations are ZX and ZY respectively. Arc ZY is the great circle arc joining X and Y. In triangles PZX and PZY:

Given:

    PX, PY (polar distances from *Nautical Almanac*)
    ZX, ZY (co-altitudes of X and Y)
    XPY (difference between R.A'S of X and Y)

To find:

    PZ and hence the observer's latitude $(90-PZ)$

In triangle PXY:

    Using PX, PY and XPY find XY (arc 1)

In triangle PXY:

    Using PX, PY and XY find PXY (arc 2)

In triangle ZXY:

    Using ZX, ZY and XY, find ZXY (arc 3)
    From arcs (2) and (3) find PXZ (arc 4)

In triangle PZX:

    Using PXZ, PX and ZX, find PZ.

Then:

    Latitude of observer $= (90-PZ)°$

METHODS OF FINDING LATITUDE

Should the stars X and Y be a 'selected pair' which appear in the Star Tables (of Lynn's or Shadwell's) arcs (1) and (2) may be found by inspection, and the labour of computation thereby effectively reduced.

An interesting method of finding latitude based on the principle of the star double altitude was given by Thomas Lynn in his *Navigation* which was published in 1825. Lynn's ingenious method is explained with reference to Fig. 9.

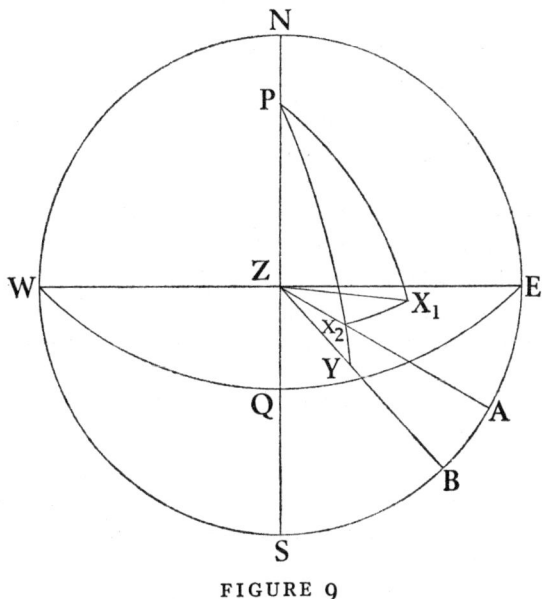

FIGURE 9

Fig. 9 represents the celestial sphere projected on to the plane of the celestial horizon of an observer whose zenith is projected at Z. WQE is the projection of the equinoctial and P that of the elevated celestial pole. X and Y are the projections of two stars whose R.A's differ by a small angle YPX, Y lying to the west of X.

The altitude of the westernmost star Y is observed and, after an interval equal to the difference between the R.A's of X and Y, the star X is observed. At the time of observation of X, the star will occupy the same hour circle as was occupied by star Y when the first observation was made. The altitude of Y at the time of the first observation is arc BY; and the altitude of X at the time of the second observation is arc AX_2.

In the spherical triangle ZX_2Y, the arcs ZX_2 and ZY are known —these being the zenith distances of the stars observed. The arc X_2Y is also known—this being the difference between the declinations of the two stars. Given the three sides of the triangle ZX_2Y the angle ZYX_2 may be calculated.

In the spherical triangle PZY, given PY, ZY and PYZ, the arc PZ—which is the complement of the observer's latitude—may be found.

The accuracy of Lynn's method, as pointed out by Commander W. R. Martin in his *Navigation and Nautical Astronomy* of 1888, is greatest when the difference of bearings of the two stars is 90°, a condition which may be approximated to by selecting stars which have a small difference of Right Ascensions and a big difference of declinations. Martin also pointed out that:

'in order that errors of altitude may affect the latitude least, it is desirable that, when practicable, one of the stars should have a small altitude near the meridian and the other a large altitude near the prime vertical circle; the altitude of the star first observed requires the usual correction for run in the interval.'

The reliability of the calculated latitude obtained from the normal double-altitude problem in which the Sun is employed depends entirely upon the change of bearing of the Sun between the times of the two observations. The Rev. James Inman is credited with being the first to point out this very important fact. The most dependable result from the Sun double altitude applies when the change in azimuth between the times of the two observations is 90°. This condition was stated by Inman in the first edition of his *Navigation* which appeared in 1826.

Many writers on navigation gave complex rules and advice on limiting conditions for the double-altitude problem. These rules and conditions were, in many cases, related quite mistakably to time from noon. Lieutenant Raper, in his review of Captain Sumner's newly-published book, which appeared in the *Nautical Magazine* of 1844, remarked:

'Projection [of Sumner lines] therefore, affords evidence of the simplest and most convincing kind that the value of a double altitude depends altogether on the difference of azimuth. This condition, first pointed out by Dr. Inman, has nothing to do with time from noon, which more popular works reiterate as the proper limiting condition of the double altitude, to the great detriment of the extensive and successful practice of this important observation.'

It is interesting to note that in a critique of Bowditch's *Practical Navigator*, which appeared in the *Nautical Magazine* of 1842, the reviewer pointed out that: 'Bowditch does not once allude to the difference of azimuth as the criterion of the value of the double-altitude observation.'
He also stated that:

'Bowditch gives no case of latitude by meridian altitude of a star, and yet he does the double altitude of a star—an observation not taken once in a whole servitude.'

Captain J. Trivett, a contributor to the *Nautical Magazine* of 1850, and one of the early Examiners of Masters and Mates, described a method for solving the star double-altitude problem by using a table of angular distances between selected stars. The method, which appears to have been popular at the time, was known as the *I O U method*. The star nearer the meridian was referred to as the Inner star. The other was referred to as the Outer star, and the angular distance between the two stars was known as U. To facilitate the computation, and at the same time render the method easy to remember, the solution was commenced by writing down:

        Inner altitude
        Outer altitude
     arc U from Star Tables [Lynn's or Shadwell's]
in this order. Hence the mnemonic I O U.

A practical seaman, L. T. Fitzmaurice, contributed a brief article to the *Nautical Magazine* of 1854 entitled 'On Finding Position by Double Altitudes with only one Latitude.' The method was described without a proof. In the following number of the magazine, a letter appeared under the signature of John Riddle, the master at the Greenwich Hospital school. Riddle wrote:

162   A HISTORY OF NAUTICAL ASTRONOMY

'I believe the writer [L. T. Fitzmaurice] to have been a pupil at Greenwich Hospital about 25 years ago. I am sorry that he [Fitzmaurice] should have forgotten his old school habit of demonstrating the propositions which he advances. The modification to which his letter refers was in full practice by our little navigators here in 1852, and since then must have spread far and wide.'

Riddle then gave a demonstration of the method which is, in effect, an adaptation of the newly-discovered Sumner method, but using only one latitude (instead of two) and calculating two hour angles and two azimuths (instead of four hour angles). (See Chapter VII.)

Staff Commander J. Burdwood, the pioneer of Azimuth Tables, described a method for finding latitude from the altitudes of two stars observed at the same instant. The description of his method appeared in the *Nautical Magazine* for 1865. It is interesting to note that Burdwood's *Tables of Sun's True Bearing (for latitudes 48°–56° N. and S.)* were published the year before, in 1864. In a review of these tables it was written:

'If J. Burdwood had really desired to leave a monument behind him of his labour for the benefit of navigation, he could have done no better than compile these valuable tables.'

8. MERIDIAN AND MAXIMUM ALTITUDES

In normal circumstances at sea it is usual to regard the meridian altitude of a heavenly body as being identical with the maximum altitude. The usual practice of measuring the meridian altitude of the Sun is to commence observing a little before noon, noting that the Sun's altitude is increasing; and then, when the Sun ceases to increase and starts to decrease his altitude—at which times the Sun is said to dip—the angle read off the sextant is taken as the meridian altitude.

A celestial body of fixed declination culminates or attains its greatest altitude when it crosses the upper celestial meridian of a stationary observer. Should, however, the observer be changing his position towards the north or south, and/or the object be changing its declination the maximum altitude attained is not the meridian altitude. In other words, a celestial body attains its maximum altitude when it lies off the meridian, that is before it crosses or after it has crossed the meridian, should the observer be moving northwards or southwards, or should the declination of the body be changing. The hour angle of a celestial body when it is at maximum altitude is dependent upon the rates of change of declination of the body and the latitude of the observer.

For practical purposes, the rate of change of a star's declination is zero, and that for the Sun, which varies (being zero at the solstices and maximum at the equinoxes) is usually very small. The rate of change of declination of the Moon, however, may be considerable and, as early as the beginning of the 19th century, the Moon, because of this, was regarded as being an unfit object for meridian altitude observations.

Chauvenet, in his *Astronomy* of 1896, demonstrates that the interval between the times of meridian and maximum altitudes of a celestial body, when observed by a stationary observer, is given by the formula:

$$T = \frac{d \sin (L-D)}{810,000 \sin 1'' \cos L \cos D}$$

where T is the interval in seconds of time,
        d is the rate of change of declination D in seconds of arc per hour and
        L is the observer's latitude.

It is easy to see that if the declination of a celestial body is changing in the direction of the bearing of the body at meridian passage, the maximum altitude will occur, to a stationary observer, after the meridian altitude, and vice versa. Should the observer be moving meridianally, his rate of change of latitude will have an effect. It is the algebraic combination of the rates of change of declination and latitude that determines whether the meridian altitude will occur before or after the time of maximum altitude.

With the advent of fast steam vessels during the last century the problem of the maximum altitude became one of significance. Latitude by maximum altitude became, in effect, a special case of the problem known as latitude by ex-meridian altitude.

To overcome the error that may have arisen through treating the maximum altitude as the meridian altitude, seamen have been advised, since the middle of the 19th century, to compute the time of meridian passage using the elements of the *Nautical Almanac* together with the longitude by account; and to observe the altitude of the body at the computed time, instead of waiting until it dips, which was, and still is, the traditional method.

A formula from which the difference between the meridian and maximum altitudes may be found is:

$$y = \frac{(15T)^2 \sin 1'' \cos L \cos D}{2 \sin (L \pm D)}$$

Consider a navigator observing the Moon at a time when this body's declination is zero and its hourly change of declination is 17' to the north. If the observer's ship is in the northern hemisphere and it is steaming due south at the rate of 20 knots, the combined movements of the observer's zenith and the Moon, at the time of meridian passage, would produce the same effect as if the observer were stationary and the declination of the Moon were changing at the rate of (20+17)', i.e. 37' per hour towards the north.

When a heavenly body's geographical position and an observer's position are opening, the time of maximum altitude occurs earlier than that of meridian altitude. But when these positions are closing, as they are in the case described above, the time of maximum altitude occurs later than the time of meridian passage. In the example, the Moon's maximum altitude would be reached at about eleven minutes after noon and it would be about $3\frac{1}{2}'$ of arc greater than the Moon's meridian altitude.

After the invention of position-line navigation following Captain Thomas Sumner's discovery in 1837 (see Chapter VII), astronomical navigation was brought to a state of excellence in 1875 by French astronomers and navigators, notably Marcq St Hilaire of the French Navy. Thereafter the problem of finding latitude at sea by astronomical navigation became merely a special case of the general method of obtaining an astronomical position line. The end product of every observation or *sight*, as the seaman calls it, is a position line which may be drawn on the navigation chart. A *position line* is defined simply as a line on the chart somewhere on which the ship's position is represented. The point of intersection of any two position lines is the projected position of the ship on the chart.

The astronomical position line obtained with the least effort is that which results from an observation of an object on the meridian. The position line, in this case, lies east–west along a parallel of latitude. The meridian altitude observation for latitude became available to European seamen, as we have demonstrated, as soon as tables of the Sun's declination, and a Regiment or Rules for the Sun were devised for the Portuguese navigators of five centuries ago. Since that far-off time the altitude observation of the culminating Sun has been a tradition amongst seamen. There is a distinct sense of lost hope, even in these enlightened days, should a cloudy sky or an indistinct horizon preclude the observation of the Sun's meridian altitude. There has always been something sacrosanct about the so-called noon position by observation; and this position cannot be obtained unless the noon-day height of the Sun is measured.

9. LATITUDE BY EX-MERIDIAN ALTITUDE

The troublesome disadvantage of the midday Sun-sight is that the Sun and the horizon vertically below him must be visible at one particular instant of time during the day. In tropical waters it is seldom that these conditions are not satisfied; but in the seas of middle and high latitudes, where rain, fog and cloudy skies are common, it is often impossible for the seaman to find his latitude by meridian altitude of the Sun. It is not surprising, therefore, that in pre-position-line days the enterprising navigator aimed to provide himself with alternative methods for finding latitude. Latitude, together with the lead, log and look-out, formed one of the four L's of the early mariner's creed.

The method for finding latitude which we are now about to describe, and which is known as the *ex-meridian method*, is still practised extensively by seamen. The history of the method, which dates from the middle of the 18th century, is full of interest, and the ex-meridian problem is almost as celebrated in the history of astronomical navigation as that of the double altitude.

Considerable attention was devoted to the ex-meridian problem during the 19th century—a period which was, in truth, a golden era of astronomical navigation. Many ingenious solutions were contrived and a diversity of ex-meridian tables were furnished, all aimed to facilitate the problem of finding latitude at sea.

It may be argued that the ex-meridian method became obsolescent as soon as it had been invented. The increasing use and reliability of chronometers—which were scarce instruments for a long time after Harrison had produced his famous timepiece in the 18th century—and the introduction of position-line navigation in the mid-19th century, brought astronomical navigation to a state of near perfection. Had logic prevailed in the chart-room, all the existing methods of astronomical navigation would have been swept away as soon as the 'New Navigation' of Marcq St

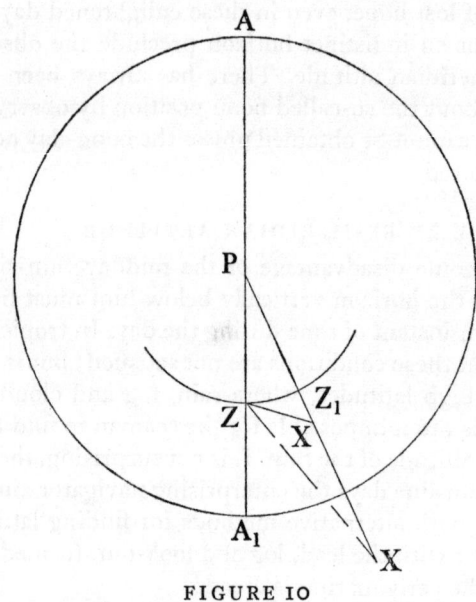

FIGURE 10

Hilaire had made its appearance. This, however, was not the case; many of the old methods of navigation remain with us, even to this day. Old-established practices die hard at sea, and modern navigators tend to manage their navigation in the same way as did their sea-faring ancestors.

The essential problem in the ex-meridian method for finding latitude is the comparison of the altitude of a celestial body at a place where the body is culminating (the latitude of the place being the same as that of the observer), with its altitude at the same instant at the observer's position.

In Fig. 10 the celestial sphere is projected on to the plane of the equinoctial. Z is the projection of the zenith of an observer, and

METHODS OF FINDING LATITUDE

Z_1 is that of the zenith of a place whose latitude is the same as that of the observer, and over whose meridian the body X is passing. P is the projection of the celestial pole and the circle is that of the equinoctial.

If the arc Z_1X can be found, the latitude of the place whose zenith is at Z_1 and, therefore, the observer's latitude, can also be found.

If l, d, z and h denote the observer's latitude, the body's declination, the body's zenith distance, and the time of the meridian passage of the body, respectively, we have, from the triangle PZX:

$$\cos h = \frac{\cos z - \sin l \sin d}{\cos l \cos d}$$

when l and d have the same name, and:

$$\cos h = \frac{\cos z + \sin l \sin d}{\cos l \cos d}$$

when l and d have different names.

When l and d have the same name:

$$\cos z - \sin l \sin d = \cos l \cos d \cos h$$

i.e.
$$\cos z - \sin l \sin d = \cos l \cos d(1 - \text{vers } h)$$

i.e.
$$\cos z - \sin l \sin d = \cos l \cos d - \cos l \cos d \text{ vers } h$$

i.e.
$$\cos z + \cos l \cos d \text{ vers } h = \cos l \cos d + \sin l \sin d$$

i.e.
$$\cos z + \cos l \cos d \text{ vers } h = \cos (l \sim d)$$

i.e.
$$\cos z + \cos l \cos d \text{ vers } h = 1 - \text{vers } (l \sim d)$$

i.e.
$$\text{vers } (l \sim d) = 1 - \cos z - \cos l \cos d \text{ vers } h$$

i.e.
$$\text{vers } (l \sim d) = \text{vers } z - \cos l \cos d \text{ vers } h$$

Similarly, when l and d have different names:

$$\text{vers } (l + d) = \text{vers } z - \cos l \cos d \text{ vers } h$$

In general:

$$\text{vers}\,(l \pm d) = \text{vers}\,z - \cos l \cos d \,\text{vers}\, h \quad . \quad . \quad (1)$$

Now $(l \pm d)$ is the meridian zenith distance of X at the place whose zenith is at Z_1. Let this be denoted by z_1. The latitude of this place and, therefore, the observer's latitude, may thus be found:

$$\text{Latitude of observer} = (l \pm d) \pm d$$

that is

$$l = z_1 \pm d$$

The above treatment is a modified form of that first given in 1754 by Cornelius Douwes, a figure famous in the history of the double-altitude problem. Douwes's investigation was modified by the Rev. James Inman D.D., whose famous nautical tables were first published in 1821. Inman's modification, used above, consists in adapting the formula to the tables of natural versines and haversines.

The ex-meridian method described above requires the use of a latitude by account, which should approximate to the observer's actual but unknown latitude. If the latitude found differs materially from that used it is necessary to repeat the computation, this time using the calculated latitude in place of the one initially used. Moreover, it is necessary for the observer to know his longitude; knowledge of this being required to find h, which figures in the computation.

It may readily be shown that:

Error in z_1 (and therefore error in calculated latitude) is proportional to cos latitude sin azimuth × error in h.

It follows, therefore, that the smaller the latitude or the nearer the azimuth to 90°, the greater will be the error in latitude consequent upon an error in h. Knowledge of correct time is all important in the ex-meridian problem.

It was early realized that when using stars for finding latitude in accordance with the ex-meridian method, those with big declinations gave the best results, this because of their relatively slow rates of change of altitude. The Pole Star, therefore, is admirably suited for the purpose; and accurate Pole Star tables have

METHODS OF FINDING LATITUDE

been available for the seaman since the early 19th century, although they were not included in the British *Nautical Almanac* until 1834.

A method for finding latitude by ex-meridian observation of the Sun, using 'direct spherics,' was given in the later editions of James Robertson's *Elements of Navigation*. This method is explained with reference to Fig. 11.

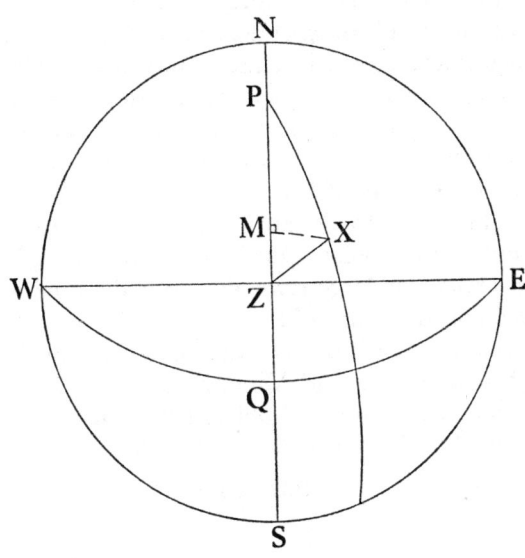

FIGURE 11

Fig. 11 is a projection of the celestial sphere on to the plane of the celestial horizon of an observer whose zenith is projected at Z. P is the projection of the elevated celestial pole; WQE that of the equinoctial; NZS the observer's celestial meridian. Arc XM is a perpendicular from X on to the observer's celestial meridian at M.

In triangle PMX:
$$\tan PM = \tan PX \cos P \quad \ldots \quad (2)$$
$$\cos PX = \cos MX \cos PM \quad \ldots \quad (3)$$

In triangle XMZ:
$$\cos ZX = \cos ZM \cos MX \quad \ldots \quad (4)$$

Divide (3) by (4) to eliminate MX:
$$\frac{\cos PX}{\cos ZX} = \frac{\cos PM}{\cos ZM}$$

i.e.
$$\cos PX \cos ZM = \cos ZX \cos PM$$
and
$$\cos ZM = \cos ZX \cos PM \sec PX \qquad . \quad . \quad (5)$$

From (2) and (5), arcs PM and ZM may be found: and these, when combined, will give arc PZ which is equivalent to the complement of the observer's latitude.

The direct method is independent of the latitude, and may be used to good effect even when the observed object has a large hour angle and azimuth, provided that the angle P is known with accuracy. A disadvantage of the method applies to cases in which the object's declination is small. In this event the arc PX is nearly 90°. Because the tangent of PX is required in (2), it is necessary, in this case, to extract the logarithm with great care. Moreover, any small error in the polar distance used will cause, in this circumstance, a relatively large error in secant PX used in (5), and this will lead to error in arc ZM.

An interesting case of the above method, described by W. R. Martin in his *Navigation and Nautical Astronomy*, applies when the declination of the observed object is less than about 1°, and the hour angle (P) is less than about half an hour. In this case:

$$PM \simeq PX$$
$$XM \simeq P$$
and
$$\cos PM \simeq 1$$

Therefore, for practical purposes:

$$\cos ZM = \cos ZX \sec P$$

The latitude is found by combining the computed arc ZM with the object's declination.

A method alternative to the direct method described above is known as the *Reduction to the Meridian Method*. This method involves the calculation of a correction to apply to the altitude of the body when out of the meridian to find its altitude when it is on the meridian. The reduction method, like that attributed to Douwes, requires the use of a latitude by account. If this proves to be materially different from the computed latitude it is necessary to repeat the computation using the computed latitude instead of the latitude by account.

METHODS OF FINDING LATITUDE 171

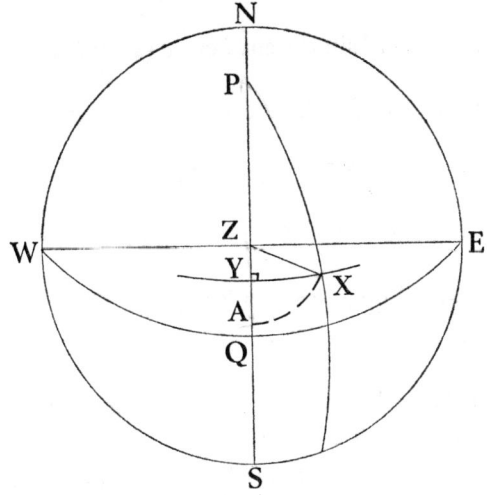

FIGURE 12

The reduction to the meridian method is attributed to the renowned French astronomer Delambre, who published it in 1814. The method is described with reference to Fig. 12.

Fig. 12 represents the celestial sphere projected on to the plane of the celestial horizon of an observer whose zenith is projected at Z. X is the projection of a celestial body whose declination is d and whose hour angle is h. Y represents the body's position when it is at meridian passage relative to the observer. The arc ZA is drawn equal to arc ZX.

$$PX = PY$$
$$ZX = ZA$$
$$YA = (ZA - ZY)$$

i.e.
$$YA = (ZX - ZY)$$

The arc YA is referred to as the reduction. Let this be denoted by r.

In triangle PZX:

$$\cos ZX = \cos PZ \cos PX + \sin PZ \sin PX \cos P$$

or
$$\cos z = \sin l \sin d + \cos l \cos d \cos h$$

Now,
$$\cos h = 1 - 2 \sin^2 h/2$$

Therefore:
$$\cos z = \sin l \sin d + \cos l \cos d\, (1 - 2 \sin^2 h/2)$$
i.e.
$$\cos z = \cos (l \mp d) - 2 \cos l \cos d \sin^2 h/2 \qquad (6)$$
Now,
$$ZX = ZA = (ZY + r)$$
Therefore:
$$\cos ZX = \cos ZY \cos r - \sin ZY \sin r$$

Since r is a small quantity and $ZY = (l \mp d)$ we have:
$$\cos z = (1 - r^2/2) \cos (l \mp d) - r \sin (l \mp d) \qquad (7)$$

By equating the values for $\cos z$ from (6) and (7) we have:
$$(1 - r^2/2) \cos (l \mp d) - r \sin (l \mp d)$$
$$= \cos (l \mp d) - 2 \cos l \cos d \sin^2 h/2$$
From which
$$r^2/2 \cos (l \mp d) + r \sin (l \mp d) = 2 \cos l \cos d \sin^2 h/2 \qquad (8)$$

The first term in (8) is small when the object is near the meridian. It may, therefore, be neglected. Hence:
$$r \sin (l \mp d) = 2 \cos l \cos d \sin^2 h/2)$$
and for practical purposes:
$$r = \frac{2 \cos l \cos d}{\sin (l \mp d)} \cdot \sin^2 h/2 \qquad (9)$$

Formula (6) may be transposed thus:
$$\cos (l \mp d) = \cos z + 2 \cos l \cos d \sin^2 h/2 \qquad (10)$$
i.e.
$$\cos z_1 = \cos z + 2 \cos l \cos d \sin^2 h/2 \qquad (11)$$

z_1 may, therefore, be found using a latitude by account provided that h is known accurately.

John Hamilton Moore, in the twelfth edition (1796) of his *Practical Navigator*, described a method for finding 'latitude by one altitude of the Sun when the time is not more distant than one hour from noon.' He gave one rule for finding the time from noon, based on formula (10); and another rule, based on formula (11), for finding the meridian zenith distance, and thence the latitude.

METHODS OF FINDING LATITUDE 173

The expression $2\sin^2 h/2$ was known as the *rising*. The rising figured in the double-altitude problem, and existing tables of *log risings*, therefore, were adapted to the ex-meridian problem.

Moore pointed out that the rule for finding time from noon should be applied to sights taken only when the Sun's altitude did not exceed 18°. He also noted that:

> 'an error in the supposed latitude can make very small difference in the change of altitude; and the nearer is the altitude taken to noon the better to find the change of altitude.'

He also warned against using the method when the time from noon exceeded one hour, and stated that in cases where the Sun's meridian altitude exceeds 60°, or when the latitude is small, good results were not to be expected unless the time from noon was much less than one hour.

Norie, in the 1814 (fourth) edition of his *Complete Epitome of Navigation*, gave the same rule as Moore's for finding latitude from a single observation of the Sun; but he did not specify any limit to time from noon. In the 1828 (ninth) edition of the same work, Norie specified that: 'in this method... the time from noon should not exceed 30 minutes.'

In the 1877 edition, the rule given in Norie's epitome was: 'The number of minutes in the time from noon should not exceed the number of degrees in the Sun's meridian zenith distance.'

This, the common rule quoted by seamen of our time, appears to have stemmed from the celebrated Raper.

After the ex-meridian method had become established among practical seamen, rules gave way to tables giving limits of time from meridian passage. In Rosser's *Self Instructor in Navigation*, published in 1885, we find a table similar to those in other textbooks of the time, giving limits of time from meridian passage (meridian distances) computed to give the number of minutes of meridian distance, when an error of half a minute in the time will produce an error of one minute of arc in the reduction. This is a reminder of Raper's definition of the term 'near the meridian.' According to Raper:

> 'The term implies a meridian distance limited according to latitude and declination, and also the degree of precision with which the time is known.'

The treatment on limits for ex-meridian by reduction given by J. Merrifield in his *Treatise on Nautical Astronomy*, first published in 1886, is interesting. After showing that:

Error in lat. = error in mer. dist. (h) sin Az. cos lat.

he concludes that:

'... as a rule, altitudes for latitude by circummeridional altitude (ex-meridian observations) should be taken within 20 minutes of the body's transit or when the object's azimuth is not more than one point.'

He then quotes the practical rule given by Norie (and others) quoted above.

Merrifield, in his discussion on ex-meridian sights, also pointed out (as others had also done) that, by finding latitude by observations of an object near the meridian both when it is east and when it is west of the meridian, and then meaning the results, '... the method is susceptible of very great accuracy.'

The books of Andrew Mackay, published during the early part of the 19th century, were among the more comprehensive works on astronomical navigation. In neither his *Theory of the Longitude*, in two volumes (1793), nor his *The Complete Navigator*, second edition 1810, do we find mention of the ex-meridian problem. Mackay did, however, describe a method for finding latitude by equal altitudes of the Sun. In this method half the elapsed interval between the times of the observations is equal to the time of either from noon (if the observer is stationary). From formula (11), putting half elapsed time equal to h, it is possible to find the meridian zenith distance, and hence the latitude.

The ex-meridian method for finding latitude became increasingly popular with the ever-increasing number of ex-meridian tables which appeared from the middle of the 19th century onwards.

If r in formula (9) is expressed in seconds of arc the formula becomes:

$$r = \frac{\cos l \cos d}{\sin (l \pm d)} \times \frac{2 \sin^2 h/2}{\sin 1''} \quad \cdot \quad \cdot \quad \cdot \quad (12)$$

Values of $(2 \sin^2 h/2)/\sin 1''$ were tabulated for suitable values of h: and by means of these tabulated values the computation of

METHODS OF FINDING LATITUDE

the reduction r may be performed with great facility. The table 'Reduction to the Meridian' which appears in Riddle's *Treatise on Navigation*, sixth edition (1855), gives values of the expression for values of h at every second of time up to twenty minutes.

An alternative solution was to use the log sine squared table and to add the constant 5·615455, which is the log of $2/\sin 1''$. Another similar solution was to use the log rising table and to add the constant corresponding to the log of $1/\sin 1''$.

It will be noticed that $2 \sin^2 h/2$ is equivalent to versine h. Inman's rule for the ex-meridian problem, which was given in his *Practical Navigation and Nautical Astronomy* (1855), was based on the formula:

$$\text{vers } z_1 = \text{vers } z - \text{vers } \theta$$

in which

$$\text{hav } \theta = \text{hav } h \cos l \cos d$$

One of the earliest of numerous ex-meridian tables was that of Captain J. T. Towson (of great-circle sailing fame), first published by the Hydrographic Office in 1849. The principles of Towson's tables are explained with reference to Fig. 11. In triangle PMX:

$$\sin \text{MX} = \sin \text{P} \sin \text{PX} \quad \text{(sine rule)}$$
$$\cos \text{PM} = \cos \text{PX} \sec \text{MX} \quad \text{(Napier's rules)}$$

In triangle ZMX:

$$\cos \text{ZM} = \cos \text{ZX} \sec \text{MX} \quad \text{(Napier's rules)}$$

From these three formulae we have:

$$\sin \text{MX} = \sin h \cos d$$
$$\cos \text{PM} = \sec \text{MX} \sin d$$
$$\cos \text{ZM} = \sec \text{MX} \cos z$$

Values of MX are tabulated in columns labelled 'Index Number.' Table 1 contains values of $(\text{PX} \sim \text{PM})$. Table 2 contains values of $(\text{ZX} - \text{ZM})$. To use the tables, Table 1 is entered using arguments d and h to extract Index Number and Augmentation 1. Augmentation 1 is added to the declination. Table 2 is then entered, using arguments altitude and Index Number, to extract Augmentation 2, which is the required reduction.

Towson's tables were designed specifically for Sun ex-meridian observations, as were others such as those of James Bairnson (c. 1880).

The ex-meridian tables of Brent, Walter and Williams, first published in 1886, provided for star sights as well as for Sun sights. When θ is small $\sin\theta = \theta$ radians nearly. Formula (12), therefore, may be expressed thus:

$$r = \frac{\cos l \cos d \ h^2 \sin^2 15'}{\sin (l \pm d) \ 2 \sin 1''} \quad \ldots \quad (13)$$

where r is in seconds of arc and h is in minutes of time. The tables of Brent, Walter and Williams were based on this formula.

It follows, from formula (13), that the change of altitude of a body when it is near meridian passage varies as the square of the meridian distance. That is:

$$r \propto h^2$$

This is the principle of an ingenious method for finding the reduction to the meridian graphically. The curve of r against corresponding meridian distance h is a parabola, which became known as Foscolo's parabola, after Professor Foscolo of Venice who devised this graphical method for solving the ex-meridian problem. Foscolo's parabola was published by the British Hydrographic Office in 1867.

'Cloudy Weather' Johnson, in his *Brief and Simple Methods for Finding Latitude and Longitude* (third edition 1895), gave a table for finding latitude by ex-meridian observations based on formula (1). He used the term 'reduced versine' for the natural versine corresponding to the log versine of h, diminished by the sum of the log secants of l and d. Three small tables—all at a single double-page opening—were provided for finding the reduced versine.

Johnson's rule for finding the meridian zenith distance was simple: 'The natural versine of the ex-meridian zenith distance diminished by the reduced versine is the natural versine of the meridian zenith distance.'

In the same work, Johnson introduced his methods for finding longitude by ex-meridian altitudes, and latitude and longitude by double ex-meridian.

Johnson's explanation of the reduction to the meridian, given

in his well-known *On finding Latitude and Longitude in Cloudy Weather*, concludes with the formula:
$$r = 2C \text{ hav } h \cos l \cos d \sec \text{ altitude}$$
where
$$C = 1 \text{ radian}$$

The upper part of Johnson's ex-meridian table in this work gives values of cos l sec altitude or N. The lower part gives values of 2 C hav $h \times$ N. The factor cos d was ignored (presumably because, when used with the Sun, cos d approximates to unity). Johnson justified himself by stating: 'A further correction (for the declination) may be applied when both the reduction and declination are considerable.'

Formula (13) may be written:
$$r = A h^2$$
where
$$A = \frac{\cos l \cos d \sin^2 15'}{\sin (l \pm d) \, 2 \sin 1''}$$

where l and d may be considered to have their meridian values.

The ex-meridian tables in present-day nautical tables are based on this relationship. The principle of the method was described by H. B. Goodwin as a kinematic principle applied to navigation. The kinematics in this case is related to the distance travelled by a body in any time t when it is accelerating at the rate of a. The same mechanical principle is used in the method for finding latitude known as the *short double altitude*.

Raper, in his *Practice of Navigation*, informs us that the first work in which a method occurs of finding the latitude by two altitudes observed near the meridian (but restricted to the same side) with an interval of a few minutes, is *Cours d' Observations Nautiques* by Ducom.

Robertson, in his *Elements of Navigation*, drew attention to the fact that during the 18th century the problem of finding latitude by three ascending or descending altitudes of the Sun '... exercised the talents of many ingenious persons.'

Numerous solutions to the problem were given, but the only case of practical value applied when the intervals between the first and second, and the second and third, observations are equal, and the observed object is near the meridian. The Abbé de la Caille is credited with giving a solution to this problem as early as 1760.

Let a be the meridian altitude and a_1, a_2 and a_3 the ex-meridian altitudes observed at intervals of t. If h is the time from meridian passage at which the middle of the three ex-meridian observations are made:

$$a = a_1 - k(h \pm t)^2$$
$$a = a_2 - kh^2$$
$$a = a_3 - k(h \mp t)^2$$

where

$$k = \frac{\cos l \cos d \sin^2 15'}{\sin (l \pm d) \, 2 \sin 1''}$$

From these relationships, the meridian altitude, and hence the latitude, may be found algebraically.

In an interesting paper by John White on the ex-meridian problem, published in the *Nautical Magazine* in 1896, the following formula for the reduction, based on a formula by Godfray, was given:

$$r = \frac{h^2}{2(\tan l + \tan d) \sin 1' \operatorname{cosec}^2 15'}$$

This formula suggested the construction of a table giving values of $2 \tan l \sin 1' \operatorname{cosec}^2 15'$, and $2 \tan d \sin 1' \operatorname{cosec}^2 15'$, against arguments h in minutes of time from unity to 60. By means of this table the denominator of the formula given above could readily be found and the solution to the ex-meridian problem thereby facilitated.

The Admiralty published, in 1895, a diagram devised by White for obtaining the reduction in an ex-meridian altitude observation.

The well-known nautical publishers, Messrs J. D. Potter of the Minories, published a diagram for solving the ex-meridian problem. This first appeared in 1897, when the inventor, F. Kitchin, a naval instructor on H.M.S. *Britannia*, described his diagram in the *Nautical Magazine* in 1901.

Notable among a profusion of ex-meridian tables published during the present century are those of Captain H. S. Blackburne. In a gentle rebuke, Captain Blackburne stated in the 1918 edition of his *Tables for Azimuth, Great-Circle Sailing, and Reduction to the Meridian*, first published in 1905, that: 'the only ex-meridian tables which are at present allowed to candidates for the B.O.T. examinations are those of Towson and Raper.' This restriction doubtless limited the use of other ex-meridian tables.

METHODS OF FINDING LATITUDE

Blackburne devised several ex-meridian tables, each designed for a special purpose. He was a great advocate of star sights and, in his *Excelsior Ex-meridian and Position-Finding Tables* first published in 1917, he gave a comprehensive set of tables giving azimuths and reductions for 29 of the brighter stars.

The ex-meridian method for finding latitude has, since its introduction, been a firm favourite with seamen and ex-meridian tables are still widely used.

When a heavenly body is on the prime vertical circle its rate of change of altitude is proportional to the cosine of the latitude. It follows, therefore, that:

$$\cos \text{latitude} \propto \text{rate of change of altitude}$$

and

$$\cos l = k \cdot \frac{\text{change of altitude } a \text{ in time } t}{t}$$

If a is in seconds of arc and t is in seconds of time, $k = 1/15$. Therefore:

$$\cos l = \frac{1}{15} \times \frac{a}{t}$$

This relationship provides a method for finding latitude. It was described in some textbooks of the late 19th century. Some writers recommended observing the time taken for the Sun, when its true bearing was east or west, to rise or fall an angle equal to its own diameter, this requiring only one setting of the sextant. The change in altitude, in this case, could be checked against the value of the Sun's semi-diameter given in the *Nautical Almanac*.

CHAPTER VI
Methods of finding Longitude

1. INTRODUCTORY

Anaximander is often venerated by geographers who regard him as being the father of their subject. Credit for the invention of the map is due to this philosopher of Ancient Greece.

Anaximander, who flourished during the 6th century BC, noted that the stars appear to revolve around the celestial pole—a manifestation of the Earth's axial rotation. He concluded that the Earth must lie at the centre of a vast sphere on the inside surface of which the stars are fixed. Anaximander, therefore, may rightly be claimed to be the father of mathematical astronomy, as well as of geography, by virtue of his invention of the celestial sphere.

The Greek philosopher Democritus (fl. 450 BC) is credited with being the first to construct a rectangular map. His map was based on his personal travels which led him to the conclusion that the habitable part of the Earth is one and a half times as long in the east–west direction as it is broad in the north–south direction. This notion, which became the common view, is perpetuated by the very names we give to the two coordinates used to describe terrestrial positions, the words *latitude* and *longitude* being derived from *latus* and *longus* signifying, respectively, breadth and length.

The longitude of a place is a measure of the angle contained between the plane of the meridian of the place and that of a standard meridian from which longitudes are measured east or west. The datum meridian commonly used is that of Greenwich, so that the meridian of Greenwich and its antipodal meridian divide the Earth into the eastern and western hemispheres in much the same way as the equator divides the Earth into the northern and southern hemispheres.

We have noted, in our discussion on the history of the latitude, that finding the latitude of a place was but a trivial problem to the

METHODS OF FINDING LONGITUDE

Greeks of antiquity. Finding longitude, on the other hand, presented considerable difficulty.

The difference between the longitudes of two places is equivalent to the difference between their local times, reckoning 15° of difference of longitude to one hour difference of local times. The problem of finding longitude, therefore, is one in which the local time at a particular instant is to be compared with the local time at the datum or standard meridian, for the same instant. If, for example, the local time is 6 a.m. at the instant when it is 8 a.m. at the Greenwich meridian, the local longitude is two hours or 30°. The name of the longitude, in this instance, is west of the Greenwich meridian, because the spin of the Earth to the east results in clock times at the Greenwich meridian being in advance of those at places in the western hemisphere and behind those in the eastern hemisphere.

Apart from the rough method of dead reckoning, in which estimations of the ship's courses and distances are made, there are two principal methods by which the longitude of a terrestrial position, east or west of a given meridian, may be found. One is an astronomical method, and the other is mechanical.

The astronomical method of finding longitude involves the use of a predicted time of an astronomical event, such as an eclipse or occultation, the predicted time being given for a particular datum meridian. If the local time of the event may be found, the longitude of the place east or west of the datum meridian may also be found by comparing the local and predicted times.

The mechanical method of finding longitude involves the use of a timepiece. If the rate of gaining or losing is known, and if the error of the timepiece on the time at some datum or prime meridian is also known, the correct standard time may be found. This, compared with the local time of any given moment, will give a measure of the local longitude east or west of the standard meridian.

The ancient philosophers of the Mediterranean world used as a datum line, from which longitude was measured eastwards, the meridian through the 'Fortunate Isles'—believed to be the Canaries. These islands were believed to be the westernmost part of the habitable Earth. Throughout classical times, and during the period of the Renaissance, the meridian through the 'Fortunate Isles' continued to be the *first* or *prime meridian*.

Following the Golden Age of discovery, when the peoples of Western Europe emerged as sea-traders, almost every European nation used as a prime meridian one that passed through its State territory. The French, for example, used the meridian of Paris; the Dutch, that of Amsterdam; the English, that of London; and so on.

It is interesting to note that as far back as the late 17th century, when so many prime meridians had already been established, we find a Spanish professor writing to the Royal Society in London proposing a new place for the first meridian. The comment on his proposal was no less plain than it was lamentable. '... a thing which never could be accomplished.'

The difficulties related to the profusion of prime-meridians—difficulties thought by many to be insurmountable—were not brought to an end until the closing decades of the 19th century, when it was decided by international agreement to adopt the meridian through Greenwich as the prime meridian from which longitudes should be measured.

2. LONGITUDE FROM ECLIPSE OBSERVATIONS

Ancient maps were constructed on the basis of observed latitudes and estimated longitudes. East-west errors, therefore, were in many cases considerable. We have noted in Chapter I the improvement which the great Hipparchus introduced in relation to mapping the Earth's surface. He is credited with being the first to suggest the use of eclipses of the Moon for finding the longitudes of places. Hipparchus explained how the difference of longitude between two places could be found from a comparison of the times at the two places of the occurrence of a lunar eclipse. It appears, however, that during the following three centuries very few eclipse observations had been made for this purpose; and we find the famous Ptolemy complaining about this in AD 150.

During the period of the Great Discoveries, Columbus, on his second voyage in 1494, had recourse to the method of finding longitude by observation of an eclipse of the Moon. It is recorded that the result of his observation, which marked the first attempt to find the position of a place in the New World using astronomical principles, was in error to the extent of about 18°. This error was due to inaccurate predictions of the time of the eclipse through imperfect knowledge of the motion of the Moon, and to the error

METHODS OF FINDING LONGITUDE 183

in the determination of the local times of the precise stages of the eclipse.

Some interesting accounts of finding longitude from eclipse observations are to be found amongst the early *Transactions* of the Royal Society of London. One writer describes how the method was employed for finding the longitude of Moscow in the year 1688. Observations of a lunar eclipse were made at Leipzig and Moscow, the local times being 8.54 p.m. and 10.40 p.m. respectively. This showed that Moscow lies 1 hr. 46 mins. of time or 26° 30′ of longitude east of Leipzig. From an earlier eclipse observation it had been shown that Leipzig is 49 mins. of time or 12° 15′ of longitude east of London. It was concluded, therefore, that Moscow is 38° 45′ east of London.

In 1719 the astronomer Edmund Halley found the longitude of the Cape of Good Hope from an observation of an eclipse made at sea in latitude 34° 23′ S., at a position 180 leagues east of the Cape. From his observation he concluded that the Cape of Good Hope lies 11° east of London.

With the improvements in accuracy of eclipse predictions concomitant with the advance of astronomical knowledge, it became possible to estimate very accurately the path of the Moon's shadow across the Earth's surface during a solar eclipse. Eclipses of the Sun, therefore, as well as those of the Moon could be used for longitude determination. Perhaps the most noteworthy of these observations was that made by the famous navigator Captain James Cook. The account, part of which is extracted from the *Philosophical Transactions* of the Royal Society, is of considerable interest.

'Mr. Cook, a good mathematician and very expert in his business, having been appointed by the Lords Commissioners of the Admiralty to survey the sea-coasts of New Found Land, Labrador, etc., took with him a very good apparatus of instruments, among them a brass telescope quadrant by Mr. John Bird.

'Being August 5th 1766 at one of the Burgeo Islands near Cape Ray Latitude 47° 36′ 19″, the south west extremity of New Found Land, and having carefully rectified his quadrant, he waited for the eclipse of the Sun.... [He] considered the eclipse to have begun at 00 h 04 m 48 s apparent time [astronomical time] and to have ended at 03 h 45 m 26 s apparent time.

'Mr. George Witchell had exact observations of the same eclipse taken at Oxford by the Rev. Mr. Hornby, and ... from the comparison the difference of longitude of the places of observation, making due allowance for parallax and the Earth's spheroidal figure, was computed....'

Cook's place of observation became known as Eclipse Island and, according to his eclipse observations, its longitude was found to be 57° 36' 30" W. In 1874 the longitude of the same spot was found by means of electric telegraph to be 57° 36' 52", which speaks highly of the accuracy of the observations made by Cook and Hornby.

We are informed by Andrew Mackay, in his *Theory of the Longitude* of 1793, that the method of finding longitude from solar eclipse observations is:

'... the most accurate of any that has hitherto been employed. The difference of the meridians of two places may be found to the nearest second of time by comparing corresponding observations of the same eclipse.'

3. LONGITUDE FROM OBSERVATIONS OF JUPITER'S SATELLITES

A second astronomical method for finding longitude employed the satellites of Jupiter. Jupiter's four principal satellites were first observed in 1610 by Galileo, the famous mathematical professor of Padua.

The orbits of the satellites of Jupiter* are very nearly co-planar with the equator of Jupiter. The length of Jupiter's shadow cast by the Sun is about 600 times his diameter, whereas the distance of the outermost satellite is a mere 13 diameters from its parent planet. Consequently the satellites, in their orbital movements, are eclipsed. Jupiter's distance from the Sun is about five times the Earth's distance from the Sun; and the plane of Jupiter's orbit makes an angle of only about $1\frac{1}{2}°$ to the plane of the ecliptic. Therefore, the times of eclipses of the satellites are almost unaffected by the location of an observer on the Earth.

* Jupiter has at least twelve satellites although only four are relatively bright ones which are plainly visible through a ship's long glass.

METHODS OF FINDING LONGITUDE

The satellite nearest to Jupiter is called the First Satellite; the next the Second Satellite, and so on. The First Satellite makes one orbital revolution in 42 hours. The periods of the Second, Third and Fourth Satellites are 85, 170, and 400 hours respectively.

There are four different effects visible from the Earth: eclipses, occultations, transits of satellites, and transits of satellites' shadows. An eclipse occurs when a satellite enters Jupiter's shadow, and an occultation occurs when Jupiter himself hides a satellite. A transit of a satellite occurs when the satellite comes between the Earth and Jupiter, in which event the satellite appears as a tiny dark spot on the face of the planet. The entrance of the spot on the disc of Jupiter is called its ingress, and its leaving the disc, its egress. The transit of a shadow of a satellite occurs when the satellite lies on the straight line joining the satellite and the Sun.

There being four principal satellites and four effects for each satellite, the frequency of occasions when observations of the satellites may be made for the purpose of finding longitude is high.

Galileo, on discovering Jupiter's satellites, was quick to realize that the orbital movements of the satellites conformed to the planetary laws that had been enunciated by his famous contemporary Johannes Kepler. This demonstration provided compelling evidence in support of the Copernican view that the Earth and the other planets revolved around the Sun, for Jupiter and his attendant satellites could be regarded as being a small-scale model of the solar system.

Galileo seized upon the idea that tables of the eclipses and occultations of Jupiter's satellites could provide a method for finding longitude—especially at sea. In response, therefore, to the handsome reward offered by King Philip III of Spain to anyone who invented a practical method for finding accurately the position of a ship when out of sight of land, Galileo set about the task of preparing suitable tables for the purpose.

Galileo's tables of predictions were insufficiently accurate for the intended purpose; and it was not until after knowledge of the perturbations of the satellites, due to their mutual interactions, had been acquired that predictions were sufficiently accurate for finding longitude.

The famous Italian astronomer J. D. Cassini applied himself

to the problem of finding longitude by astronomical means and, in 1675, he prepared tables of predicted times of occultations and eclipses of Jupiter's satellites. The method was found to give good results for finding the longitudes of places on land.

The Danish physicist Roemer (1644–1710), in observing Jupiter's satellites at the Paris Observatory with the object of drawing up eclipse tables, discovered that predicted—and observed—times disagreed. He observed that the eclipses of the satellites were early compared with predicted times when Jupiter was relatively near to the Earth, and late when he was relatively remote from the Earth. This led him to the conclusion that light travels at a finite speed: and, moreover, he was able to make an approximate estimate of this speed.

In our own country Robert Hooke devoted some attention to drawing up tables of predictions of eclipses of Jupiter's satellites; but perhaps the most important work done in this connection was that of the celebrated Flamsteed.

John Flamsteed (1646–1719) acquired an interest in astronomy at an early age. Some papers he had written on the subject attracted attention, and he was appointed a member of a committee set up to report on a proposed method for finding longitude at sea. He became the first Astronomer Royal after the establishment of the Royal Observatory at Greenwich in 1675; and he is often regarded as being the first of the great English astronomical observers, being instrumental in introducing many improvements in observing methods. His *Historia Coelestis Britannica*, in three volumes published in 1725, some six years after his death, contains Flamsteed's catalogue of 3,000 stars.

Flamsteed drew up tables of eclipses and occultations of Jupiter's satellites, and contrived an instrument

> '... whereby with the sole help of the usual catalogue and the table of parallaxes of Jupiter's orbit, their [the satellites] distance from the axis of Jupiter may be found, to any given time within the compass of the year, and for any future year by the like tables.'

Numbers 151, 154, 165, 177, 178 of the *Transactions* of the Royal Society of London appear under Flamsteed's name; and all pertain to the problem of finding longitude from observations of eclipses of Jupiter's satellites. Flamsteed confessed it as part of his design to make

'... our more knowing seamen ashamed of that refuge of ignorance, their idle and impudent assertion that the longitude is not to be found....'

He goes on to say:

'Such of them as pretend to a greater talent of skill than others, will acknowledge that it might be attained by observations of the Moon, if we had tables that would answer her motions exactly: but after 2000 years we find the best tables extant erring sometimes 12 minutes or more in her apparent place which would cause a fault of $\frac{1}{2}$ hour or $7\frac{1}{2}°$ longitude. I undervalue not this method for I have made it my business to get a large stock of lunar observations for the correction of her theory and as a groundwork for better tables, but the examination will be a work of long time and if we should afterwards attain what we seek, that it will be found much more inconvenient and difficult than that I propose by observing the eclipses of Jupiter's satellites.'

Flamsteed anticipated the seaman's objections to the method he proposed—and very real objections they were. The long telescope required for the observation would be almost unmanageable on board a lively ship at sea; and the difficulty there would be in distinguishing the satellites from one another: these were the principal faults of the method from the seaman's viewpoint. Flamsteed pointed out the success the French had accomplished using the method, and they managed with telescopes of '14 feet long at most'! He also remarked that '... difficulty cannot be known until the method is tried....' and that '... use renders many things easy which our first thoughts conceived impracticable....'

Transaction number 214 of the Royal Society is entitled *New and Exact Tables for the Eclipses of the First Satellite of Jupiter reduced to the Julian Stile and the Meridian of London*. This was the work of Cassini who remarked that the method of finding longitude from observations of eclipses of Jupiter's satellites had been used for all the principal ports of France. Cassini had employed his skill to make easy and obvious to all capacities the calculations for finding longitude by this method.

Tables of eclipses of Jupiter's satellites were provided in the very first *British Nautical Almanac*, which made its first appearance in 1765 for the year 1767. Maskelyne, in his explanation to the tables, stated that:

> 'The eclipses of Jupiter's satellites are well known to afford the readiest, and for general Practice, the best Method of settling the Longitudes of Places at land; and it is by their means principally that Geography has been so much reformed within a Century past.'

It had been hoped that means might have been found for providing a telescope suitable for use on board ship for observing the eclipses of Jupiter's satellites. Maskelyne described the trial he made, during a voyage to Barbados under the direction of the Commissioners of Longitude in 1763, of a marine chair designed by a Mr Irwin for the purpose of facilitating the observations. He wrote:

> '... but I could not derive any advantage from the use of it ... and considering the great power requisite on a Telescope for making these observations well, and the Violence as well as the Irregularities of the Motion of the Ship, I am afraid the complete Management of a Telescope on Shipboard will always remain among the Desiderata.'

Maskelyne hastened to add, however, that he would not be understood to mean to discourage attempts founded upon good principles to get over this difficulty.

Many inventors, besides Irwin, attempted to provide the means of a steady platform suitable for use on board a rolling ship from which observations of Jupiter's satellites could be made. Commander Gould, in his work on the history of the marine chronometer, mentions several of these inventions. He also describes the steam-driven gyroscopic platform which was proposed in 1858 for the *Great Eastern*—that ill-fated leviathan of Brunel's.

The inclusion of tables of 'Eclipses of Jupiter's Satellites,' and diagrams of 'Configurations of the Satellites of Jupiter' in the

Nautical Almanac was aimed to induce keen nautical observers, in the interests of geography, to ascertain the longitudes of foreign places they visited from observations of Jupiter's satellites at a time when better methods were not available.

The method for finding longitude by eclipses of Jupiter's satellites involved observing the times of immersions (signifying the instants of disappearance of a satellite on entering the shadow of Jupiter), and emersions (signifying the reappearances of satellites on emerging from Jupiter's shadow). Comparisons of the times of observation with those given in the *Nautical Almanac* yielded the longitude of the place of observation.

For practical use at sea the method suffered, not only from difficulties of observing, but also because the eclipses do not occur instantaneously, this being due to the apparent diameters of the satellites not being inappreciable. Atmospheric effects also may affect the observations. Moreover, Jupiter is often near the Sun on the celestial sphere; and for long periods the method is not available owing to the planet not being favourably placed for observation. Many writers advocated the method for sea use; but practical seamen, and others who appreciated the difficulties of observing from a ship at sea, held little esteem for the method.

The French astronomer Lagrange (1736–1813) is credited with founding the dynamical theory of Jupiter's satellites; and the famous Laplace (1749–1827) is credited with the discovery of a remarkable numerical relationship between the satellites resulting from their mutual attractions.

The Swedish astronomer Wargentin, Secretary to the Royal Academy of Sciences at Stockholm, is noted for his tables of eclipses of Jupiter's satellites. His tables were published in the British *Nautical Almanac* for 1779 and for many succeeding years. From 1824 the predictions of the eclipses of Jupiter's satellites were from Delambre's tables; and from 1836 they were derived from Damoiseau's *Tables Ecliptiques des Satellites de Jupiter*.

4. LONGITUDE FROM OBSERVATIONS OF MOON OCCULTATIONS

The next astronomical method for finding longitude at sea, which is to demand our attention, is that in which star occultations by the Moon are employed.

The Moon, because of her real motion around the Earth, appears to move across the celestial sphere at a relatively rapid rate towards the east. This retrograde motion of the Moon is, on an average, about $\frac{1}{2}°$ per hour. Her angular diameter is also about $\frac{1}{2}°$; so that a star which lies within $\frac{1}{4}°$ of the Moon's path will be hidden by the Moon—a phenomenon known as a *star occultation*. The fact that the Moon has no atmosphere results in an occultation of a star taking place instantaneously.

The interest which ancient astrologers gave to occultations, particularly of planets by the Moon, must surely have led to the suggestion of the use of occultations for finding longitude.

The accuracy of predictions of occultations of stars or planets by the Moon depends largely upon knowledge of the complex motion of the Moon. It was not until the Moon's motion had been reduced to a reliable rule that accurate predictions became possible.

The method of finding longitude from an observation of an occultation of a fixed star is reckoned as the best means of finding longitude by astronomical means. The usefulness of the method is increased by the high frequency of star occultations; but the complex computations associated with the method renders it impracticable for finding longitude at sea.

The parallels of latitude between which particular stars cannot be occulted by the Moon were recorded in the early *Nautical Almanacs*. From this information an observer could ascertain whether or not the phenomenon would occur at his position.

During the time when the Moon is waxing, that is to say from New Moon to Full Moon, a star lying in the Moon's path will be occulted on the darkened limb of the Moon, because the enlightened edge, which faces the Sun, will be directed to the west, that is in the opposite direction to that in which the Moon is moving across the sky relative to the fixed stars. Similarly, when the Moon is waning, that is during the fortnight from Full Moon to Change of the Moon, a star lying in her path will be occulted at the enlightened limb.

To find the longitude from an observation of an occultation, the latitude of the observer and the local mean time of the observation must be known. The local mean time was usually ascertained before or after the occultation observation from an observation of the Sun or other object on or near the prime

METHODS OF FINDING LONGITUDE 191

vertical circle. The G.M.T. of the observation was also to be estimated as accurately as possible.

The problem is one of finding the Right Ascension of the Moon at the time of the occultation. It is a problem of great complexity, chiefly on account of the effects of refraction, semi-diameter of the Moon, and parallax of the Moon. After the Moon's R.A. has been computed—the star's R.A. being known and facilitating the operation—it is a simple matter to find the G.M.T. of the observation by interpolation, using the nearest tabulated values of the Moon's R.A. from the *Nautical Almanac*. Should the computed G.M.T. differ materially from the estimated G.M.T. of the observation, the problem should be re-worked.

5. LONGITUDE BY LUNAR TRANSIT OBSERVATION

Another lunar method for finding longitude involved finding the local time of the Moon's transit, and comparing it with the time of transit at a prime meridian. This method appears to have been first suggested in 1678 by Herne in a book entitled *Longitude Unveiled*.

The Moon, because of her retrograde motion across the celestial sphere relative to the fixed stars, crosses the meridian of a stationary observer later each day by a variable amount of time known as the *Moon's retardation*. Predictions of the times of the Moon's meridian passage for a standard meridian would facilitate finding longitude east or west of the standard meridian if the local time of meridian passage could be found. The unsuitability of this method for finding longitude is due to the difficulty of finding the exact local time of the Moon's transit. Numerous suggestions have been made for ascertaining local time of Moon's transit, chiefly using equal altitudes on opposite sides of the meridian; but the method has never been brought to a state whereby accurate longitudes could be found.

An interesting method for finding longitude by means of combined altitudes of the Moon and a fixed star was described by John Maurice of Chicago as late as 1900. Maurice's method involved measuring the altitudes of the Moon and a fixed star and finding therefrom their local hour angles. The sum or difference of these is equivalent to the difference between their Right Ascensions. The R.A. of the star being known, that of the Moon at the time of the observation may, therefore, be found. Having found the Moon's R.A., the G.M.T. may readily be found from the table of the Moon's R.A. given in the *Nautical Almanac*.

The method proposed by Maurice is an extension of that proposed by Lemonnier in 1771. Lemonnier proposed finding the longitude from the hour angle of the Moon. The hour angle, obtained from an altitude and the latitude, enables the observer to find the sidereal time of the observation from which the Moon's R.A. may be found.

6. REWARDS FOR DISCOVERING THE LONGITUDE

The problem of the longitude, until it was solved in the 18th century, engaged the attention of many able astronomers, mathematicians and physicists. The problem of 'east–west' navigation, as it was sometimes called, was one which was kept to the fore, largely as a result of the handsome rewards that were offered from time to time to anyone who solved the problem. To stimulate competition for the prizes offered, sums of public money were disbursed in order to give financial support to experimenters and investigators of this problem of the age. We have mentioned the reward offered by King Philip III of Spain in 1498. Other rewards were offered by the governments of France and Venice; and some private individuals offered prizes for the discovery of the longitude at sea as well. The most valuable, and the most famous of the prizes, was the considerable sum of money offered by the British Parliament in 1715. This prize appears to have been the only one that was ever paid for the discovery of the longitude.

Shortly before the passing of the Act of Parliament (12 Anne, Cap. 15), in which a reward was offered to any person who should invent or discover a practical method for finding longitude at sea, a committee was set up to investigate a petition submitted to Parliament in March 1714 by 'Several Captains of Her Majesty's Ships, Merchants of London, and Commanders of Merchantmen.' The petition, which was engineered by William Whiston, a dissenting clergyman who had held the Lucasian Chair of Mathematics at Cambridge, and Humfrey Ditton, who was a teacher of mathematics at Christ's Hospital School, set forth:

'That the discovery of longitude is of such consequence to Great Britain, for the safety of the Navy, for Merchant Ships, as well as for improvement of trade, that for want thereof many ships have been retarded in their Voyages, and many lost; but if due encouragement were proposed by the public, for such as shall discover the same, some persons would offer themselves to prove the same before the most proper judges, in order to their entire satisfaction, for the safety of men's lives, her Majesty's Navy, the increase of Trade, and the shipping of these Islands, and the lasting Honour of the British Nation.'

Whiston and Ditton had proposed an impracticable method for finding longitude at sea which was set out in a pamphlet entitled *A New Method for discovering the Longitude* published in 1714. Their proposal was to use pyrotechnic sound signals fired from vessels permanently moored at precisely defined positions along the oceanic trade routes. Following the publication of their proposal, and probably in order to give it publicity, Whiston and Ditton were instrumental in causing the petition to be submitted to Parliament.

A committee, which was appointed to examine the petition, consulted the eminent mathematicians and astronomers of the day, amongst whom were Sir Isaac Newton, John Flamsteed and Edmund Halley.

Newton, in his evidence, reviewed the several methods for finding longitude at sea that had, up to the time, been proposed. He did not favour the use of Jupiter's satellites, and held out little hope that the Moon could be employed for finding longitude. He also pointed out that the proposal made by Whiston and Ditton was merely a method of keeping an account of, rather than finding, the longitude at sea.

The famous bill which was passed following the adoption of the resolution formulated by the committee states

'... that nothing is so much wanted and desired at sea as the discovery of the longitude.'

Under the terms of the Bill, commissioners for examining, trying and judging all proposals, experiments and improvements relating to the problem of finding longitude at sea, were to be appointed. The commissioners, who became known as the Board of Longitude, were empowered to grant sums of money for experiments, and to determine the degree of exactness of any proposal. The reward offered for the discovery of the longitude was:

£10,000 for any method capable of determining longitude to an accuracy of 1°.

£15,000 for a method capable of determining longitude to an accuracy of 40'.

£20,000 for a method capable of determining longitude to an accuracy of 30' or ½°.

12 Anne, Cap. 15 stimulated not only eminent men of science but also numerous cranks and crackpots who put forward impracticable, and sometimes ridiculous, proposals. The several pamphlets published in 1714 soon after the Act had been passed, on methods of finding longitude at sea, bear testimony to this. There were many who believed that the wit of man would never reach a stage rendering it possible to find the longitude at sea. The phrase 'discovery of the longitude' entered common speech at the time, and was used to describe any practical impossibility. Nevertheless the discovery was to be made. Two methods, both of which had been proposed at least as early as the beginning of the 16th century, were brought within the bounds of practical utility at about the same time during the 18th century. These methods for finding longitude at sea are referred to as the methods of 'longitude by timepiece' and 'longitude by lunar distance,' respectively.

We shall say little about the development of the mechanical timepiece which made possible the solution to the longitude-by-chronometer problem. The method of using a clock which keeps accurate time for finding longitude had first been suggested by Gemma Frisius, the famous Flemish astronomer. The proposal had been made, not as a practical suggestion, but as a theoretical possibility, in a work entitled *De Principiis Astronomiae et Cosmographiae* which was published in Antwerp in 1530. In this same work Gemma gave several nautical axioms, included amongst which was a description of a nautical quadrant. At the time Gemma proposed the use of a timepiece for finding longitude, portable watches, using a spring as a prime mover, were of recent invention. His proposal was to lie dormant for about two hundred years before the art and science of clock-making had been perfected to render possible the production of a portable watch that could keep time on a moving ship with an accuracy sufficient to find the longitude with a reasonable degree of accuracy. The credit of the invention of the marine chronometer belongs to John

METHODS OF FINDING LONGITUDE 195

Harrison, the Yorkshire carpenter who carried off (although not without a great deal of trouble on his part) the £20,000 prize offered by the British Parliament. The history and development of the marine chronometer has been studied in great detail by the late Commander R. T. Gould, the fruits of whose painstaking researches appear in a work published in 1923. More recently, in 1966, Colonel H. Quill in his book, *John Harrison: the Man who found Longitude*, has provided his readers with a penetrating study of the life of John Harrison set against the background of the fascinating story of the development of the marine chronometer.

7. LONGITUDE BY LUNAR DISTANCE
(a) HISTORICAL SURVEY

We now come to a discussion of the method for finding longitude at sea by lunar distances. To open our discussion we shall quote part of the foreword to Gould's *Marine Chronometer*, contributed by Sir Frank W. Dyson, the Astronomer Royal at the time the work was published.

'The problem of making an almanac of the Moon's position,' wrote Sir Frank, 'is most difficult, as may be seen from the fact that in spite of the attention devoted to the lunar theory by some of the world's greatest mathematicians, it was not until 1767 that the *Nautical Almanac* was able to give predictions of the Moon's place with sufficient accuracy for them to be of use for purposes of navigation. From that time to the present day, distinguished mathematicians of England, France, Germany and America have given large portions of their lives to lunar theory. More arithmetic and algebra have been devoted to it than to any other question of astronomy or mathematical physics, but, in the end, the problem has been solved so that observed positions agree very closely with those predicted. Unfortunately, even with perfect tables, it is found that the most skilful navigator cannot obtain a very accurate position of his ship in this manner. With great pains and sometimes elaborate calculation he can be correct to within 20 miles.'

Three important points emerge from Sir Frank's comments. First, the difficulty of predicting the Moon's place; second, the difculty of observation, which limits the degree of accuracy of the method even when perfect predictions are available; and third, the elaborate calculations associated with the method.

The difficulties of observation were largely overcome by the invention and development of the reflection measuring instruments which have been discussed in Chapter III. Our main purpose in the pages that follow immediately will be related to an historical account of, and the computations involved in, the lunar method for finding longitude at sea.

The Moon's motion across the background of the fixed stars is more rapid than that of any other visible celestial body. She completes her circuit of the Heavens in a sidereal period of about $27\frac{1}{3}$ days. She moves, therefore, at an average rate of 33″ of arc in every minute of time. When she is at perigee her rate of travel is greatest and is about 36″ per minute: when she is at apogee her rate of motion is slowest and is about 30″ per minute. It is the Moon's rapid motion relative to the stars which provides a method for finding longitude. The navigator may regard the Moon as the hand of a celestial clock, so that the 'mechanism of the Heavens' provides him with a method for measuring absolute time.

The principle of the method of finding longitude by lunar distance is simple. The angle at the Earth's centre between the Moon's centre and that of the Sun, planet or any star, may be predicted if the motion of the Moon is known completely. A measured angle between the Moon and Sun (or other heavenly body), or lunar distance as it is called, may be reduced to what it would have been at the Earth's centre at the time of the observation. The reduced measured distance, when compared with the predicted or geocentric distance, provides a measure of the longitude of the observer east or west of the meridian for which the predictions apply.

The principle of finding longitude from an observation of the angular distance between the Moon and Sun, planet or zodiacal star, could hardly fail to suggest itself to an astronomer possessing a clear notion of the longitude problem and an understanding of the character of the Moon's motion relative to the fixed stars. The credit for being the first to suggest the method in print is given to Johannes Werner of Nuremberg who mentioned it in the first volume of Ptolemy's *Geography* which Werner edited in 1514. Werner not only suggested the method but also suggested the use

METHODS OF FINDING LONGITUDE

of the cross-staff 'as a very proper instrument for observing lunar distances.' A century later, in 1615, a lunar observation was made by the English explorer William Baffin with the express purpose of finding the longitude of a ship at sea.

Baffin is credited with being the first Englishman to make a lunar observation at sea. Baffin sailed in 1615 as 'mate and associate' of Robert Bylot, who commanded the *Discoverie* on a voyage in quest of the North-west Passage. When the ship was beset in ice off the Greenland coast Baffin took a complete lunar observation using the Moon and Sun. The altitudes of the two bodies were measured by means of a cross-staff, and the angle between them was found by means of azimuth observations. This method of finding the lunar distance does not lend itself to great accuracy but, doubtless, was employed in the absence of a suitable instrument capable of measuring large angles. It is not unlikely that the method of longitude by lunar distance was in the minds of many of the more enlightened navigators and astronomers of the period.

Peter Apian, in the tenth chapter of his *Cosmography* published in 1524, and the famous Gemma in his *De Principiis*, mentioned above, described the method of longitude by lunar distances.

Werner, Apian, and Gemma Frisius, treated the problem of finding longitude by lunar distance as more of a speculative, rather than a practical, method. At the time when they wrote, neither lunar tables of sufficient accuracy nor suitable angle-measuring instruments were in existence to make the problem one of practicable possibility.

Gemma Frisius, in his description of the method, discussed the refraction and parallax corrections that are necessary to apply to the observed lunar distance before it can be compared with a predicted distance in order to find the longitude of the place of observation. He pointed out that the method of making allowances for refraction and parallax is a trigonometrical process far too complex for practical seamen. There is no doubt, judging from his descriptions, that Gemma had a perfectly clear understanding of the lunar problem. Two centuries, however, were to elapse before the lunar method became practicable for finding longitude at sea.

The principal factor responsible for the delay in the solution to the lunar method for finding longitude was the rudimentary state of lunar theory. This was brought to a state of near-perfection in the middle of the 18th century.

The celebrated Kepler regarded the lunar method for finding longitude in favourable light and, in his *Rudolphine Tables*, he gave directions for observing the distance between the Moon and a star. He also gave directions for making the necessary computations. The method was also recommended by Christian Severin (Longomontanus) (1562–1647), the Danish astronomer and assistant to Tycho Brahe. Master Thomas Blundeville, in his famous *Exercises ... for Young Gentlemen ...* first published in 1594, also mentioned the method of the lunar distance, attributing it to Peter Apian.

We find in 1633 the French professor of mathematics at Paris, Jean B. Morin, proclaiming with much boasting that he had found the solution to the longitude problem. Morin requested Cardinal Richelieu to appoint a commission before whom he [Morin] could have the opportunity of establishing his claim to the honour of finding the solution to the problem and also of establishing his right to any reward that might have been forthcoming. The commission was appointed and, in 1634, Morin demonstrated the observations and mathematical computations necessary for finding the longitude by lunar distance. Morin is credited with being the first to provide detailed mathematical rules requisite for the solution to the problem; but, of course, he did not bring the problem any nearer to a practical solution. The remarks made by Capitaine de Frégate M. E. Guyou, a member of the French Bureau of Longitude, are interesting and relevant:

'The solution,' wrote Guyou in 1902, 'required nothing less than the genius of Newton, Clairaut and Euler, and the mass of observations patiently collected for three quarters of a century by Flamsteed, Halley, Lemonnier, and others. ...'

Morin had simply limited himself to an investigation of the spherical trigonometrical computations necessary for the solution of the lunar problem. These calculations, although complex in character, presented no real difficulty and had, hitherto, been passed over on account of their entire want of utility.

In a work on geography by Carpenter, printed at Oxford in 1635, notice is taken of a method of finding longitude by lunar distances. This is noted by Andrew Mackay in his *Theory of the Longitude*, where he quotes Carpenter as saying:

METHODS OF FINDING LONGITUDE 199

'This way, though more difficult may seeme better than all the rest, for as much as an Eclipse of the Moon seldom happens, and a Watch, Clocke, or Hour Glasse cannot so well be preserved, or at least so well observed in so long a Voyage, whereas every Night may seeme to give occasion to this Experiment, if so bee the ayre be freed from Clouds, and the Moone show her Face above the Horizon.'

The first effectual step made to promote the solution to the problem of finding longitude at sea by astronomical methods was the establishment of the Royal Observatory at Greenwich in 1675, and the appointment of Flamsteed as the first Astronomer Royal. Flamsteed's commission enjoined him:

'. . . to apply himself with the utmost care and diligence to the rectifying of the tables of the Motions of the Heavens and the places of the fixed stars, in order to find out the so-much desired longitude at sea, for perfecting the art of navigation.'

Up to the time of the setting up of the Royal Observatory at Greenwich the lunar tables in use were compiled entirely from the results of observations, and astronomers despaired of ever being able to predict with certainty the Moon's celestial position, her motion being so complex. With the great work on celestial mechanics, rendered possible by the illustrious Newton whose famous *Principii* was published in 1687, undertaken by several eminent 17th-century astronomers, the theory of the Moon's motion was gradually brought to a state of perfection. Newton himself undertook the construction of tables of the Moon's position founded on the theory of gravitation combined with the observations of Flamsteed. And Professor Hoyle, in a recent book, *Astronomy* 1962, recalls Newton's declaration that working out the future motion of the Moon '. . . is difficult, and it is the only problem that ever made my head ache.'

Edmund Halley, who had had valuable sea-going experience and who was, therefore, admirably qualified to talk on the matter, recommended observations of the Moon as providing the most certain method of finding longitude at sea. He had found, from experience, that all other methods were impracticable; but he also pointed out to the Royal Society the defects of the lunar tables extant. Halley published an empirical method for reducing the errors in the existing tables. This method he evolved after making careful comparisons of the Moon's positions by observations with those given in the tables. Halley found, as he had expected from his application of the principles of gravitation, that the errors of the tables recurred with regularity after a period of eighteen years eleven days—a well-known eclipse cycle known to the astronomers of antiquity as the *Saros*. Halley's paper on the subject appears under No. 420 of the *Philosophical Transactions*.

Being encouraged by his discovery of the recurrence of error after a period of eighteen years eleven days, he examined what error might arise in a period of nine years less nine days, during which period 111 lunations occur. In this manner he deduced a rule for correcting the lunar tables. Unfortunately, Halley was unable to extend his lunar observations over the whole of the eighteen-year period until he succeeded Flamsteed as Astronomer Royal in 1720. The fact that he was well over sixty years of age at the time did not, in any way, deter him from undertaking the long series of lunar observations with the object of improving lunar theory. In 1731 he announced to the Royal Society that he had taken

'... with my own eye without any assistant or interruption ... 1500 observations of the Moon ... more than Tycho, Hevelius and Flamsteed had taken altogether.'

He lived to see the completion of his tremendous project.

Halley discussed in detail the lunar method for finding longitude at sea in his *Astronomical Tables* in which he gave two complete examples using the distances between the Moon and two different stars. He mentioned, in the same work, that the method is applicable to a lunar distance using the Sun, instead of a fixed star, during the First and Last Quarters of the Moon.

The French astronomer Lemonnier undertook the task of making an eighteen-year period of lunar observations from which he drew up a table of corrections for the Moon's motion based on Halley's principle.

The Abbé de la Caille recommended the lunar method as being the only practical method for sea use. In the course of a voyage to the Cape of Good Hope in 1751, de la Caille had used the method for finding longitude. During the same year a French merchant captain, d'Apres de Mannevillette, in the service of the Compagnie des Indes, employed the lunar method using lunar tables based on Halley's principles.

Halley's empirical rule for correcting the Moon's position was not entirely satisfactory; and it was generally believed by astronomers that the inequalities of the Moon's motion were so complex that perfection in the lunar theory was not possible. Some argued, falsely, that the principle of gravitation was insufficient to explain the inequalities that affected the Moon's motion.

In 1750, the Academy of St Petersburg projected a competition the object of which was related to lunar theory. The French philosopher Clairaut presented a new theory of the Moon's motion accompanied by skeleton tables. But, because the work had to be submitted before a fixed date, the tables were not only incomplete, but they had a degree of accuracy less than that for which Clairaut's theory was susceptible. The comparison of Clairaut's tabulated lunar positions with observations made by Cassini and others exhibited errors of up to as much as 5' of arc.

The famous Swiss mathematician Leonhard Euler (1707–1783), whose work on lunar theory *Theorice Motus Lunae* was published in 1753, in an appreciation of the work of Clairaut, published in the *Philosophical Transactions of the Royal Society* for 1753, wrote:

'... It is Clairaut we have to thank for this important discovery which adds fresh lustre to the theory of the great Newton, and now for the first time, we may hope to have good astronomical tables for the Moon.'

Clairaut corrected his lunar tables in 1765, the year of his death; but ten years before this date, Johannes Tobias Mayer had compiled lunar tables which were submitted to the British Parliament, the author claiming at the same time some reward which he thought he might merit.

Tobias Mayer (1723–1762) was a self-taught German mathematician who, in 1751, was appointed Professor of Mathematics at Göttingen. His fame rests primarily on his skilful development of Euler's work in connection with lunar theory. His lunar tables were communicated in 1752 to the Royal Society of Göttingen.

Three years later he submitted an amended table in manuscript to the English Board of Longitude.

Mayer's lunar tables fell into the hands of Dr James Bradley, who had succeeded Edmund Halley as Astronomer Royal in 1742. Bradley, after examining the tables and comparing them with a large number of observations of his own, was convinced of their accuracy. In his report to the Admiralty dated February 10th 1765 Bradley wrote:

'... In obedience to their Lordships' commands I have examined the same, and carefully compared several observations that have been made (during the last five years) at the Royal Observatory at Greenwich, with the places of the Moon computed by the said tables. In more than 230 comparisons which I have already made, I did not find any difference so great as $1\frac{1}{2}'$ between the observed longitude of the Moon and that which I computed by the tables; and although the greatest difference which occurred is, in fact, but a small quantity, yet as it ought to be considered as partly arising from the error of the observations, and partly from the errors of the tables, it seems probable that, during this interval of time, the tables generally gave the Moon's place true within one minute of a degree.

'A more general comparison may perhaps discover larger errors; but those which I have hitherto met with being so small, that even the biggest could occasion an error of but little more than half a degree in longitude, it may be hoped that the tables of the Moon's motions are exact enough for the purpose of finding at sea the longitude of a ship, provided that the observations that are necessary to be made on shipboard can be taken with sufficient exactness.'

Long before Bradley had submitted this report he had shown the utility of Mayer's tables from investigation of the trials of the tables carried out by Captain Campbell on board the *Royal George* cruising within sight of Cape Finisterre in 1757, and again in sight of Ushant in 1758 and 1759. Observations made on board the *Royal George* were accurate to within 0° 37' of longitude. The lunar distances on these occasions were measured by means of Hadley's quadrant.

With the completion of the trials of Mayer's tables the problem of finding longitude at sea, to a degree of accuracy within the conditions of precision laid down by the Act of Parliament, had been solved. It might, therefore, be thought that Mayer was entitled to the reward. It was argued, however, that the method of calculating the longitude by the lunar method was far too complex for practical use; and since the solution was not given in a form available for the non-mathematical seaman, the practical conditions of the Act were not fulfilled. For this reason the Board of Longitude were led to postpone the granting of the prize.

The Board of Longitude existed from 1713 until 1828. It was empowered to grant half the reward for the discovery of the longitude as soon as a majority of the Commissioners agreed that a proposed method was practicable and useful. The remainder of the reward was to be paid as soon as a vessel, on which the method was used, sailed from Britain to a port in the West Indies and back without erring in the longitude more than the amount specified in the terms for the prize. It was not the lunar method that was instrumental in gaining the reward, although those who were responsible for perfecting the method did receive rewards. The prize was carried off, as we have already noted, by John Harrison for his ingeniously-contrived chronometer.

The perfection of lunar tables in itself was not sufficient for finding longitude at sea by the lunar method. Accurate measuring instruments and, perhaps more important than even this, methods of computing which could be reduced to relatively simple rules for the seaman were essential before the method was to become practicable.

(b) MASKELYNE AND THE *NAUTICAL ALMANAC*

Before entering upon a discussion of the methods of computation of the lunar problem, we shall first describe the development of that essential instrument of nautical astronomy known as the *Nautical Almanac and Astronomical Ephemeris*, with special reference to its founder Nevil Maskelyne.

The earliest of the principal astronomical ephemerides is the *Connoissance des Temps ou des Mouvements Celestes*, founded by Jean Picard in 1679 and published at Paris under the auspices of the Bureau de Longitude. From 1761 onwards the *Connoissance des Temps* gave the celestial positions of the Moon at twelve-hourly intervals, these positions being calculated by Mayer's method. Lunar positions were given using coordinates of the ecliptic system, that is, celestial latitude and celestial longitude respectively.

A considerable amount of computation is necessary in order to find the longitude from an observation of a lunar distance when the only assistance provided by an ephemeris consists of predicted celestial latitudes and celestial longitudes of the Moon at twelve-hourly intervals. The general practical method, in which much of the computation is dispensed with by providing the seaman with a special 'nautical' almanac, was first proposed by Abbé de la Caille. This almanac gave angular distances between the Moon's centre and that of the Sun and planets and certain zodiacal stars, for short equidistant intervals of time. A specimen of this type of lunar table, specially designed for the seaman's use, appeared in the 1761 *Connoissance des Temps*, in which lunar distances were given for four-hourly intervals. Guyou, in his article of 1902, mentions this important fact, and adds:

> 'But as has too often happened in our country [France] in connection with the application of science to the art of navigation, this proposal failed to bear fruit. It was the English astronomer Maskelyne who had the honour of carrying out the idea.'

Nevil Maskelyne, the 'Father of the Lunar Observation', was born in London in 1732. He was educated at Westminster School and Trinity College, Cambridge. He graduated in 1754 and in the following year was ordained. In 1761 he was deputed by the Royal Society to observe the transit of Venus at St Helena. He took with him a reflecting quadrant of Hadley's made by the famous instrument-maker Bird with the glass ground by Dollond; and a set of Mayer's tables. During the voyage out and home he determined the longitude of the ship using the lunar method. From St Helena he sent a letter to the Royal Society in which he described his observations (*Philosophical Transactions*, Vol. 52, 1762). He considered that his observations yielded the longitude, in each case, to within $1\frac{1}{2}°$ of the truth. On his return to England he set about the task of publishing a work which was to play a prominent part in the development of practical and scientific navigation. This work, entitled *The British Mariner's Guide*, was published in 1763.

METHODS OF FINDING LONGITUDE

Maskelyne spoke in high praise of the lunar method for finding longitude at sea. He induced the Commissioners of Longitude to grant the concession that a collection of lunar distances between the Moon and certain bright zodiacal stars, along the lines suggested by de la Caille, should be prepared for the use of seamen.

In 1765 Maskelyne succeeded Nathaniel Bliss as Astronomer Royal and, in his new position of authority, was able to put his proposal into effect. Under Maskelyne's guidance the first British *Nautical Almanac and Astronomical Ephemeris* was published in 1765 for the year 1767. The method of finding longitude by lunar distance was to become the standard astronomical method for finding longitude at sea and was to remain so for the whole of the 19th century.

Manuscript tables of the Sun and Moon, drawn up by Tobias Mayer, were received by the Board of Longitude in 1763, the year following Mayer's death. These tables were more accurate and more extensive than the original tables that had been submitted in 1752. Under Maskelyne's superintendence Mayer's lunar tables were published in 1770.

The Board of Longitude awarded the sum of £3,000 to Mayer's widow for his lunar tables. The celebrated Euler, who had furnished the theorems used by Mayer, was granted a reward of £300.

At the same time as the first *Nautical Almanac* appeared, Maskelyne published a set of tables entitled *Tables Requisite to be used with the Nautical Ephemeris for finding the Latitude and Longitude at Sea*. In the *Requisite Tables*—as this work familiarly became known—two excellent methods, with examples, for finding longitude by the lunar method, were described.

(c) PRINCIPLES AND PRACTICE

We shall now describe in some detail the principles and practice of the lunar method for finding longitude at sea.

In order to take a set of lunar observations in a regular and accurate manner, three assistants, in addition to the principal observer, are required. The principal observer measures the angle between the Moon's enlightened limb and a star, planet, or the Sun's limb. The first assistant measures the Moon's altitude; the second assistant measures the altitude of the second body; and the third assistant, armed with a watch, records the times of the observations. The three required angles should be measured simultaneously: the principal observer shouting 'stand-by' a little before each observation is to be made; and 'stop' when he obtains perfect coincidence of the enlightened limb of the Moon and the star, planet, or Sun's limb. The observations should be repeated so that four or five sets are obtained, each set being observed at approximately equal intervals within the space of six or eight minutes. The mean of each particular series of observations is then found: that is to say, the sums of the lunar distances, each of the two sets of altitudes, and the times should be divided by the number of sets. By so doing, small errors of observation are eliminated or reduced.

An expert observer might be capable of making, with a tolerable degree of accuracy, all the necessary observations himself. The normal way in which a single observer would operate would be to take several altitudes of the Moon and the second object in quick succession. Then several lunar distances would be observed; and finally several altitudes of the Moon and second object observed again. The mean of each set of observations would then be found, and the mean of the altitudes reduced to the time of the mean of the distances.

The altitudes of the Moon and the second object are required in order to ascertain the exact values of refraction and parallax-in-altitude for the Moon, and refraction for the second body. Moreover, the time of the observation must be known with tolerable accuracy in order to ascertain the declination and Right Ascension of the Moon (and the Sun if he is the second body). In circumstances when it is not possible to measure the altitudes of the Moon and second object at the time the lunar distance is measured, these angles must be computed by solving the appropriate PZX triangles.

In Fig. 1, Z represents the zenith of an observer, and HO represents that part of his horizon contained between the vertical circles through the Moon and second body (assumed to be a star).

arc MO   true altitude of Moon's centre
arc ZM   true zenith distance of Moon's centre
arc SH   true altitude of star
arc ZS   true zenith distance of star
arc MS   great-circle arc through true positions of Moon's centre and star: that is true lunar distance

The effect of atmospheric refraction on light from a star is to make it appear to have an altitude greater than its true altitude. Because of the star's immense distance from the Earth its parallax-in-altitude is considered to be nil. The elevating effect of refraction causes the star S to appear to lie at s, a point on the same vertical circle as that through its true position. The apparent position of the star is represented in Fig. 1 at s.

Refraction has an elevating effect on the Moon but parallax-in-altitude of the Moon is always greater than refraction: so that the

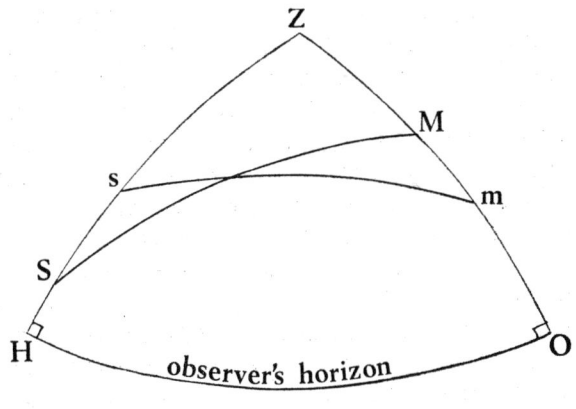

FIGURE I

depressing effect of parallax is greater than the elevating effect of refraction. Hence the apparent position of the Moon has a smaller altitude than that of the Moon's true position. The apparent position of the Moon is represented in Fig. 1 at m.

arc mO   apparent altitude of the Moon's centre
arc Zm   apparent zenith distance of Moon's centre
arc sH   apparent altitude of star
arc Zs   apparent zenith distance of star
arc ms   great-circle arc through the apparent positions of Moon and star: that is the apparent lunar distance

The apparent lunar distance of a star or planet is found by applying the Moon's semi-diameter to the measured lunar distance. The apparent lunar distance of the Sun is found by applying a combination of the augmented semi-diameters of the Moon and

Sun: the combination being dependent upon which limbs of the Moon and Sun are observed. If the adjacent limbs are observed, the apparent lunar distance is found by applying the sum of the semi-diameters of the Moon and Sun to the observed distance, etc.

(d) METHODS FOR CLEARING THE DISTANCE

When finding longitude by the lunar method, the most tedious part of the process is that which is known as *clearing the distance*. This involves reducing the apparent lunar distance to the true lunar distance or, in other words, clearing the apparent distance from the effects of parallax and refraction in order to ascertain the angle at the Earth's centre between the directions of the Moon's centre and the star, planet or Sun's centre. Many eminent astronomers and mathematicians, and many lesser men too, have given compendious rules, table or diagrams, to facilitate the problem of clearing the lunar distance. Among the numerous methods that have been popular at different times, are those given by Chevalier de Borda, Abbé de la Caille, Maskelyne, Delambre, Lyons, Witchell, Dunthorne, Krafft and Airy. As long ago as 1797 Mendoza del Rios, in a paper published in the *Philosophical Transactions of the Royal Society*, described forty different methods of clearing the lunar distance; and during the following hundred years numerous other methods were proposed, and many of the earlier methods were modified.

The multitude of methods available for clearing the lunar distance reduce themselves into two principal types; and the several rules given by different investigators are obtained by using different trigonometrical transformations of the common fundamental spherical trigonometrical formulae.

The first type to be described provides a rigorous solution to the problem and no approximations are used in the process. The solution is obtained as follows. Referring to Fig. 1, using arcs Zm, Zs and ms, all of which are deduced from observations made on the Earth's surface, the angle at the zenith contained between the vertical circles through m and s is calculated. Using the calculated angle mZs, and the true zenith distances ZM and ZS, the true distance MS is readily found. Typical of the methods based on this type of solution are those of Borda, Delambre, Krafft, Young and Airy, each of which we shall describe.

Borda's Method for Clearing the Distance

Chevalier Jean Borda, the French mathematician and astronomer, was born at Dax in 1733. He entered the French Navy and busied himself with nautical astronomical investigations for which he became well known. It was in his *On the Reflecting Circle*, published in 1787, and to which we have made reference in Chapter III, that Borda described his method of clearing lunar distances. At the time of its introduction to navigators, and for many decades afterwards, Borda's method was considered by competent authorities to be the best.

Referring to Fig. 1:
In triangle ZSM:

$$\cos Z = \frac{\cos SM - \cos ZS \cos ZM}{\sin ZS \sin ZM}$$

i.e.

$$\cos Z = \frac{\cos SM - \sin HS \sin OM}{\cos HS \cos OM} \quad \ldots \quad (1)$$

In triangle Zsm:

$$\cos Z = \frac{\cos sm - \cos Zs \cos Zm}{\sin Zs \sin Zm}$$

i.e.

$$\cos Z = \frac{\cos sm - \sin Hs \sin Om}{\cos Hs \cos Om} \quad \ldots \quad (2)$$

Let M be the true altitude of the Moon
m be apparent altitude of the Moon
S be true altitude of Sun or star
s be apparent altitude of Sun or star
D be true lunar distance
d be apparent lunar distance

Equating (1) and (2), we have:

$$\frac{\cos D - \sin S \sin M}{\cos S \cos M} = \frac{\cos d - \sin s \sin m}{\cos s \cos m}$$

Add unity to each side we have:

$$1 + \frac{\cos D - \sin S \sin M}{\cos S \cos M} = 1 + \frac{\cos d - \sin s \sin m}{\cos s \cos m}$$

i.e.
$$\frac{\cos S \cos M + \cos D - \sin S \sin M}{\cos S \cos M} = \frac{\cos s \cos m + \cos d - \sin s \sin m}{\cos s \cos m}$$

i.e.
$$\frac{\cos D + \cos (M+S)}{\cos S \cos M} = \frac{\cos d + \cos (m+s)}{\cos s \cos m}$$

i.e.
$$\frac{1 - 2 \sin^2 D/2 + \{2 \cos^2 (M+S)/2 - 1\}}{\cos S \cos M} = \frac{2 \cos (m+s+d)/2 \cos (m+s-d)/2}{\cos s \cos m}$$

From which:
$$\sin^2 D/2 = \cos^2 (M+S)/2 - \frac{\cos S \cos M}{\cos s \cos m} \cos (m+s+d)/2 \cos (m+s-d)/2$$

Let:
$$\cos^2 \theta = \frac{\cos S \cos M}{\cos s \cos m} \cos (m+s+d)/2 \cos (m+s-d)/2 \quad . \quad (X)$$

Then:
$$\sin^2 D/2 = \cos^2 (M+S)/2 - \cos^2 \theta$$

i.e.
$$\sin^2 D/2 = \tfrac{1}{2}\{1 + \cos (M+S)\} - \tfrac{1}{2}\{1 + \cos 2\theta\}$$

i.e.
$$\sin^2 D/2 = \tfrac{1}{2}\{\cos (M+S) - \cos 2\theta\}$$

i.e.
$$\sin^2 D/2 = \sin\{(M+S)/2 + \theta\} \sin\{(M+S)/2 - \theta\}$$

i.e.
$$\sin D/2 = [\{\sin (\overline{M+S}/2 + \theta)\} \sin \{(\overline{M+S}/2 - \theta)\}]^{\frac{1}{2}} \quad . \quad (Y)$$

Equations (X) and (Y) are used in the problem after being adapted to logarithmic computation as follows:

$\log \cos \theta = \tfrac{1}{2}\{\log \sec m + \log \sec s + \log \cos (m+s+d)/2 + \log \cos (m+s-d)/2 + \log \cos M + \log \cos S\}$

$\log \sin D/2 = \tfrac{1}{2}\{\log \sin (\overline{M+S}/2 + \theta) + \log \sin (\overline{M+S}/2 - \theta)\}$

A desirable feature of Borda's method is that it is simple and

METHODS OF FINDING LONGITUDE 211

direct and gives the true lunar distance without embarrassment of algebraic signs. It also has the advantage, when using it, that no special tables are required, apart from those of the common log trig functions.

The rules for solving the problem of clearing the distance using Borda's method are as follows:

1. Find M, S, m and s.
2. Correct the observed distance d for index error and semi-diameter.
3. Place under one another the apparent distance d and the apparent altitudes m and s: and take half their sum, L. From the half sum L subtract the apparent distance d. Under this place the true altitudes M and S.
4. Take from tables log secant m, log secant s and log cosines of L, $(L-d)$, M and S. Add these six quantities and divide the sum by 2. The result is the log cosine of θ.
5. Take half the sum of the true altitudes M and S. Call this ϕ. Find the sum of and difference between θ and ϕ. Add the sines of the sum and difference. Divide by 2. The result is the sine of half the true lunar distance, that is $D/2$.

A typical problem and its solution by Borda's method is Ex. 519 taken from Merrifield's *Nautical Astronomy* of 1886. Ex. Given the undermentioned data to compute the true distance between the Moon and the Sun.

Apparent Altitudes
$m = 13°\ 29'\ 27''$
$s = 31°\ 11'\ 34''$

True Altitudes
$M = 14°\ 18'\ 32''$
$S = 31°\ 10'\ 07''$

Apparent Distance
$d = 107°\ 52'\ 4''$

		°	′	″		
	d	107	52	04		
	m	13	29	27	sec	0·012151
	s	31	11	34	sec	0·067815
	2)	152	33	05		
L		76	16	32·5	cos	9·375207
$(L \sim d)$		31	35	31·5	cos	9·930337
M		14	18	32	cos	9·986314
S		31	10	07	cos	9·932295
ϕ		22	44	19·5		2)19·304119

θ	63	19	08·3	cos	9·652059
$\theta+\phi$	86	04	17·8	sin	9·998978
$\theta-\phi$	40	35	38·8	sin	9·813378
				2)	19·812356
$D/2$	53	40	43·2	sin	9·906178
		×2			
D	107	21	26·4	True Distance	

With the assistance of an auxiliary table, the operation of clearing the distance using Borda's method may be simplified. The auxiliary table alluded to gives values called *logarithmic differences* which are tabulated against arguments 'Apparent Altitude of Moon's Centre' and 'Moon's Horizontal Parallax.' The logarithm of the expression $(\cos S \cos M)/(\cos s \cos m)$ is the logarithmic difference; and, unlike some auxiliary tables used for abridging the calculations involved in clearing lunar distances, tables of logarithmic differences may be used with confidence.

Delambre's Method for Clearing the Lunar Distance

Delambre (1749–1822), the famous French astronomer who served on the French Bureau of Longitude, succeeded Lalande in 1807 as Professor of Astronomy at the Collège de France. He devoted considerable attention to nautical astronomical problems. The following method of Delambre's for clearing the distance is not dissimilar to that of Borda's.

Equating (1) and (2), as in Borda's method, we have:

$$\frac{\cos D - \sin S \sin M}{\cos S \cos M} = \frac{\cos d - \sin s \sin m}{\cos s \cos m}$$

From which:

$$\cos D = \frac{(\cos d - \sin s \sin m) \cos S \cos M}{\cos s \cos m} + \sin S \sin M$$

$$= \frac{(\cos d + \cos (m+s) - \cos s \cos m)}{\cos s \cos m} \cos S \cos M - \sin S \sin M$$

because $\cos (m+s) = \cos m \cos s - \sin m \sin s$,

$$\cos D = \left\{ \frac{\cos d + \cos (m+s)}{\cos s \cos m} - 1 \right\} \cos S \cos M + \sin S \sin M$$

$$= \left\{\frac{\cos d + \cos (m+s)}{\cos s \cos m}\right\}$$
$$\cos S \cos M - (\cos S \cos M - \sin S \sin M)$$

$$= \left\{\frac{\cos d + \cos (m+s)}{\cos s \cos m}\right\} \cos S \cos M - \cos (M+S)$$

$$= \left[\frac{2 \cos \tfrac{1}{2}\{(m+s)+d\} \cos \tfrac{1}{2}\{(m+s) \sim d\}}{\cos s \cos m}\right]$$
$$\cos S \cos M - \cos (M+S)$$

The logarithm of the first expression is calculated, and the natural number answering to this logarithm is extracted from log tables. The natural cosine of $(M+S)$ is then subtracted from this to give the natural cosine of the true distance D.

A notable feature of the above method is that tables of natural- and logarithmic-cosines are all that are necessary.

An example, taken from Professor J. R. Young's book on *Practical Astronomy etc.* published in 1856, illustrates the operation of clearing the distance using the above method of Delambre's.
Ex. Find the true distance if:

$d = 83° 57' 33''; m = 27° 34' 05''; s = 48° 27' 32''$
$M = 28° 20' 48''; S = 48° 26' 49''.$

	°	′	″			
$d =$	83	57	33	. .	Ar Comp cos	0·1783835
$m =$	27	34	05	. .	Ar Comp cos	0·0523390
$s =$	48	27	32		log 2	0·3010300
2)	159	59	10			
½ sum	79	59	35	. .	log cos	9·2399686
½ sum ~ d	3	57	58		log cos	9·9989587
M	28	20	48		log cos	9·9445275
S	48	26	49		log cos	9·8217187
					log 0·3442921	9·5369260
$(M+S)$	76	47	37	nat. cos.	0·2284595	
D	83	20	54	nat. cos	0·1158326	

It will be noticed that Ar Comps of cosines are used in the above example. These obviously correspond to secants. 'Ar Comp' is the abbreviation of 'Arithmetical Complement.' The Arithmetical Complement of a logarithm is the number it wants of 10·0000000. ... This artifice was used extensively in the practice of navigation in the 18th and 19th centuries, and was designed to facilitate problems on proportion, and for trigonometrical calculations. In the example above the divisor 'cos s cos m' is dealt with by taking the arithmetical complements of the logs of the cosines of s and m; and the solution is thereby facilitated in that the required log, viz. 9·5369260 is obtained by addition only. The problem of finding an arithmetical complement is quite easy of solution. Starting at the left hand every figure is subtracted from 9 except the last which is subtracted from 10. Thus, the Ar Comp of the cosine of 27° 34′ is 0·05233 to five places of decimals, because the log cosine of the angle is 9·94767, which, of course, is the log secant of 27° 34′.

Young's Method for clearing the distance

Young, whose work we have quoted above, gave the following method for clearing lunar distances.

From equations (1) and (2), in the development of Borda's method, we have:

$$\cos Z = \frac{\cos D - \sin S \sin M}{\cos S \cos M} \quad \ldots \quad \text{(I)}$$

and

$$\cos Z = \frac{\cos d - \sin s \sin m}{\cos s \cos m} \quad \ldots \quad \text{(II)}$$

By adding unity to each side of equations (I) and (II) we have:

$$1 + \cos Z = \frac{\cos D + \cos M \cos S - \sin M \sin S}{\cos M \cos S}$$

$$= \frac{\cos D + \cos (M+S)}{\cos M \cos S} \quad \ldots \quad \text{(III)}$$

and

$$1 + \cos Z = \frac{\cos d + \cos m \cos s - \sin m \sin s}{\cos m \cos s}$$

$$= \frac{\cos d + \cos (m+s)}{\cos m \cos s} \quad \ldots \quad \text{(IV)}$$

From (III) and (IV) we have:
$$\frac{\cos D + \cos (M+S)}{\cos M \cos S} = \frac{\cos d + \cos (m+s)}{\cos m \cos s}$$

From which:
$$\cos D = \{\cos d + \cos (m+s)\}\frac{\cos M \cos S}{\cos m \cos s} - \cos (M+S)$$

The computation of cos D, using Young's method, is shortened considerably if a table of logarithmic differences is used to evaluate (cos M cos S/cos m cos s).

Young pointed out that the altitudes of the objects are not required to such a high degree of accuracy as that for the lunar distance; and he remarks:

'This is a desirable circumstance, because, from the frequent obscurity of the sea horizon, it is more difficult to get the altitudes accurately than the distance.'

It is evident that, for a given distance d, small changes in the values of m and s, and the same changes in the values of M and S, cannot produce any sensible effect upon the value of D. It is obvious that the logarithmic difference is always nearly unity; and this is the principal reason why small errors in the altitudes do not sensibly affect the distance. Young goes on to say that it is important that the proper corrections be carefully applied to the observed altitudes to obtain the true altitudes, even though the former should not have been taken with precision—the relative values of the observed and true altitudes must still be preserved.

Krafft's Method for Clearing the Distance

Krafft's method was published in St Petersburg in 1791. Like Borda's, it is a rigorous method based on the common formulae of spherical trigonometry. As in Borda's and Delambre's methods, the spherical cosine formula is applied to triangles ZMS and Zms for finding Z; and the two expressions for cos Z are equated thus:

$$\frac{\cos D - \sin S \sin M}{\cos S \cos M} = \frac{\cos d - \sin s \sin m}{\cos s \cos m}$$

216   A HISTORY OF NAUTICAL ASTRONOMY

From which:
$$\cos D = -\cos(M+S) - \frac{\cos M \cos S}{\cos m \cos s}\{\cos d + \cos(m+s)\}$$

Let
$$\frac{\cos M \cos S}{\cos m \cos s} = 2\cos A \quad \ldots \quad (X)$$

Then:
$$\cos D = -\cos(M+S) + 2\cos A\{\cos d + \cos(m+s)\}$$

i.e.
$$\cos D = -\cos(M+S) + 2\cos A \cos d + 2\cos A \cos(m+s)$$

i.e.
$$\cos D = -\cos(M+S) + \cos(d+A) + \cos(d\sim A) \\ + \cos\{(m+s)+A\} + \cos\{(m+s)\sim A\}$$

Multiplying both sides by -1 and adding 1 to both sides we get:

$$1 - \cos D = \{1 + \cos(M+S)\} + \{1 - \cos(d+A)\} + \{1 - \cos(d\sim A)\} \\ \{1 - \cos(m+s)+A\} + \{1 - \cos(m+s)\sim A\} - 4$$

i.e.
$$\text{vers } D = \text{suvers}(M+S) + \text{vers}(d+A) + \text{vers}(d\sim A) \\ + \text{vers}\{(m+s)+A\} + \text{vers}\{(M+s)\sim A\} - 4$$

and:
$$\text{vers } D = \text{vers (sum zen. dists)} + \text{vers}(d+A) + \text{vers}(d\sim A) \\ + \text{vers}\{(m+s)+A\} + \text{vers}\{(m+s)\sim A\} - 4 \quad . \quad . \quad (Y)$$

Equations (X) and (Y) give the full solution to the problem. The rules for Krafft's method are as follows:

1. Find M, S, m, s and d.
2. Find the sum of the apparent altitudes and of the true zenith distances.
3. From the sum of log cos M, log sec m, log cos S and log sec s, subtract the log of 2; the remainder is the log cos of A.
4. Add vers $(d+A)$, vers $(d\sim A)$, vers $\{(m+s)+A\}$ vers $\{(m+s)\sim A\}$, and vers sum, $(Z+z)$, of the true zenith distances, and subtract 4 from the sum; the remainder is the versine of the true distance.

Krafft's method for solving the problem on p. 211 is given hereunder.

METHODS OF FINDING LONGITUDE 217

$\cos A = \frac{1}{2} \cos M \sec m \cos S \sec s$

$M =$	14°	18′	32″	log cos	9·986314
$m =$	13	29	27	log sec	0·012151
$S =$	31	10	07	log cos	9·932295
$s =$	31	11	34	log sec	0·067815
					9·998575
				log 2	0·301030
$A =$	60°	06′	30″	log cos	9·697545
$d =$	107	52	04		
$(m+s) =$	44	41	01		
$(Z+z) =$	134	31	12		

vers D = vers $(Z+z)$ + vers $(d+A)$ + vers $(d \sim A)$ + vers $\{(m+s)+A\}$ + vers $\{(m+s) \sim A\}$

$(Z+z) =$	134°	31′	21″	vers	1·701117	+	73
$(d+A) =$	167	58	34	vers	1·978026	+	36
$(d \sim A) =$	47	45	34	vers	0·327633	+	121
$\{(m+s)+A\} =$	104	47	31	vers	1·255165	+	145
$\{(m+s) \sim A\} =$	15	25	29	vers	0·035982	+	37
			parts for secs		412		412
$D =$	107	21	27	vers	1·298335		

The interesting feature of Krafft's method is that no logarithmic tables are required, and the true distance is deduced from addition of natural versines only

The method given by Dr James Inman in his *Navigation* of 1821, is a modified version of Krafft's method. The angle A in the equation is called, by Inman, the 'auxiliary angle' and, being pre-computed, is extracted from Inman's tables.

Although the use of the Auxiliary Table given by Inman is often called the Inman method for clearing the distance, the method is due, not to Inman, but to Mendoza del Rios.

Mendoza del Rios became well known to navigators during the early 19th century through his collection of nautical tables—a massive collection in quarto and containing over 600 pages. Some 300 pages of these tables were designed for facilitating the clearing of lunar distances. Rios undertook the considerable amount of labour of calculating values of vers $\{(m+s)+A\}$ + vers $\{(m+S)$

$\sim A\}$; and vers $(d+A) +$ vers $(d \sim A)$, for every minute of arc. The Rev. William Hall, a well known naval instructor remarked, in an essay written in 1902:

> 'It was a matter of 300 pages quarto containing 150,000 figures. I know nothing to equal it except a folio volume by the Board of Longitude containing the correction of the distances for every degree of distance, and of the two apparent altitudes with differences for each and corrections for parallax.'

We shall have occasion later to refer to the folio volume mentioned by Hall.

Airy's Method for Clearing the Distance

Sir George Bidell Airy (1801–1892) was born at Alnwick. He was educated at Hereford and Colchester before entering Cambridge where he had a brilliant academical career. He was appointed Astronomer Royal in 1835 in succession to Pond. To seamen Airy is well known for his work on ship magnetism and the corrections for deviations in iron ships. His work on lunar theory was of great importance; and it is interesting to note that Hansen's *Tables of the Moon* are dedicated to Airy.

Airy's method for clearing a lunar distance dates from 1882, from which time it became customary to place a copy of the method in the chart boxes supplied to H.M. ships. The method is direct, simple in application, requires logarithms to five figures only, and gives results with an accuracy sensibly perfect. It is described by Airy as follows:

> 'The characteristic circumstance upon which this treatment depends is the use, in the factors of corrections, not of each apparent element nor of the corresponding correcting element, but of the mean between the two.
>
> 'The elements which we require are—the apparent altitude and the corrected altitude of the Moon, the apparent altitude and the corrected altitude of the Sun, and the apparent and corrected distance. The first five of these are known accurately. The last (the corrected distance between the Sun and the Moon) must be estimated. There is no difficulty in doing this with accuracy abundantly sufficient for this investigation. With

METHODS OF FINDING LONGITUDE 219

Greenwich time by account, the distance may be rudely computed from the distances in the *Nautical Almanac*. Or, without time or calculation, a navigator accustomed to lunar distances may form a shrewd guess of the probable amount of correction. (The effect of a possible error will be exhibited hereafter.) We have now all the six elements required for the investigation.

'Let Moon's corrected altitude + Moon's app. alt. = $2A$
Moon's corrected altitude − Moon's app. alt. = $2a$
Sun's apparent altitude + Sun's corr. alt. = $2B$
Sun's apparent altitude − Sun's corr. alt. = $2b$
Corrected distance + apparent distance = $2C$
Corrected distance − apparent distance = $2c$

'Then:
Moon's apparent altitude = $A-a$
Corrected altitude = $A+a$
Sun's apparent altitude = $B+b$
Corrected altitude = $B-b$
Apparent distance = $C-c$
Corrected distance = $C+c$

'The essential circumstance which directs the further investigations is the equality of the zenithal angles and consequently of the cosines of the zenithal angles. The corresponding equation is:

$$\frac{\cos(C-c) - \sin(A-a)\sin(B+b)}{\cos(A-a)\cos(B+b)} = \frac{\cos(C+c) - \sin(A+a)\sin(B-b)}{\cos(A+a)\cos(B-b)}$$

or, multiplying out the denominators:

(First side) $\cos(C-c) \cdot \cos(A+a) \cdot \cos(B-b)$
$\quad -\sin(A-a) \cdot \sin(B+b) \cdot \cos(A+a) \cdot \cos(B-b)$
= (Sec'd side) $\cos(C+c) \cdot \cos(A-a) \cdot \cos(B+b)$
$\quad -\sin(A+a) \cdot \sin(B-b) \cdot \cos(A-a) \cdot \cos(B+b)$

'For development of these terms it must be remembered that:

$\sin(A+a) = \sin A \cos a + \cos A \sin a$
$\sin(A-a) = \sin A \cos a - \cos A \sin a$

$$\cos(A+a) = \cos A \cos a - \sin A \sin a$$
$$\cos(A-a) = \cos A \cos a + \sin A \sin a$$
and similarly for B and C.

'We now proceed to develop the first side. Making the substitutions just stated, the first large product gives:

Line 1, not containing $\sin a$ or $\sin b$ or $\sin c$:
$\quad +\cos A \cos B \cos C \cos a \cos b \cos c$

Line 2, containing simply $\sin a$, or $\sin b$, or $\sin c$:
$\quad +\cos A \cos B \sin C \cos a \cos b \sin c$
$\quad -\sin A \cos B \cos C \sin a \cos b \cos c$
$\quad +\cos A \sin B \cos C \cos a \sin b \cos c$

Line 3, containing $\sin a \sin b$, or $\sin b \sin c$, or $\sin a \sin c$:
$\quad -\sin A \sin B \cos C \sin a \sin b \cos c$
$\quad +\cos A \sin B \sin C \cos a \sin b \sin c$
$\quad -\sin A \cos B \sin C \sin a \cos b \sin c$

Line 4, containing $\sin a \sin b \sin c$:
$\quad -\sin A \sin B \sin C \sin a \sin b \sin c$.

'And the second large product gives:

Line 7:
$\quad -\sin A \cos A \sin B \cos B$

Line 2:
$\quad -\sin A \cos A \sin b \cos b + \sin B \cos B \sin a \cos a$

Line 3:
$\quad +\sin a \cos a \sin b \cos b$.

There is no line 4.

'We now examine the second side of the equation:

'The difference between the first side and the second side is this: that in every place where a occurs in the equation, or $\sin a$ in the development, of the first side, $-a$, or $-\sin a$ occurs on the second side; and similarly for b, $\sin b$; c, $\sin c$. And, moreover, these changes occur simultaneously; so that wherever $\sin a . \sin b$ occurs on the first side there will be $-\sin a . -\sin b$ on the second side; and where $\sin a . \sin b$, $\sin c$ occurs on the

METHODS OF FINDING LONGITUDE

first side there will be $-\sin a$. $-\sin b$. $-\sin c$ on the second side. And thus we see that:

'For line 1 the first side and the second side are the same.
'For line 2 the first side and the second side are equal but have opposite signs.
'For line 3 the first side and the second side are the same.
'For line 4 the first side and the second side are equal but have opposite signs.

'There transferring the second side with sign changed to the first side, the equation becomes the following:

$$\left.\begin{array}{l} +2.\cos A.\cos B.\sin C.\cos a.\cos b.\sin c. \\ -2.\sin A.\cos B.\cos C.\sin a.\cos b.\cos c. \\ +2.\cos A.\sin B.\cos C.\cos a.\sin b.\cos c. \\ -2.\sin A.\sin B.\sin C.\sin a.\sin b.\sin c. \\ -2.\sin A.\cos A.\sin b.\cos b. \\ +2.\sin B.\cos B.\sin a.\cos a. \end{array}\right\} = 0$$

'This equation is rigorously accurate.

'We will now consider what simplification it will admit, preserving the character of practical accuracy of the highest order.

'a, which is half the correction of the Moon's altitude, can never exceed 30'. Cosine of a can never differ from one by 1/20,000 part, sin a/a can never differ from one by 1/60,000 part; and therefore for cos a and sin a we may put 1 and a. The same applies to b and c. In the product sin a.sin b.sin c the factor of sin c can rarely or never amount to 1/50,000, and that term may be neglected. The equation now becomes:

$$\left.\begin{array}{l} +\cos A.\cos B.\sin C.2.c. \\ -\sin A.\cos B.\cos C.2.a + \cos A.\sin B.\cos C.2.b. \\ -\sin A.\cos A.2.b. + \sin B.\cos B.2.a. \end{array}\right\} = 0$$

'Remarking that $2a$, $2b$, $2c$ are the corrections of Moon's altitude, Sun's altitude, and distance, the result of this equation is:

Corr'n of Distance $\begin{cases} (+\tan A.\cot C - \sec A.\sin B.\operatorname{cosec} C) \\ \quad \times \text{correction of Moon's altitude} \\ (-\tan B.\cot C. + \sin A.\sec B.\operatorname{cosec} C) \\ \quad \times \text{correction of Sun's altitude} \end{cases}$

'The only opening in error in this formula is in the estimated value of C, as depending on error in the estimated *Nautical Almanac* distance, or in the estimated correction to the observed distance. Suppose that the time by account was 4 m. in error (implying error in longitude of 1°). The approximate correction of distance would be taken out about 2' in error, and C would be about 1' in error. If the value of the distance was about 60°, and error of 1' would produce in cot C an error of about 1/1400 of that term of the computed correction, and in cosec C the error would be 1/6000. These would be hardly sensible. But if, with C corrected by this approximation, the calculation be repeated (requiring only a few minutes), the error of result wll be totally insensible.

'The following is offered as a form proper to be used with this method:

'Prepare this table, inserting numbers instead of the printed words.

See Table 1.

'Then proceed with the following calculations, using 5-figure logarithms:

See Table 2.

'This form supposes that C is less than 90°. When C exceeds 90°, the supplement to 180° is to be taken, the cosec and cotan of that supplement are to be used, and the signs of the first and fourth numbers, which are produced by cotan C, are to be changed; the first number will become subtractive, and the fourth number additive.

'The second approximation will very rarely be required. If, however, the final "correction to apparent distance" differs from that assumed at the beginning by 2' or 3' it may be satisfactory to use the second approximation; it is very easy.'

Airy's method has been given in full to illustrate the remarkably clever manner in which the lunar problem was analysed and the ingeniously-contrived solution made possible.

The Approximate Method of Clearing Lunar Distances

The alternative type of solution to the problem of clearing a lunar distance is often referred to as the approximate method, of

Correct. to Moon's apparent alt. (additive)	Correct. to Sun's apparent alt. (subtractive)	First Approximation	Second Approximation (if necessary)
Moon's app. alt. Moon's cor. alt.	Sun's app. alt. Sun's app. alt.	Assumed cor. to app. distance. (add've) (sub've)	Assumed cor. to app. distance (add've) (sub've)
Sum A = half sum	Sum B = half sum	App. distance Cor. distance	App. distance Cor. distance
		Sum C = half sum	Sum C = half sum

TABLE 1

First Approximation	Second Approximation (if necessary)
Additive terms	Additive terms
log tan A	(Repeat) log tan A
log cot C	log cot C
log cor. to Moon's alt.	(Repeat) log cor. to Moon's alt.
Sum and number	Sum and number
log sin A	(Repeat) log sin A
log secant B	(Repeat) log sec B
log cosec C	log cosec C
log cor. to Sun's alt.	(Repeat) log cor. to Sun's alt.
Sum and number	Sum and number
Sum additive terms	Sum additive terms
Subtractive terms	Subtractive terms
log sec A	(Repeat) log sec A
log sin B	(Repeat) log sin B
log cosec C	log cosec C
log cor. to Moon's alt.	(Repeat) log cor to Moon's alt.
Sum and number	Sum and number
log tan B	(Repeat) log tan B
log cot C	log cot C
log cor. to Sun's alt.	(Repeat) log cor. to Sun's alt.
Sum and number	Sum and number
Sum subtr. terms	Sum subtr. terms
Combination of additive and subtractive = correction to apparent distance	Combination of additive and subtractive = correction to apparent distance

TABLE 2

which there are many varieties. The principles of the approximate method are described in relation to Fig. 2.

Perpendiculars are drawn from M and S, the true places of the Moon and Sun (or star or planet) respectively, on to the arc joining the apparent positions m and s. The difference between the apparent- and true-distances is found from the small triangles Mma and Ssb, either by successive approximations (hence the name approximate method), or from a development of an algebraic series of which the smaller terms are neglected.

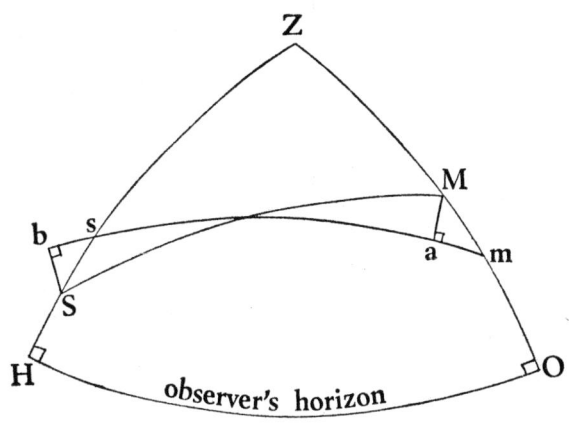

FIGURE 2

The invention of the approximate method for clearing a lunar distance is attributed to the English astronomer Israel Lyons.

Israel Lyons (1739–1775) was born at Cambridge. He showed great aptitude for mathematics; and the Master of Trinity, Dr Robert Smith, offered to provide for his education. He studied botany as well as mathematics, and read a course of lectures on botany at Oxford in 1764. Lyons was engaged by the Board of Longitude to calculate the *Nautical Almanac*—a task for which he received £100 per annum. In 1773, he was appointed by the Board of Longitude as astronomer to Captain Phipps during a polar voyage of discovery.

Lyons's method for clearing lunar distances was described in the *Nautical Almanac* of 1767. An alternative method for clearing lunar distances was given in the same work: this being Mr Dunthorne's method.

Dunthorne (1711–1775) of Cambridge, although not fortunate enough to have had an academic education, became an expert in many branches of learning, and particularly in astronomy. Both Lyons and Dunthorne received £50 each as a reward from the Commissioners of Longitude for their methods for clearing lunar distances.

In the *Nautical Almanac* for 1772 Maskelyne and George Witchell each contributed a method for clearing lunar distances. An improved version of Dunthorne's solution was also given. The same methods were described in the second edition of Maskelyne's *Requisite Tables*, which were prepared by William Wales and published in 1781.

Among the many successful attempts made to shorten the necessary calculations for clearing lunar distances, a notable effort was that made by Witchell, the mathematical master of the Royal Naval Academy at Portsmouth. Witchell conceived the idea of devising a table of corrections, based on the approximate method of clearing a lunar distance. In 1765 the Commissioners of Longitude awarded him £100 to enable him to complete and print 1000 copies of the tables. Later, they advanced a further £200. The work was continued by the Plumian Professor of Astronomy at Cambridge, and they thus became known as the *Cambridge Tables*. These are the tables referred to by the Rev. William Hall, mentioned above. They were published in 1772 and formed a ponderous folio volume containing the corrections for every degree of lunar distance, and for the two apparent altitudes, with differences for each degree, and corrections for parallax. It was a costly production—over £3,000 having been spent on it—and was ill-adapted for sea use.

Those varieties of the approximate method for clearing lunar distances, which are based upon accurate formulae, require a considerable amount of computation. More usually tables were employed. Such tables are based on the use of a mean refraction—no method being available for correcting the actual refractions which affect the observations. The approximate method, however, is capable of giving correct results; and, in fact, for the same amount of computation, the approximate method gives better results than the rigorous method.

Referring to Fig. 2: in the small triangles Mma and Ssb, which are right-angled at a and b respectively, the angles at m and s may

METHODS OF FINDING LONGITUDE 227

be calculated using three given sides in triangle mZs. The arcs Mm and Ss are also known, these being the corrections respectively to the Moon's and Sun's (or planet's or star's) altitudes; the triangles are small, the arc Mm never being more than about 1°, so that they may be treated as being plane right-angled triangles. In this case:

arc ma = correction to Moon's alt. × cos m
arc sb = correction to second body's alt. × cos s

When the angles at m and s are acute, as they are in the figure, ma is subtractive from, and sb is additive to, the arc sm in order to obtain arc SM. The contrary applies when the angles are obtuse. This principle is sometimes modified by calculating the effects of refraction and parallax-in-altitude of the Moon separately.

We shall now describe a selection of the approximate methods for clearing a lunar distance, commencing with a relatively modern one to best illustrate the method.

Merrifield's Method for Clearing a Lunar Distance

Dr John Merrifield, Ll.D., was headmaster of the Navigation School at Plymouth for many years during the last century. It is interesting to note that his son W. V. Merrifield, B.A., became headmaster, in succession to Mr J. Gill, of the Navigation School at Liverpool which had been established in the closing decade of the last century.

John Merrifield was a well-known author of works on navigation and, at the time of publication of his *Treatise on Nautical Astronomy*, published in 1886, the author had been engaged in teaching navigation for nearly a quarter of a century. In his treatise, Merrifield described a method for clearing lunar distances which he had invented, and which had appeared in the *Monthly Notices* of the Royal Astronomical Society for April 1884. The method is direct in its application, requires no special tables, and is claimed to be a very close approximation well adapted for sea use. The method is described in relation to Fig. 3.

In Fig. 3:

M and S are the true positions of the Moon and Second body respectively; and m and s are their apparent positions.

Arcs MS and ms are the true- and apparent-distances respectively.

Q is the point of intersection of arcs MS and ms.
With centre Q describe arcs Mp and Sq. SqQ and QpM are, therefore, right angles.

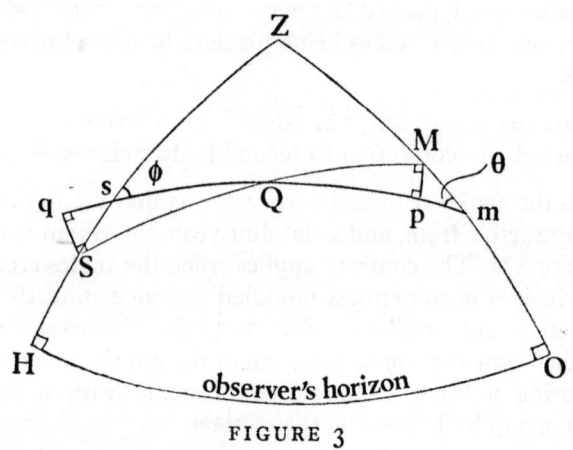

FIGURE 3

Let d be the apparent distance
D be the true distance
m the Moon's apparent altitude
s the second body's apparent altitude
y the Moon's apparent zenith distance
z the second body's apparent zenith distance

$$S' = \frac{y+z+d}{2} \quad \text{and} \quad S = \frac{m+s+d}{2}$$

C = Moon's correction for altitude
 = parallax in altitude − refraction
c = second body's correction
 = parallax in altitude − refraction
$\theta = Zms$, and $\phi = Zsm$

Then:
$$\begin{aligned}
D &= \text{arc MS} \\
&= \text{arc qp} \\
&= sm - pm + sq \\
&= d - C \cdot \cos\theta + c \cdot \cos\phi \\
&= d - C(1 - 2\sin^2\theta/2) + c(1 - 2\sin^2\phi/2)
\end{aligned}$$

i.e.
$$D = d - (C-c) + 2(C \cdot \sin^2\theta/2 - c \cdot \sin^2\phi/2) \qquad . \quad (1)$$

Now:
$$\sin^2 \theta/2 = \frac{\sin(S'-d).\sin(S'-y)}{\sin d.\sin y}$$
$$= \frac{\sin(y+z-d)/2.\sin(z+d-y)/2}{\sin d.\sin y}$$
$$= \frac{\sin(90-(m+s+d)/2).\sin(m+d-s)/2}{\sin d.\cos m}$$

i.e.
$$\sin^2 \theta/2 = \frac{\cos S.\sin(S-s)}{\sin d.\cos m}$$

Similarly:
$$\sin^2 \phi/2 = \frac{\cos S.\sin(S-m)}{\sin d.\cos s}$$

Therefore:
$$C\sin^2 \theta/2 - c\sin^2 \phi/2$$
$$= \frac{\cos S}{\sin d}\left\{\frac{C.\sin(S-s)}{\cos m} - \frac{c.\sin(S-m)}{\cos s}\right\}$$
$$= \operatorname{cosec} d.\cos S(M-N)$$

where $M = C \sec m \sin(S-s)$,
and $N = c \sec s \sin(S-m)$.

Then, from formula (1):
$$D = d - (C-c) + 2\operatorname{cosec} d.\cos S(M-N)$$

Merrifield's rule is as follows:

1. Place the Moon's apparent altitude, the second body's apparent altitude and the apparent lunar distance under one another, and take half their sum (S); from which subtract the second body's apparent altitude ($S-s$), and the Moon's apparent altitude ($S-m$). Under these place the Moon's correction for altitude (C) and the correction for altitude of the second body (c).
2. Add together log secant Moon's apparent altitude, sin ($S-s$) and Moon's correction reduced to seconds; the sum is log the number of seconds in M.

$$M = C \sec m.\sin(S-s)$$

230   A HISTORY OF NAUTICAL ASTRONOMY

3. Add together log secant second body's apparent altitude, sin $(S-m)$, and second body's correction reduced to seconds; the sum is log the number of seconds in N.

$$N = c \sec s \cdot \sin (S-m)$$

4. Find difference between M and N, and add together log cosecant apparent distance, cos S, and $(M-N)$; the sum is log $C \cdot \sin^2 \theta/2 - c \cdot \sin^2 \phi/2$ in seconds.
5. Double the result in (4), and add to the apparent distance d, from which subtract the difference of corrections for altitudes; the remainder is the true distance D.

The solution of the problem given on p. 211, using Merrifield's method, is as follows:

	Apparent Altitude	True Altitude	Apparent Distance
Moon m =	13° 29′ 27″	m' = 14° 18′ 32″	d = 107° 52′ 04″
2nd s =	31° 11′ 34″	s' = 31° 10′ 07″	

Here C = correction for Moon's altitude
    = $(m-m')$
    = 49′ 05″
    = 2495″

and c = correction for second body's altitude
    = $(S-s')$
    = 01′ 27″
    = 87″

$$D = d - (C-c) + 2(C \cdot \sin^2 \theta/2 - c \cdot \sin^2 \phi/2)$$

```
            m'     13°   29'   27"      sec 0·012151
            s'     31    11    34       sec 0·067815
            d'    107    52    04                        cosec 0·021470
                2)152    33    05
            S      76    16    32·5
(S-s')             45    04    58·5     sin 9·850113    cos 9·375207
(S-m')             62    47    05·5     sin 9·949046
            C                  29    45  log 3·469085
            c                        87  log 1·939519
```

METHODS OF FINDING LONGITUDE 231

$$
\begin{array}{lrr}
M \quad 2144\cdot6 & & \log 3\cdot331349 \\
N \quad 90\cdot4 & & \log 1\cdot966380 \\
\hline
(M-N) & & 2054\cdot2'' \log 3\cdot312642 \\
C.\sin^2 \theta/2 - c.\sin^2 \phi/2 & & 512\cdot2 \quad \log 2\cdot709319 \\
\hline
& = \quad 8' \quad 32\cdot1'' \\
& \quad\quad \times 2 \\
\hline
& 17' \quad 04\cdot2'' \\
(C-c) & -47 \quad 38 \\
\hline
\text{Correction for app. dist.} & -30 \quad 33\cdot8 \\
d = & 107° \quad 52' \quad 04'' \\
\hline
\text{True Distance } D & 107° \quad 21' \quad 30\cdot2''
\end{array}
$$

Dunthorne's Method for Clearing a Lunar Distance

Dunthorne's method for clearing lunar distances was reproduced in many navigation manuals of the last century, and in particular in Norie's *Epitome*, the firm favourite of seafaring men of the Merchant Service for almost the whole of the 19th century. In the earlier editions of Norie's *Epitome*, of the four solutions for clearing a lunar distance, that of Dunthorne's formed Method 1.

It is interesting to note that tables of 'Logarithmic Differences' appeared in textbooks of nautical astronomy as early as the end of the 18th century. Mackay gave them in his *Theory of the Longitude* published in 1793; and so did Norie in the earlier editions of his famous *Nautical Tables*. Norie, in explaining these tables refers to their use in connection with Mr Dunthorne's method for clearing a lunar distance. We shall state Dunthorne's method as given by Norie. No demonstration of the rule was attempted, this being a common feature of works such as Norie's, which were aimed for the instruction of the unfortunate navigational practitioners who had no desire to understand the methods they used, being content merely to work 'according to the rule.' Compared with many of the rules for clearing lunar distances, that of Dunthorne's is about the least complex.

Rule: 1. To the correction of the Moon's altitude, add the correction of the Sun's or star's altitude; their sum, added to the difference of the apparent altitudes when the Moon's altitude is greater, or subtracted from it when the Moon's altitude is less than the Sun's or star's, will give the difference of their true altitudes.

2. From the natural cosine of the difference of the apparent altitudes subtract the natural cosine of the apparent distance, when the apparent distance is not less than 90°; but when it is greater add together the natural cosines; and to the logarithm of their sum or difference add the logarithmic difference (Table 31): then the difference between the natural number of this sum and the natural cosine of the difference of the two altitudes will be the natural cosine of the true distance, when the natural number is less than the natural cosine of the difference of the true altitudes; otherwise the remainder will be the natural cosine of the supplement of the true distance or the natural sine of the excess of the true distance above 90°.

William Hall's Method for Clearing a Lunar Distance

What is probably one of the last methods proposed for clearing a lunar distance is that described by the Rev. William Hall in a paper printed in the *Nautical Magazine* for the year 1903.

Hall wrote at a time when the lunar problem clearly had outlived the limits of its usefulness. It had been announced at about this time that the 1906, and subsequent *Nautical Almanacs* would not contain tables of lunar distances. A considerable amount of correspondence appeared in the pages of the *Nautical Magazine* over a period of several years during the early part of the 20th century, the writers being divided in their opinions of the lunar method for finding longitude at sea. Captain Lecky, in his famous *Wrinkles*, had been accused—not without justification—of burying the lunar in an unbefitting manner, in view of the great service it had performed to seamen for well over a hundred years. A handsome 'obituary notice of the lunar', to quote Hall, had prompted him to write on the subject in the August issue of the *Nautical Magazine* for 1903. Hall, in common with numerous other so-called 'lunarians', were genuinely sorry that the time had come when the lunar problem for finding longitude was beyond resurrection.

Hall's interesting method is described in relation to Fig. 4. Referring to Fig. 4:

Mm = Moon's correction in altitude
    = parallax in altitude—refraction

METHODS OF FINDING LONGITUDE 233

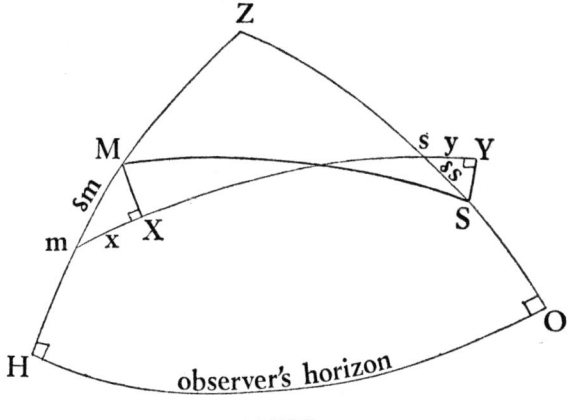

FIGURE 4

Ss = second body's correction in altitude
 = refraction—parallax in altitude
Mm is plus, and Ss is minus to apparent altitude.
MX and SY are arcs perpendicular to ms produced, so that corrections to apparent distance are $-mX$ and $-sY$.

Let:
m = apparent altitude of Moon
$\delta m+$ = correction to apparent altitude of Moon
s = apparent altitude of second body
$\delta s-$ = correction to apparent altitude of second body
d = apparent distance
$sY - mX = \delta d+$
 = correction to apparent distance
P = Moon's horizontal parallax
p = second body's horizontal parallax

Now:
$$\delta d = \delta s \cdot \cos YsS - \delta m \cos MmX$$

Applying the fundamental cosine formula to triangle Zms:

$$\cos s = \frac{\cos Zm - \cos Zs \cdot \cos ms}{\sin Zs \cdot \sin ms}$$

i.e.
$$\cos s = \frac{\sin m - \sin s \cdot \cos d}{\cos s \sin d}$$

Similarly:
$$\cos m = \frac{\sin s - \sin m \cdot \cos d}{\cos m \cdot \sin d}$$

Considering only that part of the correction due to δs we have:

δs = Refraction—parallax in altitude
 = Ref. $-p \cdot \cos s$
 = $\cos s(\text{Ref. sec}.s - p)$

Hall tabulated the quantity Ref. sec alt. It is obtained by adding to the ordinary tables of refraction the following correction to obtain what may be called 'Horizontal Refraction' on the analogy of 'Horizontal Parallax.'

To obtain Hor. Ref. add to Tabulated Refraction

App. Alt.	0°	1°	2°	3°	4°	5°	6°	7°	8°	9°
0°	0	0	1	1	2	2	3	3	4	4
10	5	5	6	7	7	8	8	9	9	10
20	10	11	11	12	12	13	13	14	14	15
30	15	16	17	17	18	18	19	19	20	21
40	21	22	22	23	23	24	25	25	26	26
50	27	27	28	29	29	30	31	31	32	33
60	34	34	35	36	36	37	38	38	39	39
70	40	41	42	43	44	45	45	46	46	47
80	48	49	49	50	51	52	53	54	55	56

Take the Hor. Ref. and its difference from the Hor. Parallax. Call this difference y for second body.

$$\delta s = y \cdot \cos s$$

and

Correction to distance due to

$$\delta s = \delta s \cdot \cos s$$
$$= \delta s \left(\frac{\sin m - \sin s \cdot \cos d}{\cos s \cdot \sin d} \right)$$

Correction due to $\delta s = \dfrac{y(\sin m - \sin s \cdot \cos d)}{\sin d}$ . . . . (I)

Now consider that part of the correction due to δm.

Again:

Take the Hor. Ref. and its difference from the Hor. Parallax of Moon. Call this difference x for Moon.

$$\delta m = x \cdot \cos m,$$

and
Correction to distance due to $\delta m = \delta m \cdot \cos m$

$$= \delta m \left(\dfrac{\sin s - \sin m \cdot \cos d}{\cos m \sin d} \right)$$

Correction due to $\delta m = \dfrac{x(\sin s - \sin m \cdot \cos d)}{\sin d}$ . . (II)

By combining (I) and (II) we may find the correction to distance δd.

Now,

$$\delta d = \delta s \cdot \cos s - \delta m \cdot \cos m$$

$$\delta d = \dfrac{y(\sin m - \sin s \cdot \cos d)}{\sin d} - \dfrac{x(\sin s - \sin m \cos d)}{\sin d}$$

i.e.

$$\delta d \sin d = y(\sin m - \sin s \cdot \cos d) - x(\sin s - \sin m \cdot \cos d)$$

From which, because δd is small:

$$\sin (\delta d) = \sin y(\sin m \cdot \operatorname{cosec} d - \sin s \cdot \cot d)$$
$$- \sin x(\sin s \cdot \operatorname{cosec} d - \sin m \cdot \cot d)$$

This formula, for a lunar distance, according to Hall, is fairly simple. The footnote to the paper in which he described his method is interesting and relevant to the position of the lunar method at the beginning of the 20th century.

'Footnote: Since this article was penned, an official memorandum has been issued in which notice is given that the lunar is no longer to be included in the course of instruction for junior officers of the Navy. . . . Hence my contribution must be read in the light of pure mathematics, and the lunar can no longer be included in the sphere of practical navigation. Nevertheless I am unable to cancel my expression of opinion that the lunar must remain the supreme test of powers of observation and of computation. . . . The midshipman of the future will doubtless find some other problem to trouble him in place of the lunar.

'I have a certain satisfaction in the thought that this article is probably the last which will be written on a problem which has engaged the attention of mathematicians and navigators for hundreds of years . . . which in the hands of early pioneers of exploration was the one and only means of fixing longitude, which will be remembered in the future by the coming race, who—smile as they may at our rough methods—can hardly grudge a modified approval of the industry and ingenuity displayed by oldfashioned "Lunarians."

'The Lunar is dead; let us bury it with due respect!'

Several graphical methods were devised for clearing lunar distances. The Abbé de la Caille devised a graphical method for this purpose as early as 1759. De la Caille's method, by which it was claimed that distances could be cleared graphically with an accuracy of 20" of arc, was published in *Connoissance des Temps* for 1761. It also appeared in the later editions of Bouguer's *Traité de Navigation*—a well-known French manual of navigation which the Abbé revised.

Andrew Mackay, in his *Theory of the Longitude*, not only refers to a 'Parallactic Rotula' invented by James Ferguson and constructed on the same principles as de la Caille's 'Chassis de Reduction,' but he also describes a graphical method of his own invention for clearing lunar distances. Mackay's method is similar to de la Caille's, and consists of four scales which provide the means for finding the various quantities used in the process of clearing the distance. By means of Mackay's (and similar) graphical method, lunar distances could be cleared accurately and expeditiously without resort to tables or calculations.

In 1790, the English mathematician Margetts published tables expressed graphically in curves. Margetts's graphical method for clearing lunar distances was similar in plan to that of the Abbé de la Caille.

John William Norie devised and published a set of 'linear

7. Hadley Octant. By Benjamin Martin, c. 1760.

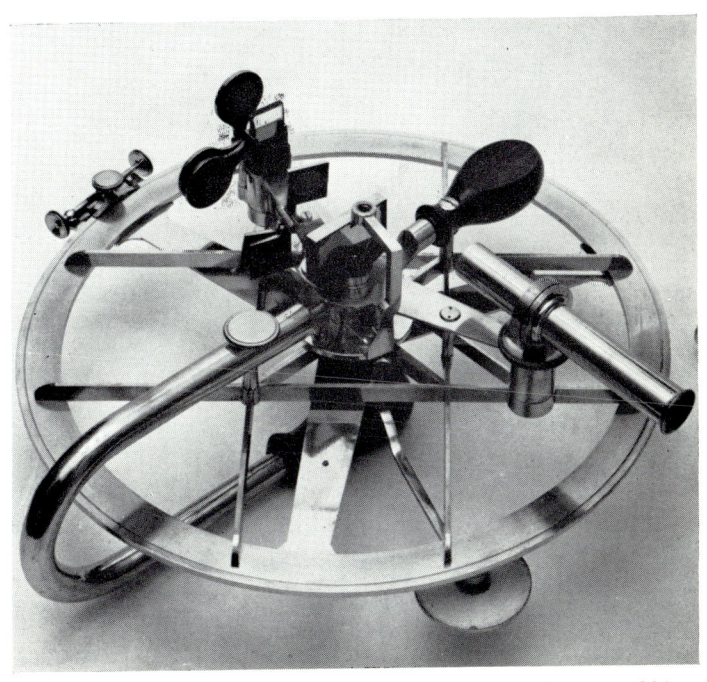

8. Reflecting Circle. By Edward Troughton, c. 1800. (See page 83.)

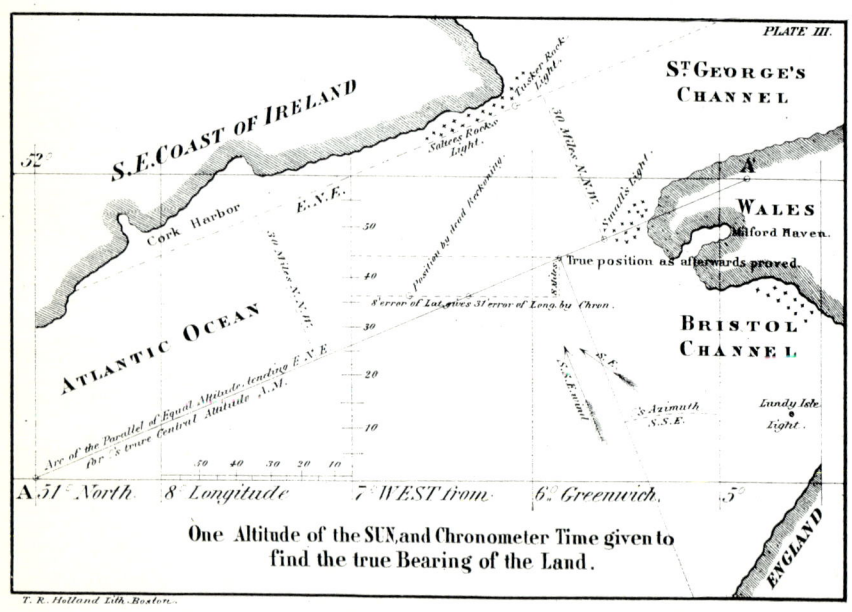

9. Sextant. By Kelvin Hughes, 1967.

10. Facsimile of Plate 3 of Sumner's Pamphlet. (See page 277.)

DISTANCES of MOON's Center from SUN, and from STARS EAST of her.

Stars Names.	Days	Noon. D. M. S.	III^h. D. M. S.	VI^h. D. M. S.	IX^h. D. M. S.	Midnight. D. M. S.	XV^h. D. M. S.	XVIII^h. D. M. S.	XXI^h. D. M. S.
α Arietis.	1	78. 51. 29	77. 24. 50	75. 58. 9	74. 31. 24	73. 4. 36	71. 37. 44	70. 10. 50	68. 43. 52
	2	67. 16. 51	65. 49. 46	64. 22. 39	62. 55. 28	61. 28. 15	60. 0. 59	58. 33. 40	57. 6. 20
	3	55. 38. 57	54. 11. 32	52. 44. 5	51. 16. 39	49. 49. 13	48. 21. 49	46. 54. 26	45. 27. 7
	4	43. 59. 50							
Aldebaran.	4	73. 12. 40	71. 39. 19	70. 5. 40	68. 31. 50	66. 57. 43	65. 23. 19	63. 48. 38	62. 13. 39
	5	60. 38. 23	59. 2. 48	57. 26. 53	55. 50. 39	54. 14. 5	52. 37. 11	50. 59. 57	49. 22. 22
	6	47. 44. 26	46. 6. 8	44. 27. 28	42. 48. 25	41. 9. 1	39. 29. 14	37. 49. 5	36. 8. 33
	7	34. 27. 39	32. 46. 21	31. 4. 41	29. 22. 40	27. 40. 18	25. 57. 35	24. 14. 34	22. 31. 14
	8	20. 47. 35							
Pollux.	8	64. 59. 42	63. 14. 53	61. 29. 40	59. 44. 3	57. 58. 0	56. 11. 32	54. 24. 41	52. 37. 26
	9	50. 49. 48	49. 1. 47	47. 13. 24	45. 24. 40	43. 35. 37	41. 46. 15	39. 56. 35	38. 6. 37
	10	36. 16. 24							
Regulus.	10	72. 0. 35	70. 9. 10	68. 17. 26	66. 25. 24	64. 33. 5	62. 40. 30	60. 47. 40	58. 54. 37
	11	57. 1. 20	55. 7. 51	53. 14. 13	51. 20. 25	49. 26. 28	47. 32. 25	45. 38. 18	43. 44. 7
	12	41. 49. 54	39. 55. 40	38. 1. 29	36. 7. 21	34. 13. 19	32. 19. 22	30. 25. 35	28. 32. 2
	13	26. 38. 44	24. 45. 43	22. 53. 4	21. 0. 49	19. 9. 2			
Spica ♍.	13	- - -	- - -	- - -	- - -	73. 7. 29	71. 14. 43	69. 22. 11	67. 29. 55
	14	65. 37. 56	63. 46. 15	61. 54. 53	60. 3. 52	58. 13. 13	56. 22. 58	54. 33. 7	52. 43. 41

Stars Names.	Days	Noon. D. M. S.	III^h. D. M. S.	VI^h. D. M. S.	IX^h. D. M. S.	Midnight. D. M. S.	XV^h. D. M. S.	XVIII^h. D. M. S.	XXI^h. D. M. S.
Spica ♍.	15	50. 54. 41	49. 6. 8	47. 18. 3	45. 30. 28	43. 43. 23	41. 56. 50	40. 10. 50	38. 25. 26
	16	36. 40. 37	34. 56. 23	33. 12. 50	31. 29. 59	29. 47. 53	28. 6. 35	26. 26. 7	24. 46. 31
	17	23. 7. 50							
Antares.	17	68. 30. 37	66. 48. 34	*65. 6. 58	63. 25. 51	61. 45. 12	60. 5. 1	58. 25. 18	56. 46. 3
	18	55. 7. 15	53. 28. 55	51. 51. 2	50. 13. 36	48. 36. 38	47. 0. 7	45. 24. 3	43. 48. 26
	19	42. 13. 16	40. 38. 33	39. 4. 18	37. 30. 29	35. 57. 8			
The Sun.	16								120. 25. 9
	17	118. 49. 31	117. 14. 21	115. 39. 37	114. 5. 20	112. 31. 30	110. 58. 7	109. 25. 10	107. 52. 39
	18	106. 20. 33	104. 48. 53	103. 17. 38	101. 46. 47	100. 16. 20	98. 46. 18	97. 16. 40	95. 47. 22
	19	94. 18. 29	92. 49. 58	91. 21. 47	89. 53. 58	88. 26. 29	86. 59. 20	85. 32. 30	84. 5. 59
	20	82. 39. 47	81. 13. 54	79. 48. 17	78. 22. 57	76. 57. 54	75. 33. 6	74. 8. 31	72. 44. 12
	21	71. 20. 7	69. 56. 17	68. 32. 38	67. 9. 13	65. 46. 0	64. 22. 59	63. 0. 8	61. 37. 27
	22	60. 14. 57	58. 52. 36	57. 30. 22	56. 8. 17	54. 46. 21	53. 24. 33	52. 2. 51	50. 41. 16
	23	49. 19. 48	47. 58. 26	46. 37. 9	*45. 15. 58	43. 54. 51	42. 33. 49	41. 12. 50	39. 51. 54
	24	38. 31. 2							
α Arietis.	29					64. 11. 21	62. 43. 38	61. 15. 54	59. 48. 8
	30	58. 20. 22	56. 52. 36	55. 24. 50	53. 57. 6	52. 29. 23	51. 1. 41	49. 34. 3	48. 6. 30
	31	46. 39. 2	45. 11. 40	43. 44. 28	42. 17. 26	40. 50. 38	39. 24. 5	37. 57. 49	36. 31. 54
	F.1	35. 6. 16							

* Read thus in first Edition.

AUGUST, 1897.

MEAN TIME. LUNAR DISTANCES.

Day	Star's Name and Position		Noon.	P.L. of diff.	IIIh.	P.L. of diff.	VIh.	P.L. of diff.	IXh.	P.L. of diff.
1	SUN	W.	34 50 16	2798	36 24 41	2789	37 59 23	2782	39 34 15	2774
	Spica	E.	38 15 16	2498	36 34 0	2492	34 52 35	2486	33 11 2	2480
	Saturn	E.	70 10 8	2521	68 29 20	2514	66 48 30	2508	65 7 28	2502
	Antares	E.	83 58 4	2480	82 16 33	2473	80 34 42	2466	78 51 41	2459
2	SUN	W.	47 31 6	2737	49 6 57	2730	50 42 57	2723	52 19 7	2717
	Jupiter	W.	16 40 56	2615	18 10 39	2588	19 58 41	2565	21 38 4	2543
	Saturn	E.	56 40 16	2475	54 58 28	2470	53 16 33	2467	51 34 33	2463
	Antares	E.	70 20 5	2411	68 37 5	2417	66 53 55	2411	65 10 36	2405
3	SUN	W.	60 22 3	2684	61 59 9	2679	63 36 1	2673	65 13 28	2667
	Jupiter	W.	30 2 2	2482	31 43 44	2472	33 25 59	2463	35 8 4	2455
	Mars	W.	26 15 2	2632	27 53 13	2621	29 31 59	2610	31 11 11	2601
	Saturn	E.	43 33 11	2453	41 52 11	2461	39 38 50	2457	37 56 31	2455
	Antares	E.	56 31 55	2375	54 47 43	2370	53 3 25	2365	51 19 0	2360
4	SUN	W.	73 21 47	2641	74 59 24	2635	76 37 49	2630	78 16 3	2626
	Jupiter	W.	43 48 11	2419	45 32 19	2414	47 16 46	2408	49 1 34	2403
	Mars	W.	39 27 6	2561	41 6 55	2554	42 46 53	2548	44 27 0	2542
	Antares	E.	42 34 59	2334	40 51 19	2329	39 7 23	2325	37 23 17	2320
	α Aquilæ	E.	96 39 58	2954	95 18 47	2945	93 57 37	2940	92 36 5	2938
5	SUN	W.	86 28 47	2603	88 7 38	2599	89 46 58	2595	91 25 36	2591
	Jupiter	W.	57 49 32	2378	59 13 56	2373	60 58 11	2369	62 43 31	2364
	Mars	W.	52 49 25	2515	54 30 0	2510	56 11 5	2504	57 52 30	2501
	Antares	E.	28 30 36	2299	26 44 35	2296	24 58 9	2292	23 12 17	2288
	α Aquilæ	E.	84 26 50	2920	82 55 2	2920	81 23 13	2920	79 51 19	2920
6	SUN	W.	99 42 0	2574	101 21 31	2572	103 1 5	2569	104 40 43	2566
	Jupiter	W.	71 25 33	2346	73 10 25	2344	74 55 27	2341	76 40 20	2338
	Mars	W.	66 19 22	2482	68 1 20	2482	69 42 42	2477	71 24 28	2474
	Spica	W.	31 37 21	2293	33 23 29	2289	35 9 45	2285	36 56 7	2281
	α Aquilæ	E.	72 13 46	2995	70 41 49	2985	69 9 21	2981	67 36 37	2977
	Fomalhaut	E.	96 2 31	2648	94 22 17	2645	92 41 41	2643	91 0 44	2640
7	SUN	W.	112 59 42	2556	114 39 37	2555	116 19 36	2554	117 59 37	2553
	Jupiter	W.	85 26 9	2328	87 11 27	2328	88 56 48	2326	90 42 11	2326
	Mars	W.	79 54 11	2464	81 36 15	2464	83 18 22	2461	85 0 30	2460
	Spica	W.	45 49 10	2266	47 35 59	2265	49 22 50	2263	51 9 44	2262
	α Aquilæ	E.	60 15 45	3125	58 41 19	3158	57 6 56	3179	55 32 17	3194
	Fomalhaut	E.	82 58 1	2635	81 19 54	2638	79 41 50	2641	78 3 50	2643
	α Pegasi	E.	104 30 13	2427	102 47 17	2424	101 4 16	2420	99 21 12	2418
8	SUN	W.	126 19 33	2553	127 59 39	2556	129 39 39	2556	131 19 26	2557
	Mars	W.	93 31 59	2459	95 13 28	2462	96 55 35	2463	98 37 41	2465
	Spica	W.	60 4 32	2260	61 51 31	2260	63 38 28	2260	65 25 25	2261
	Saturn	W.	28 52 17	2344	30 36 21	2356	32 20 23	2371	34 4 40	2362
	Fomalhaut	E.	69 35 50	2654	67 58 28	2696	66 20 42	2729	64 42 38	2752
	α Pegasi	E.	90 45 6	2414	89 1 51	2415	87 18 38	2417	85 35 35	2419
9	Spica	W.	74 19 34	2273	76 6 13	2277	77 52 47	2280	79 39 16	2283
	Saturn	W.	42 48 21	2341	44 33 19	2341	46 18 13	2341	48 2 58	2342
	Antares	W.	28 28 54	2265	30 15 43	2270	32 2 27	2273	33 49 6	2277

XIV. AUGUST, 1897.

MEAN TIME. LUNAR DISTANCES.

Day	Star's Name and Position		Midnight.	P.L. of diff.	XVh.	P.L. of diff.	XVIIIh.	P.L. of diff.	XXIh.	P.L. of diff.
1	SUN	W.	41 9 17	2766	42 44 30	2759	44 19 52	2751	45 55 24	2744
	Spica	E.	31 29 21	2476	29 47 34	2472	28 5 41	2468	26 23 43	2465
	Saturn	E.	63 26 17	2496	61 44 58	2490	60 3 31	2485	58 21 57	2480
	Antares	E.	77 10 30	2451	75 28 8	2445	73 45 37	2438	72 2 56	2431
2	SUN	W.	53 55 25	2710	55 31 52	2704	57 8 27	2697	58 45 11	2691
	Jupiter	W.	23 18 32	2523	24 59 3	2520	26 39 56	2516	28 21 13	2492
	Saturn	E.	49 52 28	2461	48 10 19	2458	46 28 6	2456	44 45 50	2453
	Antares	E.	63 27 9	2399	61 43 33	2393	59 59 48	2387	58 15 55	2381
3	SUN	W.	66 50 52	2661	68 28 4	2656	70 6 2	2651	71 43 49	2646
	Jupiter	W.	36 50 20	2447	38 32 48	2440	40 15 26	2433	41 58 14	2426
	Mars	W.	32 49 34	2591	34 28 41	2583	36 7 59	2576	37 47 27	2568
	Saturn	E.	36 14 14	2457	34 32 0	2461	32 49 52	2467	31 7 53	2475
	Antares	E.	49 34 27	2354	47 49 46	2349	46 4 58	2344	44 20 2	2339
4	SUN	W.	79 54 23	2621	81 32 50	2616	83 11 23	2612	84 50 2	2607
	Jupiter	W.	50 47 34	2396	52 32 16	2391	54 17 16	2387	55 59 49	2382
	Mars	W.	46 7 15	2536	47 47 38	2530	49 28 5	2525	51 8 46	2520
	Antares	E.	35 39 8	2315	33 54 42	2312	32 10 20	2307	30 25 50	2303
	α Aquilæ	E.	90 34 16	2927	89 12 15	2924	87 50 20	2921	86 28 34	2920
5	SUN	W.	93 4 3	2588	94 43 15	2584	96 22 36	2581	98 2 3	2577
	Jupiter	W.	64 26 57	2361	66 11 28	2357	67 56 5	2353	69 40 47	2350
	Mars	W.	59 33 42	2497	61 15 23	2494	62 56 21	2490	64 37 49	2486
	Antares	E.	21 26 0	2283	19 39 38	2281	17 53 13	2278	16 6 39	2275
	α Aquilæ	E.	78 19 32	2937	76 47 57	2937	75 16 20	2950	73 44 57	2954
6	SUN	W.	106 20 25	2564	108 0 10	2561	109 39 58	2559	111 19 49	2558
	Jupiter	W.	78 25 24	2336	80 10 31	2334	81 55 43	2332	83 40 54	2331
	Mars	W.	73 6 12	2471	74 48 8	2469	76 30 8	2467	78 12 8	2465
	Spica	W.	38 42 34	2277	40 29 7	2274	42 15 44	2272	44 2 25	2269
	α Aquilæ	E.	66 4 1	2974	64 31 22	3047	62 58 45	3070	61 25 59	3096
	Fomalhaut	E.	89 30 39	2637	87 52 31	2635	86 14 21	2633	84 36 11	2633
7	SUN	W.	119 39 32	2553	121 19 32	2553	122 59 32	2553	124 39 32	2553
	Jupiter	W.	92 27 29	2335	94 12 47	2334	95 58 1	2335	97 43 38	2336
	Mars	W.	86 42 39	2460	88 24 49	2460	90 6 59	2460	91 49 9	2460
	Spica	W.	52 56 40	2261	54 43 37	2261	56 30 34	2260	58 17 33	2260
	α Aquilæ	E.	54 29 23	3279	52 45 41	3339	51 1 41	3384	50 18 33	3445
	Fomalhaut	E.	76 25 56	2651	74 48 10	2657	73 10 32	2665	71 33 5	2674
	α Pegasi	E.	97 38 1	2416	95 54 49	2415	94 11 36	2414	92 28 21	2414
8	SUN	W.	132 59 20	2558	134 39 1	2561	136 19 0	2564	137 58 45	2567
	Mars	W.	100 19 44	2466	102 1 45	2469	103 43 47	2471	105 25 36	2475
	Spica	W.	67 12 20	2264	68 59 13	2266	70 46 3	2268	72 32 50	2270
	Saturn	W.	35 49 9	2356	37 33 47	2351	39 18 33	2346	41 3 15	2344
	Fomalhaut	E.	63 4 25	2778	61 25 38	2758	60 18 15	2779	58 43 19	2782
	α Pegasi	E.	83 52 31	2422	82 9 15	2425	80 26 16	2429	78 43 3	2436
9	Spica	W.	81 25 40	2288	83 12 4	2293	84 58 7	2298	86 44 10	2302
	Saturn	W.	49 48 18	2343	51 33 5	2345	53 18 0	2348	55 2 59	2351
	Antares	W.	35 35 39	2281	37 22 5	2284	39 8 25	2292	40 54 37	2297

tables' for use with Lyon's method of clearing lunar distances. 'These tables,' claimed Norie, 'render Lyons' method one of the easiest and shortest [methods] that have been proposed.'

Norie, in the preface to the sixth edition of his treatise on practical navigation refers to the generous and valuable assistance given to him by Mr George Coleman, F.R.A.S., who for many years had been an officer in the Honourable East India Company's ships. This gentleman, a former Christ's Hospital boy, who succeeded Norie as teacher at the Nautical Academy, published a set of *Nautical and Lunar Tables*. The lunar tables followed the pattern of Norie's, and were designed for use with an improved Lyons's method for clearing distances. Arthur B. Martin, who succeeded Coleman at Norie's Nautical Academy, and who edited the twenty-first edition of Norie's *Epitome* and *Tables* gave, as the third of five methods for clearing the distance 'Lyons's method improved by Mr Coleman.'

8. FINDING G.M.T. FROM A LUNAR OBSERVATION

Having discussed some of the numerous methods that have been used for clearing lunar distances, we shall now turn our attention to the final part of the lunar problem, in which the G.M.T. of the observation is found.

Predicted angular distances between the Moon and Sun and the nine bright zodiacal stars—α *Aquilae, Fomalhaut,* α *Arietis, Aldebaran, Pollux, Regulus, Antares, Spica,* and α *Pegasi*—were given for Noon, III hrs., VI hrs., IX hrs., etc., G.M.T.

The stars whose distances from the Moon were tabulated in the *Nautical Almanac* are, with one exception, those whose declinations do not exceed the maximum declination of the Moon. The exception to the rule is *Fomalhaut*, whose declination is slightly greater than the Moon's maximum declination of $28\frac{3}{4}°$. *Fomalhaut* was chosen for it is a bright star located in a large part of the heavens in which no other bright star having a smaller declination is to be found.

The angular distances given in the *Nautical Almanac* for every third hour of G.M.T. are described as *geocentric distances*: that is to say, they are angular distances between the true direction of the Moon's centre and that of 'second bodies' at the Earth's centre. After a lunar distance has been observed and reduced to the Earth's centre by clearing it of the effects of refraction and parallax, it remains for the observer to compare the cleared observed distance with predicted angular distances given in the *Nautical Almanac*.

The arrangement of predicted lunar distances to be found in the earlier *Nautical Almanacs* is illustrated in the accompanying plate (11), which is a facsimile of a page of lunar distances from the 1797 *Nautical Almanac*.

In later almanacs, the arrangement of predicted distances is different from that in the earlier almanacs. A facsimile page from the 1897 *Nautical Almanac* is reproduced (Plate 12). It will be seen from this reproduction that the predicted distances are arranged from west to east commencing each day with the object which is at the greatest angular distance west of the Moon, in the order in which they appear—W indicating that the object is west, and E, east of the Moon.

The rule for finding the G.M.T. is as follows:

For the given date take from the *Nautical Almanac* the distances between which the calculated distance lies, and find their difference. Find the difference between the distance tabulated for the earlier time and the observed distance. This is the change in distance for the required interval, assuming the Moon's motion to be uniform. This 'change in distance' may be calculated by simple proportion as follows:

Let Δ = difference of distances between which the computed distance lies

δ = difference between the earlier of these and the true distance found from observation

t = time elapsed since the time corresponding to the earlier distance

Then:

$$\frac{\Delta}{\delta} = \frac{3 \text{ hrs.}}{t}$$

$$t = \frac{3 \cdot \delta}{\Delta} \text{ hrs} \qquad \ldots \quad \text{(I)}$$

$$= 10800 \cdot \frac{\delta}{\Delta} \text{ secs.}$$

The laborious calculation entailed in finding the interval t, using this formula, led to the invention of an artifice in the form

METHODS OF FINDING LONGITUDE 239

of a *Table of Proportional Logarithms*. Proportional logarithms are the invention of Dr Nevil Maskelyne, and they were first given in his *Tables Requisite to be used with the Nautical Almanac*.

The proportional logarithm (prop. log) of a time interval t is defined as the common log of 10800 diminished by the common log of t seconds (less than 10800). Thus the prop. log of 1 hr. is log 10800 − log 3600, i.e.

$$4 \cdot 0334 - 3 \cdot 5563 = 0 \cdot 4771$$

which is tabulated as 4771.

Similarly, the prop. log of 2 hrs. is log 10800 − log 7200, i.e.

$$4 \cdot 0334 - 3 \cdot 8573 = 0 \cdot 1761$$

which is tabulated as 1761.

Proportional logarithms are given to four figures, this being sufficiently accurate for their purpose. It will be noticed that prop. logs decrease as the numbers of which they are logarithms increase.

From equation (I) above, we have:

$$t \cdot \Delta = 3 \cdot \delta$$

and

$$\log t + \log \Delta = \log 3 \text{ hrs.} + \log \delta$$

or

$$\log t = \log 3 \text{ hrs.} - \log \Delta + \log \delta$$

from which

$$\log 3 \text{ hrs.} - \log t = (\log 3 \text{ hrs.} - \log \delta) - (\log 3 \text{ hrs.} - \log \Delta)$$

That is:

$$\text{prop. log } t = \text{prop. log } \delta - \text{prop. log } \Delta \quad . \quad . \quad \text{(II)}$$

In deducing the time corresponding to a given true lunar distance by simple proportion, four entries in log tables are necessary for finding the elapsed time from a tabulated time given in the almanac. By using proportional logs, as will be evident from equation (II), two entries only are sufficient—prop. log δ being given in the almanac.

Users of the earlier *Nautical Almanacs* normally provided themselves with Maskelyne's *Requisite Tables*, included amongst which, as we have noted, was the table of proportional logarithms. This table eventually found its way into other collections of nautical tables, the latter gradually superseding the *Requisite Tables*, the last (third) edition, being published in 1811.

In later *Nautical Almanacs*, tables of lunar distances gave, abreast of each tabulated distance, the corresponding prop. log of difference. These figures, which appear in columns labelled *prop. log. of diff.* (see facsimile of page from 1897 *Nautical Almanac*) correspond to prop. log in equation (II) above.

An example, taken from Ainsley's *Extra Master's Guide* of 1888, illustrates how the G.M.T. of a lunar observation was found using prop. logs.

Example: 1888, January 20th. Required to find the Greenwich Mean Astronomical Time at which the true distance between the Moon and α *Pegasi* was 40° 18′ 40″.

From *Nautical Almanac* (an extract appears below) it is seen that on January 20th the star α *Pegasi* was west of the Moon, and its distance from that object as under:

Day of month.	Star's name and Position.	Noon	P.L. of Diff.	III^h	P.L. of Diff.	VI^h	P.L. of Diff.
20	α *Pegasi* W	38° 38′ 19″	4107	39° 48′ 14″	4045	40° 59′ 09″	3990

It appears, by inspecting the distances in the *Nautical Almanac*, that the given true distance 40° 18′ 40″ lies between the two consecutive distances viz. 39° 48′ 14″ the distance at III hrs., and 40° 59′ 09″, the distance at VI hrs.: whence it is evident that the time must be January 20th between III hrs. and VI hrs., the nearer distance preceding in order of time, the given distance is at III hrs.

The given distance is	.	40°	18′	40″	
At III hrs. the distance is		39	48	14	⟶ P.L. = 4045
Difference	. . . .		30	26	⟶ P.L. = 7719
Interval t	. . . .	01^h	17^m	15^s	⟵ P.L. = 3674

Time corresponding to
    distance at . . .   03   00   00

    G.M.T. January 20   04   17   15   (approximate)

Owing to want of uniformity in the motion of the Moon, it is necessary to apply a correction for the 'inequality of the Moon's motion,' to the interval t found by simple proportion as in the preceding example. It is for this reason that the G.M.T. found in the above example is designated (approximate). The correction for the Moon's inequality is referred to as the 'correction for second differences.'

In extreme cases an error of 50′ of longitude results when the correction for second differences is ignored. For any given angular distance the correction is least when the direction of the 'second body' from the Moon is perpendicular to the line of the cusps of the Moon.

Second differences are found by taking the difference between successive or (first) differences of any varying function. By taking the difference between successive 'second differences,' [third differences] are found. An example will make this clear.

If a, b, c, d, etc. are the successive values of a varying function, we get:

Successive values of varying function	First Diffs.	Second Diffs.	Third Diffs.	Fourth Diffs.
a				
	$a-b$			
b		$a-2b+c$		
	$b-c$		$a-3b+3c-d$	
c		$b-2c+d$		$a-4b+6c-4d+e$
	$c-d$		$b-3c+3d-e$	
d		$c-2d+e$		
	$d-e$			
e				

It will be found in practice that if differences be carried far enough they tend to equality. When this is attained the value of the function may be found for any given intermediate quantity. The so-called 'method of differences' was of importance in interpolating between tabulated elements such as the Moon's, planets' or Sun's declination or Right Ascension given in the *Nautical Almanac* for uniform time intervals—the time being referred to as the 'independent variable.'

The Moon's rate of motion towards, or away from, a second body used in the lunar problem varies with time; and, therefore, the exact position of the Moon relative to a 'second body' at any instant between the time given for those in the almanac cannot be found exactly by simple proportion. It will be found, however, that by carrying the interpolation to second differences, a result sufficiently accurate for practical purposes is obtained. The second differences, therefore, are the rates at which the change of motion varies.

A table for finding the correction for second differences is included in the later *Nautical Almanacs*, and also in the popular collections of nautical tables such as those of Norie, Raper and Inman. The arguments in the table are: interval t; and mean differences of prop. logs (P.L. of Diff.). In the example given above, the prop. log following the distance taken from the *Nautical Almanac* at III hrs. is 4045: the prop. logs to the right and left of this are 3990 and 4107 respectively. The differences between these last two prop. logs and that at III hrs. are 55 and 62; and the mean difference is, therefore, $\frac{1}{2}$ (55 + 62), i.e. 58; and, it will be noticed, the prop. logs are decreasing. Entering the table of corrections for second differences, with $t = 01$ hr. 17 mins. 15 secs., and mean difference P.L. 58, the correction for second differences is 18 secs. This is to be added to the approximate G.M.T. found above, because the prop. logs are decreasing.

Proportional logs served to indicate which of the tabulated stars, etc., used in the lunar problem, was most favourably placed for furnishing the most exact measurement of the lunar distance. That object was to be preferred which had the least proportional log opposite to it, because the greater velocity of the Moon towards, or away from, the second body, the greater is the reliance to be placed on an observation of the distance. Because proportional logs decrease as the natural numbers increase, the smaller is the proportional log abreast of a tabulated distance, the greater is the Moon's velocity towards, or away from, the body whose distance is tabulated, and the greater the change of distance between Moon and body in the interval. It is upon this change of distance that the value of the observation depends.

We now take leave of the lunar problem in order to discuss the 'longitude by chronometer' method which superseded the 'longitude by lunar' method.

METHODS OF FINDING LONGITUDE 243

9. LONGITUDE BY CHRONOMETER

(a) METHODS

Although the marine chronometer was perfected by John Harrison in the middle of the 18th century, the instrument never came into popular use at sea until the middle of the 19th century.

Finding longitude by means of a chronometer requires knowledge of the observer's latitude and certain astronomical data in order to define the observed object's celestial position at the time of the observation. The exact error and daily rate of the chronometer must also be known. We have already discussed, in Chapter III, the problems relating to estimating the chronometer error in the days before radio time signals became available. During a voyage the careful navigator missed no opportunity of checking his chronometers by means of time-ball observations used at places the exact longitudes of which were known, or by calculating his longitude very accurately when in port using the exact latitude of the place of observation. At sea the lunar problem was often used in the latter part of the 19th century, not as a means for finding longitude, but as a means for checking the chronometer error and rate.

In all astronomical methods for finding longitude it is necessary to compare local time with a standard time for a particular instant. The difference between these times is a measure of the difference of longitude between the local and standard meridians. In the *longitude by chronometer* method the standard time is provided by the chronometer the accumulated error of which is known. The local time may be found after observing the altitude of a heavenly body—or *taking a sight* as the seaman would say—and solving a celestial triangle known as the *astronomical* or *PZX triangle*. The method of finding local time of sight is explained with reference to Fig. 5.

The circle in Fig. 5 represents the celestial horizon of an observer whose zenith is projected on to the plane of his horizon at Z. N, E, S and W are the projections of the cardinal points of the celestial horizon. P is the projection of the celestial pole and WQE is that of the equinoctial. dd_1 is the projection of a parallel of declination on which an observed object, denoted by X in the figure, is located. NZS is the projection of the observer's celestial meridian. ZXA is the projection of the vertical circle through X, and PXB is that of the hour circle through X.

arc NP = observer's latitude = l
arc AX = true altitude of X = a
arc BX = declination of X = d

In the spherical triangle PZX:

arc PZ = co-latitude of observer = $(90-l)$
arc ZX = zenith distance of X = $(90-a)$
arc PX = polar distance of X = $(90 \pm d)$

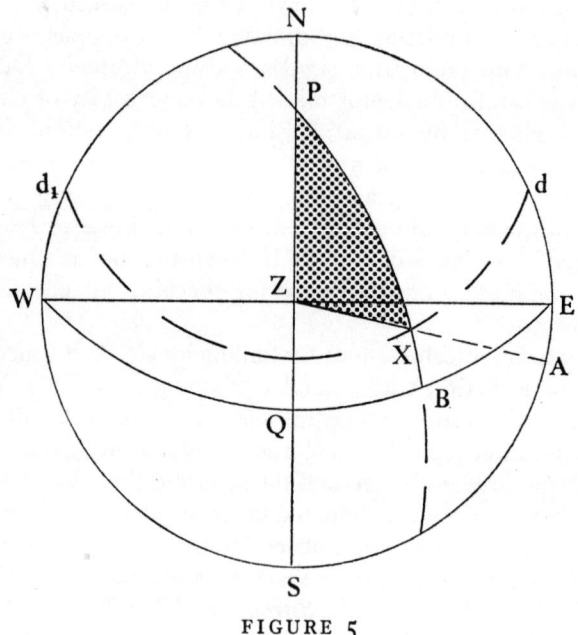

FIGURE 5

It will be noticed that the arc PX is $(90+d)$ or $(90-d)$ according as to whether latitude and declination have different or equivalent names respectively.

If the three sides of the spherical triangle PZX are known, it is an easy matter to find the angle ZPX. This angle, which is referred to as the 'meridian distance' of the body X, the 'hour angle' of the body or, more familiarly, angle P, is a measure of the time that will elapse before, or has elapsed since, the body X will be, or was, at meridian passage.

METHODS OF FINDING LONGITUDE 245

If the body observed is the Sun, the angle P is a measure of the time interval from apparent noon, so that if P can be found the Local Apparent Time (L.A.T.) may also be found. To the L.A.T. the equation of time (given in the *Nautical Almanac*) is applied to give the Local Mean Time (L.M.T.).

If the body observed is a star or celestial body other than the Sun, the problem of finding the L.M.T. of the observation is more complex than with the case for the Sun. By applying to the local hour angle of the observed object its Right Ascension (R.A.) the Local Sidereal Time (L.S.T.) may be found. The L.S.T. is equivalent to the Right Ascension of the observer's meridian (R.A.M.). Having found this, the Sun's Right Ascension is applied to it to find the L.A.T., and to this the equation of time is applied to give the L.M.T.

Because of the relative complexity involved when finding the L.M.T. of an observation of a star, Moon or planet, the Sun was normally employed for finding longitude. To a large extent this is true even for today when Sun observations for longitude are still regarded, in many instances, as providing the best fixes.

If, at the time of a sight, the chronometer time is recorded, the G.M.T. of the observation is known; and this, when compared with the L.M.T. of the observation, enables the observer to find his longitude, for:

$$\text{Longitude} = \text{G.M.T.} \sim \text{L.M.T.}$$

Longitude is named West when G.M.T. exceeds L.S.T. and East when L.M.T. exceeds G.M.T. Hence the well-known aid to memory:

Longitude west, Greenwich time best.
Longitude east, Greenwich time least.

Given the three sides of a spherical triangle, any of the three angles may be calculated by using the fundamental cosine formula of spherical trigonometry. Thus, in the astronomical triangle of nautical astronomy, we have:

$$\cos P = \frac{\cos ZX - \cos PZ . \cos PX}{\sin PZ . \sin PX}$$

from which:

$$\cos P = \frac{\cos z - \sin l . \sin d}{\cos l \cos d}$$

when l and d have the same name, and

$$\cos P = \frac{\cos z + \sin l . \sin d}{\cos l . \cos d}$$

when l and d have different names.

The cosine formula is not suited to logarithmic computation, so that the calculation of angle P by its use is tedious. This defect prepared the way for other formulae, derived from the fundamental cosine formula but which do not suffer its principal disadvantage.

There is a singularly wide variety of trigonometrical rules for finding an angle in a spherical triangle, and many of these have been used for solving the PZX triangle. A layman well may be surprised that seamen were not provided with a standard method for solving the PZX triangle.

The methods used at different times for solving the astronomical triangle have seldom provided the shortest, or the simplest, or even the most accurate solution. The method employed often depended upon which set of nautical tables a seaman was accustomed to using, and the number of collections of tables available during the 19th century was legion. Moreover, once a specified method of sight reduction had been accepted, mastered and committed to memory, a seaman tended to use it, to the exclusion of all others, throughout his sea-going career. This applies generally to the mathematically-minded seaman of today, and doubtless applied more so to the non-mathematical mariners of the past. Seamen of all times have been content to work according to the rule, caring little for the derivation of the rule. We shall discuss some of the many methods used for calculating angle P in the astronomical triangle.

Since $\cos P = 1 - 2 . \sin^2 P/2$, by substitution in the spherical cosine formula for solving P in the PZX triangle, we have:

$$1 - 2 \sin^2 P/2 = \frac{\cos ZX - \cos PZ . \cos PX}{\sin PZ . \sin PX}$$

i.e.

$$1 - 2 \sin^2 P/2 = \frac{\sin a - \sin l \cos PX}{\cos l \sin PX}$$

and

$$2 \sin^2 P/2 = 1 - \frac{\sin a - \sin l \cos PX}{\cos l \sin PX}$$

METHODS OF FINDING LONGITUDE

$$= \frac{\cos l \sin PX + \sin l \cdot \cos PX - \sin a}{\cos l \sin PX}$$

$$= \frac{\sin (PX + l) - \sin a}{\cos l \cdot \sin PX}$$

Now,
$$\sin A - \sin B = 2 \cdot \cos \tfrac{1}{2}(A+B) \cdot \sin \tfrac{1}{2}(A-B)$$

so that
$$2 \sin^2 P/2 = \frac{2 \cdot \cos \tfrac{1}{2}(PX + l + a) \cdot \sin \tfrac{1}{2}(PX + l - a)}{\cos l \cdot \sin PX}$$

Therefore:
$$\sin P/2 = \sqrt{\sec l \cdot \operatorname{cosec} PX \cdot \cos \tfrac{1}{2}(PX + l + a) \sin \tfrac{1}{2}(PX + l - a)}$$

let
$$s = \tfrac{1}{2}(PX + l + a)$$

Then:
$$\sin P/2 = \sqrt{\sec l \cdot \operatorname{cosec} PX \cdot \cos s \cdot \sin (s - a)}$$

This formula was derived by the French naval officer Jean Charles de Borda (1733–1799), who published it in his *On the Reflecting Circle* in 1787.

Borda's method for finding angle P was popular during the first half of the 19th century. It was described in many manuals of the period, including those of Andrew Mackay (1793), and Edward Riddle (1824).

Henry Raper improved Borda's method by adapting it for use with haversines. Now:
$$\operatorname{hav} \theta = \tfrac{1}{2} \operatorname{vers} \theta = \tfrac{1}{2}(1 - \cos \theta)$$

Also,
$$\cos \theta = 1 - 2 \sin^2 \theta/2$$

Therefore:
$$\operatorname{hav} \theta = \sin^2 \theta/2$$

so that
$$\operatorname{hav} P = \sqrt{\sec l \cdot \operatorname{cosec} PX \cdot \cos s \cdot \sin (s - a)}$$

Raper provided a table of logarithms of $\sin^2 \theta/2$, which he called the *Table of Log Sine Squares*. By using this table the method he suggested, for solving P in the PZX triangle, was shortened compared with Borda's method. The method suffered the disadvantage in that five tables of different log trig functions were required.

The trigonometrical function the versine was adapted to the needs of nautical astronomy during the latter part of the 17th century. Now versine θ is equal to $(1 - \cos \theta)$, so that the spherical cosine formula may be reduced to:

$$\text{vers P} = \frac{\text{vers ZX} - \text{vers (PX} \pm \text{PX)}}{\sin \text{PZ} . \sin \text{PX}}$$

The quantity $(PZ \pm PX)$ is equivalent to the Meridian Zenith Distance (M.Z.D.) of the body, disregarding any change in the body's declination during the period of time equal to P. Thus:

$$\text{vers P} = \frac{\text{vers ZX} - \text{vers M.Z.D.}}{\sin \text{PZ} . \sin \text{PX}}$$

Now,

vers ZX − vers M.Z.D. = $(1 - \cos \text{ZX}) - (1 - \cos \text{M.Z.D.})$
 = cos M.Z.D. − cos ZX
 = sin mer. alt. − sin alt.

Therefore:

vers P = (sin mer. alt. − sin a) sec l . sec d

or,

vers P = (sin mer. alt. − sin a) sec d . cosec co l

This is the basis of the rule given by Nevil Maskelyne in his *Requisite Tables*. Included in these tables are those of natural and artificial- or logarithmic-versines, the latter being referred to as *log risings*. The layout of the solution to the longitude problem recommended by Maskelyne is illustrated in the following example which is extracted from the second edition of the *Requisite Tables*.

Example: Sun's declination 22° 23½′ N. Latitude 34° 55′ N. Corrected altitude 36° 59¾′. Find the Sun's meridian distance.

Dec.	22° 23½′ N.	→ log sec	0·03405	
Co-lat.	55° 05′	→ log cosec	0·08619	
Mer. alt.	77° 28½′	—————————→	nat. sin	97620
Cor'd. alt.	36° 59¾′	—————————→	nat. sin	60176
			4·57338→log diff.	37444
		Log rising	4·69363	
		Mer. dist.	3 hrs. 58 mins. 23 secs.	

METHODS OF FINDING LONGITUDE 249

It will be noticed that in the example above, log- and natural-functions are given to no more than five places of decimals. It was Maskelyne's considered opinion that, for purposes of astronomical navigation, this was abundantly sufficient.

Andrew Mackay, in his *Theory of the Longitude* of 1793, gave a method for finding longitude by chronometer based on the spherical cosine formula. His method required the use of three special tables (numbered in his collection 27, 28 and 29 respectively), in addition to tables of natural- and logarithmic-sines. Mackay's rule is:

'Enter Table 27 with latitude and declination, and take out corresponding number to which prefix the index 4; and add to it the log sine of the corrected altitude. Find the natural number answering thereto, to which apply the number from Table 28, by subtracting or adding, according as latitude and declination are of same or contrary name. Now find the above difference or sum in Table 29, and the corresponding time will be the distance of the object from the meridian.'

The rule appears to have been derived in the following way:

$$\cos P = (\cos ZX - \cos PZ \cdot \cos PX) \operatorname{cosec} PZ \cdot \operatorname{cosec} PX$$

From which:

$$\cos P = \sin a \cdot \sec l \cdot \sec d \pm \tan l \cdot \tan d$$

By adding unity to each side, we have:

$$1 + \cos P = 1 + \sin a \cdot \sec l \cdot \sec d \pm \sin l \cdot \tan d$$

Now,

$$1 + \cos \theta = \text{suversine } \theta$$

Therefore:

$$\text{suvers } P = 1 + \sin a \cdot \sec l \cdot \sec d \pm \tan l \cdot \tan d$$

and

$$\underbrace{\text{suvers } P - 1} = \sin a \cdot \underbrace{\sec l \cdot \sec d} \pm \underbrace{\tan l \cdot \tan d}$$

Table 27 gives values of $\sec l \cdot \sec d$ multiplied by 10,000.
Table 28 gives values of $\tan l \cdot \tan d$ multiplied by 10,000.
Table 29 gives values of suvers $P - 1$ multiplied by 10,000.

The rule of Mackay's would appear to provide for a shorter

solution than either Borda's or Maskelyne's. This, however, is not the case, because Tables 27 and 28 give tabulated values for each whole degree of latitude and declination; the necessary tedious interpolation, therefore, made the method unwieldy.

It is an easy matter to derive the following four formulae from the fundamental cosine formula by expressing cos P in terms of sin P, sin P/2, cos P/2 and tan P/2 respectively:

$$\sin P = \frac{2/\sin PZ \sin PX}{\sqrt{\sin s . \sin (s-ZX) . \sin (s-PZ) . \sin (s-PX)}}$$

$$\sin P/2 = \sqrt{\frac{\sin (s-PZ) . \sin (s-PX)}{\sin PZ . \sin PX}}$$

$$\cos P/2 = \sqrt{\frac{\sin s . \sin (s_2-ZX)}{\sin PZ . \sin PX}}$$

$$\tan P/2 = \sqrt{\frac{\sin (s-PZ) . \sin (s-PX)}{\sin s . \sin (s-ZX)}}$$

All of these formulae are adapted for logarithmic computation. To which of the four preference ought to be given over the others should depend on the value of angle P. It can be demonstrated that the first is suitable for cases in which P is near 90°. It is expedient to use the second or fourth when P is acute; and to use the third when P considerably exceeds 90°.

Rules based on these formulae for solving angle P in the PZX triangle are given in many of the 19th-century navigation manuals. Mrs Janet Taylor, a prominent teacher of navigation who had an academy in the Minories in London gave, in the 1837 edition of her book *The Principles of Navigation Simplified*, rules based on all four: and, in addition, she gave seven additional rules, making twelve in all, for solving an angle in a spherical triangle. This celebrated lady, in common with most other manual writers of the time, made no attempt to advise seamen on the relative methods of the several rules given. A rule was given, invariably without demonstration, and the seaman was expected to accept it without question. Navigation manuals of the 18th and 19th centuries are crowded with rules—many of them extremely complex and often shrouded in mystery. It was not until the dawn of the present century that the written rules gave way to the more

METHODS OF FINDING LONGITUDE 251

comprehensible trigonometrical formulae to be found in the manuals of the present day.

Of the half-angle formulae suitable for solving angle P, those giving sin P/2 and cos P/2 were most commonly used. John Hamilton Moore's single rule for finding meridian distances was based on the formula for cos P/2. The formula for sin P/2 was a favourite during the latter part of the 19th century seemingly because sines alone were employed in the computation.

The principal stumbling-block in most of the above formulae, from the seaman's point of view, is related to the difficulty of handling trigonometrical functions of angles over 90°. An interesting method, designed to overcome this difficulty, was suggested in 1860 in a letter to the editor of the *Nautical Magazine*. The proposed method involved working from the nadir of the celestial concave instead of from the zenith, and using the object's nadir distance instead of its zenith distance.

The formula:

$$\sin A/2 = \sqrt{\frac{\sin(s-b)\sin(s-c)}{\sin b \sin c}}$$

may be simplified by squaring both sides and substituting haversine A for $\sin^2 A/2$, thus:

$$\text{hav } A = \frac{\sin(s-b).\sin(s-c)}{\sin b . \sin c}$$

By using this formula the need for dividing the sum of the logs by two and then having to double the angle A/2 to find A—which is unavoidable when using the formula for sin A/2—is eliminated.

An interesting method used for finding P is derived from the cosine formula as follows:

$$\cos P = \frac{\cos ZX - \cos PZ . \cos PX}{\sin PZ . \sin PX}$$

from which:

$$2.\sin^2 P/2 = 1 - \frac{\cos ZX - \cos PZ . \cos PX}{\sin PX . \sin PZ}$$

$$= \frac{\cos(l \sim d) - \cos ZX}{\cos l . \cos d}$$

$$= \frac{2.\sin \tfrac{1}{2}(ZX + \overline{l \sim d}).\sin \tfrac{1}{2}(ZX - \overline{l \sim d})}{\cos l . \cos d}$$

Therefore:
$$\text{hav } P = \sec l \cdot \sec d \sqrt{\text{hav}\{ZX+(l\sim d)\} \text{ hav }\{ZX-(l\sim d)\}}$$

This was the formula used by British naval officers during the closing decades of the last, and the opening decades of the present, century.

Tables of versines from 0° to 180°, log haversines, and half log haversines were provided in Inman's tables. These tables were used almost exclusively in the Royal Navy during the greater part of the last century.

James Inman D.D., who was Senior Wrangler at Cambridge in 1800, became Professor of Mathematics at the Royal Naval College at Portsmouth. During a long and distinguished career Inman devoted considerable attention to promoting nautical astronomy. Although versines and haversines had been used for navigational purposes during the 18th century, it is chiefly to Inman that we owe the popularity of these functions which has lasted to the present time.

The significant feature of the versine is that it has a unique value for every angle between 0° and 180°: so that angles in the second quadrant present no difficulty in respect of algebraic sign when versines, instead of the basic trigonometrical functions cosine, tangent, cotangent and secant, are used.

We have noted above that the cosine formula for angle P in the PZX triangle modified for the use of versines is:

$$\text{vers } P = \frac{\text{vers } ZX - \text{vers } (PZ \pm PX)}{\sin PZ \sin PX}$$

or

$$\text{vers } P = \frac{\text{vers } z - \text{vers } (l \pm d)}{\cos l \cos d}$$

i.e.

$$\text{vers } P = \{\text{vers } z - \text{vers } (l \pm d)\} \sec l \sec d$$

i.e.

$$\text{vers } P = \text{vers } \theta \sec l \sec d$$

where

$$\text{vers } \theta = \text{vers } z - \text{vers } (l \pm d)$$

H. B. Goodwin, who edited the second edition of William Hall's adaptation of Inman's tables in 1918, proposed, in 1899, the above formula for finding angle P. Goodwin pointed out that:

'... since in Inman we have no logarithmic versines, but only log haversines, we must divide each side by 2.'

If follows, therefore, that:

$$\text{hav } \theta = \text{hav } z - \text{hav } (l \pm d)$$

and

$$\text{hav } P = \{\text{hav } z - \text{hav } (l \pm d)\} \sec l \sec d$$

Goodwin demonstrated that all practical astronomical problems of navigation could be solved by the application of versines (or haversines) to the spherical cosine formula.

The formulae for the Ex-meridian and the Marcq St Hilaire problems are respectively:

$$\text{hav M.Z.D.} = \text{hav ZX} - \text{hav } P \cos l \cos d$$
$$\text{hav ZX} = \text{hav } P \cos l \cos d + \text{hav M.Z.D.}$$

Goodwin's comments are interesting in view of the general belief that Percy L. H. Davis was responsible for the introduction of the so-called *cosine-haversine method* of sight reduction commonly employed at the present time. It is true that Davis, in his *Requisite Tables* of 1905, published the haversine table in its present form for the first time: a form in which natural and logarithmic haversines are tabulated in adjacent columns to facilitate the solution of the PZX triangle by the cosine-haversine method. The following statement made by Goodwin in the paper referred to above is, however, important:

'... The conclusion to which a comparison of the various methods employed today in practical nautical astronomy appears to point may be summed up briefly as follows:

'Firstly that a table of natural versines and another of log haversines, or better still, versines, should be included in every collection of nautical tables....'

It is interesting to note that in Goodwin's edition of Inman's tables, Davis's haversine table appears with a note by Davis dated 1932. The note clearly indicates the jealousy with which Davis regarded his table and the contempt he held for those whom he regarded as being pirates of his copyright.

Had Goodwin's recommendation been carried out it is

doubtless that present-day navigators who use the so-called 'longitude by chronometer' method in preference to the superior intercept method, would use Goodwin's formula:

$$\text{vers } P = \text{vers } \theta \sec l \sec d$$

In view of this, tables of haversines would have fallen into disuse.

The importance of the chronometer for finding longitude at sea using astronomical methods was, in olden times, paramount. At the present time the costly and delicate chronometer, although carried on almost every ocean-going craft, no longer ranks as a vitally necessary aid to navigation. A good pocket watch with a sweep second hand is sufficient for navigational purposes; this on account, not only of the availability of good and inexpensive time-keepers but also because of the frequency of radio time-signals by means of which the error and rate of a watch may be checked.

(b) OBSERVATIONS FOR CHECKING THE CHRONOMETER

Before the days of radio time-signals the longitudes of numerous shore stations were determined by electric telegraph. Before the invention of the telegraph, chronometric differences and other methods were used to establish longitudes. The celebrated Henry Raper played a prominent rôle in compiling tables of maritime positions with the principal aim of providing the means whereby a seaman could check his chronometer. Raper's first table of maritime positions was published in the *Nautical Magazines* of 1839 and 1840, and gave details of upwards of 200 maritime positions. Raper's communication to the *Nautical Magazine* consisted of five sections as follows:

1. An abstract of the principal scientific voyages and surveys, from which he obtained his data.
2. Remarks on the different modes of determining longitudes.
3. Remarks on the necessity of adopting a uniform method of placing on record chronometric determinations in order to be immediately available for the construction and examination of charts.
4. Remarks on the necessity of adopting secondary meridians.
5. A discussion of the principal maritime positions and the methods employed to determine their positions.

METHODS OF FINDING LONGITUDE 255

In the first section, the abstracts begin with Cook's voyages:

'Cook. 1st voyage. H.M.S. Endeavour, left Plymouth Aug. 26. 1768. Madeira, Rio, Straits le Maire, Tierra del fuego, Lagoon I., Bow I., Bird I. Discovered Chain I., Society I., Otaheite, where he observed the Transit of Venus. New Zealand, which he sailed around. Coast of Australia, Rotte I., Sava, Java Head, Batavia, Cape, St. Helena.

'2nd voyage. In the Resolution and Adventure. Capt. Furneaux took 4 chronometers (one of these was the first chronometer made by Kendall upon Harrison's description, and Cook's favourable report procured to Harrison the additional £10,000 from Parliament. The first trial of Harrison's own watch had been made on a voyage to Jamaica in 1761, and his son's voyage to Barbados in 1764). Left Plymouth July 13th 1772. Madeira, C. Verds, C.G. Hope. Searched for Southern Continent. New Zealand (Dusky Bay) Resolution I., Doubtful I., Marguesa, Society I., Friendly I., New Hebrides, N. Caledonia discovered. Discovered Norfolk I. Corrected his former position of N. Zealand. Tierra del fuego, Georgia, Fernando, Azores.

'3rd voyage. Resolution and Discovery. Capt. Clarke Sailed from Plymouth July 12th 1776. Teneriffe, C. Verds, Cape, Prince Edward I., Kerguelens I. V.D. Land, New Zealand, Discovered Mangeea I and Wateoo. Friendly Is. Otaheite. Bolabola, Discovered Christmas I. Sandwich I. New Albion, Nootka Sound, Coast of Oonalashka and thence to northwards. C. North, Discovered Owhyee where he was killed. Capt. Clarke Kamschatka in search of N.W. passage died. Coast of Japan. Sulphur I. Discovered, Prata, Macao.'

Other abstracts included those of French, Spanish, German, Russian, Danish, Dutch, as well as British voyages.

In his discussion on methods of determining longitude, Raper pointed out that, although the most common astronomical observation employed for the purpose was that of the lunar distance, it was well known that no results differed more widely from each other than those of lunars taken at different times, and often at the same time. He quoted the Rev. G. Fisher, who was the Astronomer to Captain Parry's second voyage (1821–1823). Fisher stated that:

'... the mean of 2500 observations in December differed 14' from the mean of 2500 observations in the following March; and that the mean of a great many observations made on board both ships (Griper and Hecla) in the following summer differed 10' from the last or 24' from the first.'

Most experienced navigators of the time considered the result of a lunar observation as serving only to correct gross errors in the chronometer.

Raper discussed the deficiencies in the methods of finding longitude from observations of Jupiter's satellites, eclipses of the Sun, Moon culminating stars, and lunar transits. He concluded his remarks by suggesting the use of steam vessels carrying several chronometers (fifty or sixty) to measure chronometric differences for ascertaining the longitudes of places.

Captain (later Admiral) Charles F. A. Shadwell, R.N., followed up Raper's labours and produced, in the middle of the 19th century, a classic work entitled *Notes of the Management of Chronometers and the Measurement of Meridian Distances*.

In his preface to this work, Shadwell mentions the important services rendered by chronometers in the ordinary course of navigation at sea, and then points out that:

'... they are susceptible, when placed in intelligent hands, of being applied to higher scientific uses; and, when rightly employed, are capable of affording valuable contributions towards the gradual perfection of Maritime Geography.'

The treatises on navigation used by seamen of the last century contained ample rules and directions for using chronometers for determining longitude at sea, but little or no instruction was normally to be found on how to measure 'chronometric differences' or 'meridian distances.' Shadwell's aim was to overcome this defect of the seaman's manual.

At the time he wrote, the invention of the electric telegraph and the successful accomplishment of its submarine connection, had already been used for astronomical purposes of measuring differences of longitude. Shadwell, however, never fully realized that this marvellous device was soon to prove of inestimable value in settling terrestrial longitudes. In this connection the American

METHODS OF FINDING LONGITUDE   257

Government, from the year 1874, played a prominent part. Under the direction of the Bureau of Navigation, longitude chains of meridian distances connecting many parts of the world were measured. One of the first of these, which was determined in 1878-1879, embraced Greenwich, Lisbon, Madeira, St Vincent, Pernambuco, Rio, Montevideo, Buenos Aires and Para (Brazil), and was later continued to Cordoba. It was found from this telegraphic investigation that the longitude of the observatory at Lisbon was in error to the extent of no less than 2'.

In 1883-1884, another telegraphic chain embraced Vera Cruz, (previously connected telegraphically to Washington), Lima, West Africa, Valparaiso and closing, as did the 1879 chain, in Cordoba. The two independent longitudes of Cordoba, found in this way, differed by no more than about 0·2' of arc.

The longitudes of numerous maritime positions having been established, it became possible for mariners to check their chronometers by time signals which were made at most of the principal harbours of the world for the express use of shipping. The time-signal at the Royal Greenwich Observatory, for example, consisted of the dropping of a black ball from the mast on the Eastern turret. The ball was hoisted half-mast as preparatory at five minutes to 1 p.m. G.M.T. At two-and-a-half minutes to 1 it was hoisted close up, and dropped at 1 p.m. G.M.T. precisely. At Dover, a gun was fired at noon G.M.T. precisely. At Hobart (Tasmania), a ball was dropped simultaneously with the firing of a gun at 1 p.m. local time (15 hrs. 10 mins. 39·6 secs. Astronomical G.M.T.) precisely.

(c) ABSOLUTE-ALTITUDE AND EQUAL-ALTITUDES OBSERVATIONS

In the event of there being no local visual or audible time-signal, the navigator who wished to check his chronometer had to fall back on astronomical methods. The problem involved is simply one of finding the G.M.T. by either absolute-altitude, or equal-altitudes, observations.

The term *absolute altitude*, in this context, means observations on one or both sides of the meridian worked out as ordinary longitude by chronometer problems. In contrast, the term *equal altitudes* applies to a method in which the times when the observed body has equal altitudes east and west of the meridian are noted, the G.M.T. of meridian passage of the object being computed from these times.

The three kinds of observations which came in the category of absolute altitudes were:

1. Absolute altitudes of the Sun or a star on one side of the meridian.
2. Mean of the results of absolute altitudes of the Sun taken before and after noon.
3. Mean of the results of absolute altitudes of stars taken east and west of the meridian.

The normal method of observation included the use of an artificial horizon ashore at a place whose exact latitude and longitude are found from the chart or sailing directions. At many harbours special 'observation spots' were provided for this purpose. The position of an observation spot was accurately known, the spot having been used in the survey of the area, and was to be found in the title of the chart.

A series of an odd number—usually seven, nine or eleven—of altitude observations were made and the chronometer times of each observation noted. The mean of the altitudes together with the known latitude were used to compute the apparent time, and hence the G.M.T. of the observation at the position of observation. In the case of the Sun

$$G.M.T. = L.M.T. \pm \text{longitude}$$

where

$$L.M.T. = L.A.T. \pm \text{equation of time}$$

In the case of a star:

$$G.M.T. = L.H.A.^* + R.A.^* - R.A.M.S. \pm \text{long}$$

The equation of time or the Right Ascension of the Mean Sun, as the case may be, was extracted from the *Nautical Almanac*. The comparison of the calculated G.M.T. of observation with the chronometer time gave the chronometer error for the time of the observation.

The observations were made ashore in preference to on board, because of the unreliability of altitudes measured above the sea horizon on account of the unknown refraction effects on the dip of the sea horizon.

The errors involved in absolute altitudes taken on one side only of the meridian included instrumental error, shade error, roof error, error due to irradiation, error due to abnormal refraction, and personal error.

Shade error was due to the two rays from the horizon glass and mercury surface, respectively, passing through different-coloured shades before entering the observer's eye, and the possibility of these shades having different errors. Shade error was obviated by using the coloured eyepiece, with which most sextants were provided, instead of shades, when making observations with an artificial horizon.

Roof error was due to want of parallelism of the faces of the glass used in the roof of the artificial horizon. This error was eliminated by reversing the roof half-way through the set of observations.

Irradiation error, due to the optical illusion in which bright objects viewed against a dark background appear to be larger than they really are, was eliminated, when observing the Sun, by taking two sets of observations, one of the Sun's upper limb, and the other of his lower limb, working each set separately and taking the mean of the results.

Instrumental error, and error due to abnormal refraction cannot be eliminated in observations of absolute altitudes observed on one side only of the meridian. They do, however, tend to cancel out when absolute altitudes are taken on both sides of the meridian when the Sun has the same, or nearly the same, altitude at both observations.

The most accurate results are obtained from observations of absolute altitudes of two stars, one east and the other west of the meridian, the two stars having approximately the same altitude, and the interval between the observations being made as short as possible.

The second, and more accurate, method of finding the chronometer error is by equal altitudes.

If the times T_1 and T_2, marked by a chronometer, be noted when a star has the same altitude east and west of the meridian, then it is clear that, provided that the observer is stationary, the time of the star's meridian passage is $\frac{1}{2}(T_1 + T_2)$. If this be known, the error of the chronometer may readily be found. This is the basis of the method of equal altitudes for finding the chronometer error.

One of the advantages of the equal-altitudes method over the method of absolute altitudes is that any error of graduation of the sextant will have no effect on the result, because the two altitudes are taken at the same graduation.

The method of finding the time of meridian passage of a heavenly body by equal altitudes was employed in astronomical observations ashore as early as the 17th century. It appears that the French astronomer Jean Picard was the first to employ the method at sea in 1671.

When the Sun's centre is on the meridian of any place, the local apparent time (L.A.T.) is noon (or midnight). To obtain the Local Mean Time (L.M.T.) of meridian passage, it is necessary to apply the equation of time to the L.A.T. After this has been done the G.M.T. of meridian passage may be found by applying the longitude in time to the L.M.T.

For a stationary observer, the error of his chronometer may be verified from equal altitudes of the Sun. Half the interval between the observations of the Sun with equal altitudes east and west of the meridian, added to the time by chronometer of the first observation, will be the time by chronometer when the Sun is at meridian passage, assuming that the Sun's declination does not change in the interval. The comparison between chronometer time and G.M.T. of meridian passage gives the error of the chronometer at the time of meridian passage.

When using the Sun for finding the chronometer error by equal-altitudes observations, a correction must be made to $\frac{1}{2}(T_1 + T_2)$ on account of the Sun's declination not being invariable. This correction, which normally amounts to no more than about a few seconds of time, is called the equation of equal altitudes. The equation of equal altitudes is least at the time of the solstices, and it is greatest when the Sun's declination is changing most rapidly, that is, during March and September.

In its original form the formula for finding the equation of equal altitudes consisted of two terms involving latitude, polar distance, elapsed time, and change of declination in elapsed time. In dealing with it, attention had to be given to a variety of details often perplexing in the extreme to the practical navigator. The two quantities making up the equation ϵ were denoted by A and B (or M and N) and the usual form of the relationship, probably introduced by the brilliant French astronomer Delambre, is:

$$\epsilon = c\,(\tan \text{lat}\, \text{cosec}\, h - \tan \text{dec}\, \cot h) \quad . \quad . \quad (I)$$

where

$c =$ half change in polar distance in interval,
$h =$ half elapsed time.

Separating the two parts in equation (I), we have:

$$A \text{ (or M)} = c \tan \text{lat}\, \text{cosec}\, h$$
$$B \text{ (or N)} = c \tan \text{dec}\, \cot h$$

From which:

$$\epsilon = A + B \text{ (or } M + N\text{)}$$

To combine A and B correctly, the observer has to consider: whether the polar distance of the observed body is increasing or decreasing; whether the declination and latitude are of the same or different names; whether the elapsed time is greater or less than twelve hours; and so on. Goodwin, writing in the *Nautical Magazine* ('New Applications of the Burdwood, Davis, and Johnson's Azimuth Tables,' 1894) described the formula in the following terms:

'The expression is open to many objections. It is a troublesome one to establish, and it is not easy to retain it permanently in the memory. It has two terms, each of which may have either of two signs: the determination of the appropriate sign is by no means a simple matter, and commonly involves continual reference to the more or less intricate precepts furnished for the purpose.'

Mrs Janet Taylor, in the third edition of her *Principles of Navigation Simplified*, made use of an ingenious method for finding the equation of equal altitudes. She employed an auxiliary angle θ, and obtained the formula:

$$\epsilon = c \cot h\, \text{cosec}\, p\, \text{cosec}\, \theta \sin(p - \theta)$$

where

$$\tan \theta = \cot l \cos h$$

in which l is the latitude, h half the elapsed time and p the polar distance.

The principal advantage of this method over the usual method of finding the equation of equal altitudes, is the simplicity of the rule for determining the sign of ϵ. Mrs Taylor's rule is: 'Additive to middle time when polar distance is increasing, and subtractive when it is decreasing.'

John Riddle, in his *Navigation and Nautical Astronomy* of 1871, also introduced a formula for finding the equation of equal altitudes consisting of one term only, but using the parallactic angle, or angle of position, PXZ in the PZX triangle.

In 1874, W. H. Bolt of Norie's Nautical Academy adapted the A and B Tables used for finding error in longitude consequent upon error in latitude (see Chapter 8) for use with the equal-altitudes problem.

The *Nautical Magazine* of 1889 contained an interesting paper by H. B. Goodwin entitled 'On the Connection between the Equation of Equal Altitudes and the Angle of Position.' Goodwin defined the equation of equal altitudes as the change in hour angle due to the change in the value of the Sun's polar distance during one half of the elapsed time between the two observations. He then proceeded to find an expression for this change. He showed that the error in hour angle resulting from a small error in the co-latitude (PZ) is given by:

$$\text{Error in hour angle} = \frac{1}{\sin PZ \tan PZX} \cdot \text{error in co-lat.}$$

Goodwin then argued that: by symmetry, error in hour angle due to a small error of, or change in, polar distance (PX) is given by:

$$\text{Error in hour angle} = \frac{1}{\sin PX \tan PXZ} \cdot \text{change in p.d.}$$

This is the expression sought for the equation of equal altitudes. Expressed in seconds of time instead of arc, the equation ϵ is:

$$\epsilon = \sec dec \cot PXZ \cdot \frac{\text{change in p.d.}}{15}$$

The beauty of this expression, like that of Janet Taylor's, is the simplicity of the rules for determining the sign of ϵ. The equation is additive for an increasing polar distance, and subtractive for a decreasing polar distance.

Goodwin pointed out that the value of PXZ, which may be solved by trigonometry or lifted from azimuth tables, is required only to the nearest degree. On the other hand, the change in the polar distance during the interval must be found very accurately, since any error in this amount will frequently cause a much larger error in the value of the equation ϵ deduced from it.

It is interesting to note that Goodwin's form for the equation of equal altitudes was used in the first *Admiralty Manual of Navigation* of 1914.

When using the equal-altitudes method for finding the error of the chronometer, a small error in the latitude is not important. W. R. Martin, in his *Navigation and Nautical Astronomy* of 1888, points out that in the most unfavourable circumstances an error of 5' in the latitude will affect the value of M (tan lat cosec h) by only the hundredth part of its value. This provided one good reason why equal altitudes should be employed for finding the error of the chronometer, in preference to the absolute-altitude method, when the latitude of the place of observation is only approximately known.

In high latitudes the value of M—which depends on tan latitude—is not so accurate as in low altitudes.

Lecky, in his *Wrinkles in Practical Navigation*, first published in 1881, declared that the method of equal altitudes of the Sun was open to many objections. Amongst these he cited:

1. The operation cannot be completed at one time.
2. During the interval the conditions which existed in the morning may be considerably changed in the afternoon.
    (a) Refraction may be different on account of a shift of wind, etc.
    (b) Personal equation may be different at the two times of observation.
    (c) Effect of heat or cold may cause divisions on sextant to alter.
3. Setting in of cloudy weather in the afternoon may render the morning's work useless.
4. Inconvenience of double journey to and from ship and the repetition of chronometer comparing.

Better than Sun equal altitudes were star equal altitudes; but these observations were not without their disadvantages. Best of all methods, perhaps, was the observation of stars east and west of the meridian, within a few minutes of each other. In this method all systematic errors, whether atmospheric, instrumental or personal, were practically neutralized in the mean result.

For rating a chronometer, in contrast to finding its error on G.M.T., the common and useful method was to note the chronometer time of successive diurnal disappearances of a fixed star behind a straight-edged board set up truly vertically. The observer's eye was to be at the same point for the observations, either to the north or south of the board, depending on which star is observed.

The interval between the successive instants when a fixed star disappears behind the vertical board is a sidereal day. This is shorter than a mean solar day by 3 mins. 55·9 secs. of mean solar time. If the difference between the times by chronometer noted at the instants of disappearance is not equal to this, the chronometer is gaining or losing.

The method of equal altitudes observed from a ship at sea provided a method for finding longitude. The principle of the method involves using a chronometer for finding the G.M.T. of meridian passage of the observed object. The G.M.T. of this event, when compared with the L.M.T. of the same event, gives the longitude of the ship at the time of the meridian passage of the object.

If the observed object is the Sun, and the observer is stationary and the change in the Sun's declination during the interval between the times of observation is ignored, the mean of the chronometer times of observation will be the chronometer time of apparent noon at ship. If the chronometer error is applied to this, the result will be the G.M.T. of apparent noon. The difference between this and the L.M.T. of apparent noon will be the ship's longitude at noon. This is equally true if the movement of the ship in the interval is due east or due west. By taking the equal-altitude sights shortly before and after noon the necessity for applying a correction for the change in the Sun's declination in the interval is obviated, since any such change will be trifling.

Had the course of the ship been northerly or southerly it would be necessary to apply a correction for the change in latitude in the interval between the times of the equal-altitudes observations. Taking into account the meridianal motion of the ship and the declination motion of the observed object, the object may be assumed to be at its maximum altitude at the mean of the chronometer times of the observations. The interval between the times of meridian- and maximum-altitudes, therefore, is the correc-

METHODS OF FINDING LONGITUDE 265

tion to apply to the mean of the chronometer times to find the chronometer time of meridian passage, and hence the G.M.T. and longitude of the ship.

The general expression for this interval t in seconds, is:

$$t = 15\cdot 28 y(\tan l \pm \tan d)(1 + x/450)$$

where y = combined movements of ship in latitude and object in declination in minutes of arc per hour,

x = rate of change of longitude towards the west in minutes of arc per hour.

Tables giving values of 15·28 tan lat (or dec) were provided for facilitating finding longitude by equal altitudes.

The value of this method of finding longitude rested in the speed with which the longitude could be found, the interval elapsing between the times of the observations being small. Moreover, as a result of the small interval of time between observations, refraction effect, in all likelihood, is the same at both observations. The method is particularly valuable in low latitudes when the Sun, near the meridian, is almost always suitable for observation.

Navigational textbooks of the last century usually ignored the problem as not being susceptible of great accuracy, and too complex for the practical seaman.

The Rev. William Hall, well known for his writings on navigation and nautical astronomy, proposed, in 1902, a solution to the problem having extreme brevity and a theoretical accuracy of 1% in the correction used. The investigation of Hall's rule is as follows:

Let h = half elapsed time
l = latitude
d = declination
δd = correction in seconds of time to reduce the mean of the chronometer times to the chronometer time of meridian altitude

Then, the equation of equal altitudes, δh, is:

$$\delta h = \frac{1}{15} h \delta d(\tan l \cos h - \tan d \cot h)$$

Hall assumed, instead of δd accounting for change in declination alone, for δd to account for meridianal motion of the ship as well. He then assumed (with justification when h is small):

$$\cos h = \cot h = \frac{1}{h \sin 1''}$$

which is true to within 1% for values of h up to thirty-two minutes of time. He then argued thus:

'Let $(V \pm v)$ be the relative velocity of Sun and ship: V being change in latitude and v variation in declination per hour expressed in seconds of arc. Then, making the substitutions and cancelling, we have:

$$\delta h = \frac{1}{15}(V \pm v)(\tan l + \tan d) \operatorname{cosec} 1''$$

'A most remarkable result in that it is independent of h.'

Hall proposed the construction of a table giving values of tan lat (or dec) cosec $1''$.

In 1847, Commander Weston of the Indian Navy proposed a method for finding longitude at sea from an observation of the time of sunrise or sunset.

Weston's method was announced by the Editor of the *Nautical Magazine* in 1848. The Editor remarked that:

'All that was wanted was to know at the instant when the Sun's centre was in the horizon of the observer what is the altitude of either limb, and then to make the observation. Now, as this depended merely on the height of eye and refraction, it became a mere matter of calculation; and, accordingly, Captain Weston has presented to the Admiralty his method accompanied by the necessary tables.

'The seaman has only to place on his sextant the minutes of altitude of either limb depending on his latitude and the Sun's declination: watch for the observation and note the time by chronometer. The tables give him the ship's time and thence the longitude.

'The tables are preparing for publication by the Admiralty,* and we shall soon have the pleasure of congratulating our nautical readers on the important addition to their astronomical resources.

* They were published in 1851.

METHODS OF FINDING LONGITUDE 267

'Captain Weston has been rewarded by the Admiralty for his ingenuity.'

In commenting upon Weston's method, W. R. Martin mentions that it is very unreliable in latitudes greater than 45°; and even in low latitudes the method is liable to a probable error of half a minute of time. In low latitudes the Sun rises out of the horizon at a large angle—90° at the equator—and the interval of time taken for him to rise out of the horizon is, therefore, comparatively short. In consequence of this the method is susceptible of accuracy. In high latitudes, however, where the angle between the Sun's diurnal path at the time of sunrise or sunset and the horizon is relatively small, the time taken for the Sun to rise out of the horizon is relatively long; so that a small error in the altitude of the Sun's limb at the time his centre is on the rational horizon results in a large error in the time.

The method of finding longitude from an observation of the Green Flash which occurs in favourable meteorological conditions when the upper limb of the Sun sinks below the visible horizon has been suggested on many occasions during the present century. This method, like that of Weston's, suffers from the uncertain effects of refraction on the angle of dip of the sea horizon. Moreover, the timing of the event is liable to error, especially in high latitudes, on account of the relatively long interval which the phenomenon may occupy. An interesting article by R. E. G. Simmons on 'Observations of the Green Flash' appears in Vol. 4 (1951) of the *Journal of the Institute of Navigation*.

CHAPTER VII

Position-line navigation

1. INTRODUCTORY

The common method of describing a position on a surface, either plane or spherical, is by means of a system of coordinates related to a point of reference on the surface. To define a position on the celestial sphere three systems are in general use. These are the *ecliptic system*, in which the coordinates are *celestial latitude* and *celestial longitude*; the *equinoctial system*, in which the coordinates are *declination* and *Right Ascension* (or *hour angle*); and the *horizon system*, in which the coordinates are *altitude* and *azimuth*. To define a terrestrial position, the coordinates most commonly used are latitude and longitude. Associated with the latitude and longitude of a particular terrestrial position are two lines on the surface of the globe which intersect at right angles. One of these lines is a parallel of latitude, which defines the position relative to the equator; and the other is a meridian, which defines the position relative to some datum- or prime-meridian.

A method of describing the position of a ship at sea, of great practical utility to the seaman, is one in which the coordinates used are related to circles of equal altitudes which intersect at the ship's position.

The centre of a circle of equal altitude lies at a point which is now known as the *geographical position* of the observed body. The geographical position of the Sun is sometimes called the *sub-solar point*, and that of a star as the *sub-stellar point*. The geographical position of a heavenly body is a point on the Earth's surface at which the body occupies the zenith; that is to say, the point at which the altitude of the body is 90°.

Because the Earth rotates, the geographical position of every heavenly body changes with time. It may readily be shown that the latitude and longitude of the geographical position of a celestial body at any time is equal to the declination and Greenwich Hour Angle of the body at the instant. Fig. 1 serves to illustrate this.

In Fig. 1, the smaller circle represents the Earth, and the larger circle the celestial sphere; p represents the Earth's North Pole

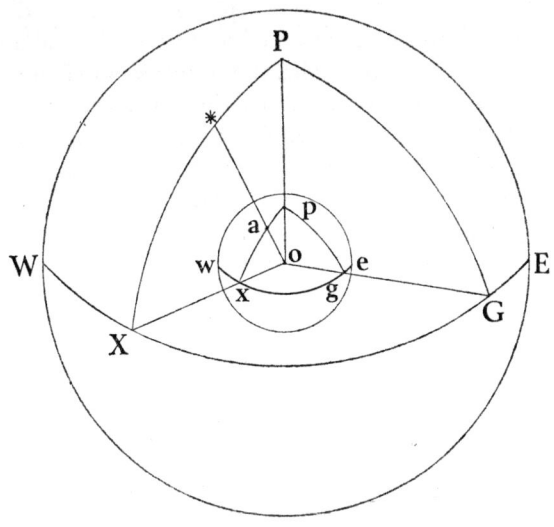

FIGURE 1

and P the celestial pole; wxge and WXGE represent the equator and equinoctial respectively. pg and PG represent the Greenwich meridian and the Greenwich celestial meridian respectively; * is any celestial body whose geographical position is at a.

$$xa = X*$$

Therefore:

$$\underline{\underline{\text{Lat. of G.P.} * = \text{dec.} *}}$$

$$gpx = GPX$$

Therefore:

$$\underline{\underline{\text{Long. of G.P.} * = \text{Greenwich Hour Angle.} *}}$$

A circle of equal altitude of a celestial body is a circle on the Earth, at every point on which the altitude of the body, at any given instant of time, is the same. Fig. 2 illustrates a circle of equal altitude related to a celestial body *.

In Fig. 2, O represents the Earth's centre and P the North Pole. A_1 represents the geographical position of celestial body *. At every point on the circle ABC the altitude of * at the instant for which the diagram applies, is $\alpha°$. It will readily be seen from the figure that the radius of a circle of equal altitude which passes through an observer's position is equal to the zenith distance of the body; and that the direction of the circle of equal altitude at any point on it is at right angles to the direction of the observed body, or that of its geographical position, at that point.

A small part of a circle of equal altitude in the vicinity of an observer's position forms, when projected on a chart, a position line. If a navigator is able to project two arcs of circles of equal altitude as position lines, he clearly is able to fix his ship's position at the intersection of the two position lines. This method of fixing a ship forms the basis of position-line navigation.

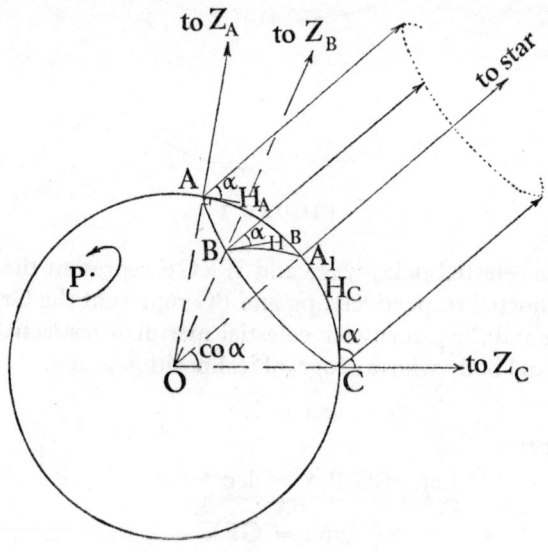

FIGURE 2

The discovery of position-line navigation rightly belongs to the American sea captain Thomas H. Sumner, who is credited with being the first to systematize the problem of finding position at sea by means of chronometer, sextant, and *Nautical Almanac*. Although Sumner is to be honoured for the discovery, the idea of position-line navigation evidently had been in men's minds for a long period before the time of Sumner's discovery. As early as the beginning of the 19th century, officers of the British Royal Navy used a method of fixing from two altitudes of the Sun which was known as the method of *Cross Bearings of the Sun*. This, and other early methods, was developed with the growing realization that azimuth is a fundamental element in every navigational problem.

2. HISTORICAL DEVELOPMENT OF POSITION-LINE NAVIGATION

Position-line or intersection-navigation, as it was sometimes called, stemmed from the method of finding latitude known as the *double altitude*. The germ of position-line navigation is contained in a work entitled *The Theory and Practice of Longitude at Sea* written by Samuel Dunn and published in 1786. In this work, which was dedicated to 'The Honourable the Court of Directors of the United Company of Merchants of England trading to the East Indies,' Dunn introduced a problem entitled:

'Of a general method whereby the latitude may be found having any two altitudes of the Sun and the time elapsed between the observations.'

By assuming two latitudes differing about 1° or less, and not widely different from the latitude by D.R., Dunn showed that the two altitudes give four hour angles, two of which appertain to each of the assumed latitudes. He then proceeded to make the following statement:

'As the difference of the elapsed times computed from the assumed latitudes is to the difference of those latitudes: so is the difference between the true elapsed time to a number of minutes which, added to or subtracted from the corresponding assumed latitude, as the case requires, gives the true latitude required, when the latitudes are assumed near enough for the truth.'

Chronometers were scarce in the days when Dunn introduced his novel problem. Had they been common, it is likely that Dunn would have extended his method for finding longitude as well as latitude. But, at the time, the only practical general method for finding longitude at sea was by the lunar distance method.

Although Dunn's resolution of the double-altitude problem had been discussed, in 1786, in all its aspects by Lalande in his *Astronomie* and *Abrégé de Navigation*, there is every justification for stating that the development of position-line navigation from

Dunn's time was carried out on the basis of his double-altitude method.

In 1812, Commander Owen, R.N., published a pamphlet in which he described a practical method for finding latitude and longitude based on Dunn's method for the double altitude for latitude. Another publication of importance in the same connection is the pamphlet published in 1833 by Commander Thomas Lynn of the East India Company. Lynn's pamphlet is entitled 'Practical Methods by Trial and Error for finding Latitude and Time at sea by the Observed Altitudes of Sun, Moon, and Stars, and the Elapsed Time between the Observations, originally submitted by Samuel Dunn.' At the time Lynn's pamphlet first appeared, chronometers had come into general use, and the time was ripe for the advent of position-fixing by astronomical position lines.

The principle of the method of making observations serve for longitude at the same time as for latitude is as follows. If observations for longitude are made, either of different heavenly bodies made at the same time, or of the same body at two different times —the correction for any run made in the interval enabling the navigator to consider both observations as having been made at the same place—then, if the two hour angles calculated with an assumed latitude agree, the ship's actual latitude is the assumed latitude. Alternatively, if the L.M.T. at each observation is calculated, both results being reduced to the place of the second observation, and it is found that the difference between the two L.M.T's corresponds with the difference between the chronometer times of the observations, the ship's latitude is the assumed latitude.

If, therefore, the intervals between chronometer times and computed times do not agree, the latitude used in the computations is NOT the ship's latitude. In this event, the method of trial and error was adopted in an attempt to find the ship's true latitude. If successful, the ship's longitude could be found provided that G.M.T's of observations were known.

The earliest solution to this valuable problem, which was known as the *double-chronometer problem*, was given by Lalande. Lalande's rule is based on the effect of an error in latitude on the hour angle of a body deduced from it, viz:

$$\Delta h = \Delta \phi \cot Z \sec \phi$$

POSITION-LINE NAVIGATION

in which h, ϕ, Z indicate hour angle, latitude and azimuth respectively. If Δh is expressed in minutes of time and $\Delta\phi$ is expressed in minutes of arc, then:

$$\Delta h = \frac{\Delta\phi \cot Z}{15 \cos \phi}$$

Suppose that two altitudes of a heavenly body, both west of the meridian, are observed at chronometer times C_1 and C_2, at which instants their corresponding azimuths are Z_1 and Z_2 (both less than 90°). Suppose that the calculated L.M.T's, using D.R. latitudes ϕ_1 and ϕ_2, are T_1 and T_2. If the correction for run in longitude be applied to T_1 to give T_c, we have T_c and T_2 the two results of the observations reduced to the instant when the second observation was made.

If:

$$(T_2 - T_c) = (C_2 - C_1)$$

then the ship's true latitude at the time of the second observation is ϕ_2.

The effect of an error $\Delta\phi$ in ϕ_1 will be, in this case, to alter the L.M.T. reduced to the time of the second observation to:

$$T_c + \frac{\Delta\phi \cot Z_1}{15 \cos \phi_1}$$

and that of the L.M.T. of the second observation to:

$$T_2 + \frac{\Delta\phi \cot Z_2}{15 \cos \phi_2}$$

Therefore:

$$\left(T_2 + \frac{\Delta\phi \cot Z_2}{15 \cos \phi_2}\right) - \left(T_c + \frac{\Delta\phi \cot Z_1}{15 \cos \phi_1}\right) = (C_2 - C_1)$$

From which:

$$\frac{\Delta\phi}{15} = \frac{(C_2 - C_1) - (T_2 - T_c)}{\cot Z_2/\cos \phi_2 - \cot Z_1/\cos \phi_1}$$

Had both observations been made when the heavenly body was east of the meridian, the quantity on the right-hand side of the equation would have a minus sign. Had the observations been made when the body was on opposite sides of the meridian or prime vertical circle, the denominator would be a sum instead of a difference. In general, therefore:

$$\frac{\Delta\phi}{15} = \pm\frac{\{(C_2-C_1)-(T_2-T_c)\}}{\cot Z_2/\cos\phi_2 \pm \cot Z_1/\cos\phi_1}$$

This is a complex formula, and the problem is one in which there are so many cases, that it is small wonder that the non-mathematical seaman found little practical use for the method.

In 1847, M. Louis Pagel of the French Navy simplified the method by considering the effect on the error in longitude due to an error in latitude of 1' of arc. Pagel published tables to simplify the problem. So also did A. C. Johnson, R.N., who improved Pagel's method. Johnson's double-chronometer method, which became exceedingly popular in both the British Royal Navy and the Merchant Service, was published in the famous pamphlet *On Finding Latitude and Longitude in Cloudy Weather*, a title which earned for its author the name 'Cloudy Weather Johnson.' Johnson's pamphlet was first published in about 1880.

As late as 1892 we find Captain Philip Parker, R.N., proposing an improvement on the existing methods of the double-chronometer problems. By this time, however, although the double chronometer in its modified form was still in use at sea, it was fast becoming redundant, to be replaced by the improved methods of astronomical navigation.

Admiral Charles Bethune, writing in 1871, stated that in the 1820's the method of finding latitude and longitude using position-line principles was regularly in use:

'... by those who felt the responsibility devolving on the person charged with the safe conduct of a ship and who were well aware that latitude by D.R. was subject to error.'

The Admiral also suggested that:

'It should be a standing order that sights be obtained as early as possible (for choice when the Sun bears due east), and again at 6 bells [11 a.m.]. Getting the time from these with two latitudes, the calculated intervals of time may be compared with the observed, and thence the latitude be determined either by simple proportion or from the altitude nearest noon. You then become independent of the noon observation.'

Henry Raper, in his *Practice of Navigation*, pointed out that Captain Sullivan, R.N., on entering the River Plate in 1843, found a line of position and got a sounding on it, and then shaped a course up river. It is certain that the concept of position-line navigation had been in a state of evolution for many years before the idea became crystallized by Captain Sumner.

3. SUMNER'S DISCOVERY

Sumner's discovery was made in 1837, and the details of his discovery are given in his famous pamphlet on the subject, which was first published in Boston in 1843.

Sumner pointed out that when knowledge of latitude is uncertain, there are only two instants in a day at which the Sun's altitude can be taken to find the ship's longitude by chronometer; and that there is only one instant in a day when a single observation of the Sun may be used to find latitude, unless the apparent time at ship is accurately known. At all times when the bearing of the Sun is not north, south, east or west, errors of latitude or longitude, proportional to the angular distance of the Sun from these points, may be great. Sumner goes on to say:

'To remedy this defect, and render a single altitude of the Sun, taken at any angle from the meridian, or from the east or west points, available, when the latitude and longitude are uncertain (the time by chronometer being given) the method of observation affords a substitute for a parallel of latitude or a meridian of longitude; namely, a line diagonal to either of these, and which is called a Parallel of Equal Altitude, which, when projected on a Mercator's chart, shows a ship to be on such projected line, corresponding to the observed altitude, ... consequently, the projected line shows the bearing of the land, in a similar way as a parallel of latitude.'

Before the introduction of position-line navigation the general method of finding position at sea was to observe the Sun on the prime vertical (or as near thereto as latitude and declination permitted) for longitude using a D.R. latitude; and to observe the Sun on the meridian to obtain the latitude at noon.

When the Sun (or other heavenly body) bears due east or west; that is, when he is on the prime vertical circle, his altitude, declination, and azimuth may be used to compute his hour angle;

from which the longitude of the observer may be found, provided that he has a chronometer, the error on G.M.T. of which is known. In other words, the longitude obtained from an observation of the Sun on the prime vertical is independent of the latitude. The longitude obtained from an observation of the Sun when his bearing is not 090° or 270° will be in error if the latitude used in the computation is not the ship's true latitude. Herein rested the importance of using a latitude which was as near as possible to the ship's true, but unknown, latitude, for calculating the hour angle of the Sun from an a.m. observation.

The latitude used in solving the longitude by chronometer from a Sun observation was generally obtained by dead reckoning from some previously known position, and may have been, therefore, greatly in error. In many cases the error was ascertained from the noonday Sun observation, after which the chronometer problem was re-worked using the correct latitude. If the latitude at noon could not be found, because of a cloudy or overcast sky, the sights for longitude taken in the morning were often discarded as being useless.

Sumner discovered that a single observation of the Sun (or other celestial object), even if the latitude is uncertain, is of value. He demonstrated that a single altitude taken at any time is available for determining a line on a chart at some point on which the ship's position lies. It seems, from what Sumner subsequently wrote, that his discovery was something of an accident. He relates how, after having sailed from Charleston, South Carolina, in November 1837 and bound for Greenock that, after passing the meridian of 21° W., no observations were made until near the land. He goes on to say that:

'... arriving about midnight, 17th December within 40 miles by dead reckoning of Tasker light, the wind hauled S.E. true, making the Irish coast a leeshore: the ship was then kept close to the wind, and several tacks made to preserve her position as nearly as possible until daylight; when, nothing being in sight, she was kept on E.N.E. under short sail with heavy gales: at about 10 a.m. an altitude of the Sun was observed, and the chronometer time noted; but, having run so far without any observation, it was plain that the latitude by D.R. was liable to error, and could not be entirely relied on.

'Using, however, this latitude, in finding the longitude by chronometer, it was found to put the ship 15' of longitude E. from her position by D.R.; which in latitude 52° N. is 9 nautical miles; this seemed to agree tolerably well with the dead reckoning; but feeling doubtful of the latitude, the observation was tried with a latitude 10' further north; finding this placed the ship E.N.E. 27 miles of the former position, it was tried again with a latitude 20' north of the dead reckoning; this also placed the ship still further E.N.E., and still 27 nautical miles. These three positions were then seen to lie in the direction of Small's light. It then at once appeared, that the observed altitude must have happened at all the three points, and at the Small's light, and at the ship, at the same instant of time; and it followed that Small's light must bear E.N.E. if the chronometer was right. Having been convinced of this truth, the ship was kept on her course E.N.E., the wind bearing still S.E., and in less than an hour, Small's light was made, bearing E.N.E. ½ E. and close aboard.'

Sumner illustrated his discovery with a diagram which forms plate 3 of his pamphlet. This historic diagram is reproduced (Plate 10). The 'Arc of the parallel of Equal Altitude tending E.N.E. for Sun's true central altitude A.M,' as Sumner described it, forms the first position line found from an astronomical observation using Sumner's method.

Captain Sumner summed up the advantages of his method of projection in the following way:

'1. When the latitude etc., are uncertain, one altitude of the Sun, at any hour, with the chronometer time, is available in a similar manner as a meridian observation, which can be taken only once a day.
'2. The errors of longitude by chronometer, consequent to any error in the latitude, are shown by inspection.
'3. The Sun's azimuth is found at the same operation.
'4. In addition to these results, found by one altitude, two similar altitudes give the true latitude, and also the longitude by chronometer. By the common methods of double altitudes, the longitude must be found by a subsequent calculation, which circumstance renders this method much the shortest.

'5. The usual simple calculation for finding the apparent time at ship is known and daily practiced by every shipmaster who uses a chronometer. No other formula is used.

'6. Double altitudes of the Sun are, therefore, within the reach of all persons who use chronometers, and who are unacquainted with the various formulas laid down in the books.'

In addition to describing how to use position lines—or Sumner lines as they later became known—obtained from Sun sights, the Captain demonstrated the application of the principles of his method to the fixed stars, the planets and the Moon.

Sumner published his important pamphlet *A New and Accurate method of Finding a Ship's Position at Sea by Projection on Mercator's Chart* in 1843. Sumner's method was first described in print in England in the following year when Lieutenant Harry Raper, R.N., communicated a detailed account of the method to the Editor of the *Nautical Magazine*. Raper described the method as '... highly ingenious and very useful when the ship is near land.'

After describing Sumner's method of laying down a position line on a chart, Raper stated:

'... a line perpendicular to the above-mentioned line [position line] towards the side on which the Sun is, shows the true azimuth of the Sun. This is so because the several latitudes and longitudes constitute a curve of equal altitudes.'

Raper pointed out that it is shorter to compute the hour angle of the Sun and thence the longitude by chronometer using the latitude by account; and then, instead of working out a second hour angle for a second latitude, to find the azimuth by means of the sine formula. He also remarked that:

'The effect of an error in altitude is easily shown by considering the place of any point of the circle of equal altitude on the chart which moves one mile for each 1' of error of altitude, and thus the corrected position of the line will be parallel to that already laid down and the distance from it the amount of the error in altitude.'

This contains the germ of the intercept method resulting from the remarkable French investigations into nautical astronomy made at a later date.

Raper also included in his extended notice of Sumner's work, an observation to the effect that it is obvious that the ship's place may be obtained by projecting two Sumner lines instead of employing the calculations of the double altitude (double chronometer).

'It is obvious,' he wrote, 'that the place of the intersection is more decisively marked as the two lines lie more nearly at right angles to each other, and as each line is perpendicular to the direction of the Sun at each observation, they will cross at near 90° when the Sun's azimuths at the two observations differ by near 90°. Projection, therefore, affords evidence of the simplest and most convincing kind, that the value of a double altitude depends altogether on the difference of azimuth. This condition was first pointed out by Dr Inman in his *Navigation* of 1826, and has nothing to do with time from noon, which more popular works reitcrate as the proper limiting condition of the double altitude to the great detriment of the extensive and successful practice of this important observation.'

It is interesting to note that the review (in all probability written by Raper) of Bowditch's book on navigation, which appeared in the *Nautical Magazine* of 1843, contains the statement:

'In the double altitude Bowditch does not once allude to the difference in azimuth as the criterion of the value of the observation. Yet this was pointed out by Dr Inman many years ago.'

In reply to an article by L. T. Fitzmaurice entitled, 'On Finding Position by Double Altitudes with only one Latitude,' which was printed in the *Nautical Magazine* of 1854, John Riddle, the Master at the Greenwich Hospital school, demonstrated the rules given by Fitzmaurice.

Fitzmaurice's rules are:

'Assume a latitude less than latitude in and compute hour angle and azimuth (calling that furthest from meridian A and the other B). Find the longitude by hour angle farthest from meridian. Also the difference between the computed and true intervals, which difference reduce to miles. When both altitudes are on the same side of the meridian, take difference azimuths, and when on the opposite sides take sum. To the difference or sum add sine azimuth B and the common log of difference of intervals. Place this sum down twice. Then for correction for latitude to this sum add sine azimuth A and cosine latitude assumed, and the sum will be the common log of the correction, always positive: and to the sum of the three logs add cosine azimuth A and sum will be log of correction of longitude. When azimuth A is less than 90° and the least altitude is east of the meridian add in east longitude and subtract in west; and when greater than 90°, subtract in east and add in west. When altitude is west of the meridian apply correction in the reverse way.'

This rule, which is typical of the numerous rules of navigation to be found in manuals of the period, was demonstrated by John Riddle. Riddle rebuked Fitzmaurice (who had been a former pupil at the Greenwich Hospital a quarter of a century earlier when John Riddle's father Edward had been the headmaster) for not demonstrating the propositions he advanced. (See Chapter V, p. 162.)

The proof given by Riddle is as follows:

Referring to Fig. 3.

Let MN be assumed parallel of latitude and m and n represent computed places upon it. Through m and n draw mQ and nQ in directions of computed bearings and draw mP and nP perpendicular to mQ and nQ. Then P is the place of the ship. PR, the meridian difference of latitude between P and m, and mR, the d. long. between m and P.

Proper difference of latitude between m and P is PR cos lat m.

Pmn and Pnm are bearings of observed object. Let these angles be denoted by θ and ϕ.

In mPn:
$$PR = mn \sin \phi \sin \theta \csc (\theta + \phi)$$

Therefore correction of assumed latitude, viz. that of point m, which is PR cos lat, is:

$$mn \sin \theta \sin \phi \csc (\theta + \phi) \cos lat \quad . \quad . \quad (1)$$

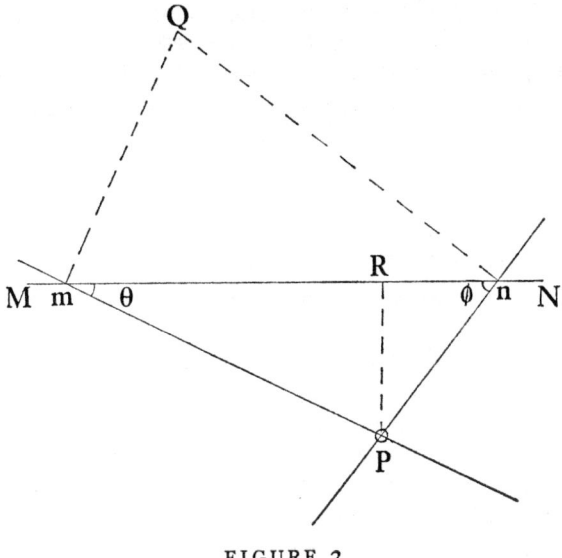

FIGURE 3

Correction of longitude of m, viz. mR is:

$$mR = PR \cot \theta$$
$$= PR \frac{\cos \theta}{\sin \theta}$$
$$= mn \cos \theta \sin \phi \cosec (\theta + \phi) \quad . \quad . \quad . \quad (2)$$

Bearings supposed one east and one west.

When both east or west $(\theta + \phi)$ must be changed to $(\theta \sim \phi)$ in formulae (1) and (2).

A graphical solution to this problem, using a simple plotting sheet, is considerably simpler and speedier than the trigonometrical solution given by Fitzmaurice and Riddle. Despite this, the method of 'figuring' instead of by drawing seems to have been preferred by seamen until relatively recent times. Even Captain S. T. L. Lecky, that great merchant seaman of the last century, seems to have had a predilection for tables and mathematics, and frequently ignored practical graphical methods. This, often to the detriment of navigators who were not aware of the existence of simple geometrical solutions to some of their relatively complex astronomical problems.

Sumner's method applied to fixing a ship by crossing position lines was often referred to as the *chord method*, because a straight line joining two points on a circle is a chord of that circle. Sumner's method was the popular method of obtaining a position line and, to a large extent, it was employed in the Merchant Service until relatively recent times.

As pointed out above, a position line may be obtained by assuming one latitude and computing a longitude and the azimuth of the observed body, instead of using two latitudes and calculating a longitude for each. This method became known as the *tangent method*, as opposed to the chord method. The generality of seamen used the chord method in preference to the tangent method, until the beginning of the present century.

About five years after the publication of Sumner's famous book, a naval instructor of the Royal Navy, J. N. Laverty, published a pamphlet apparently for private circulation, in which he discussed, in some detail, a method of finding a ship's position by means of position lines. Laverty's pamphlet is entitled *A Few Remarks on the Finding of Latitude and Longitude from Observations at Sea*. In the opening chapter, Laverty defines his position as follows:

> 'If two sets of sights be taken at sea with an interval of time of about an hour or more between them, but at such times as both may serve for finding the longitude with a chronometer—say for instance, the first at eight or half-past eight, and the other at ten, even as late as half-past ten in the morning—the latitude, and hence the longitude may be found by following a simple rule.'

Laverty's development of the problem follows along lines similar to those followed by Sumner, although he was hampered by the restriction of times of observations. He pointed out the valuable use of a single position line when combined with a terrestrial bearing or a sounding, and included among his examples, one involving simultaneous altitudes of Sun and Moon:

> 'Example: Ship's time 8 a.m. April 26, 1848. Latitude by account 49° 38′ N. True altitude Moon's centre 20° 49′ bearing S. 29° W. True altitude Sun's centre 29° 21′ bearing S. 85° E.

POSITION-LINE NAVIGATION

Longitude from altitude of Moon latitude
$$49° \ 30' \ N. = 4° \ 21' \ W.$$
Longitude from altitude of Moon latitude
$$49° \ 40' \ N. = 4° \ 50' \ W.$$
Longitude from altitude of Sun latitude
$$49° \ 30' \ N. = 4° \ 00' \ W.$$
Longitude from altitude of Sun latitude
$$49° \ 40' \ N. = 3° \ 50' \ W.$$
True Position of the ship: $49° \ 23' \ N. \ 4° \ 02' \ W.$'

H. B. Goodwin, in describing Laverty's pamphlet, pointed out that Sumner's method, on first being introduced, was confined to longitude observations. Goodwin suggested that, in the above example, the observation of the Moon, whose azimuth was only $2\frac{1}{2}$ points, would have given better results had the ex-meridian method been used.

Laverty, in his pamphlet, made the following interesting statement:

'Note: These remarks were already in the press when the writer received from a friend, a Master in the Royal Navy, to whom he had communicated them, a paper by which it appears that an American Captain had, in 1843, published "A Method of Determining the Position of a Ship". The writer must waive his claim to priority of invention, but he hopes, by the present publication, to lead many to adopt a method which, with the addition of the remarks attached to it, will be found of great utility at sea.'

Laverty did not confine himself to remarks on the position-line problem only. He included in his small pamphlet an ex-meridian table and also a table of corrections from which the error in longitude due to an error in latitude could be lifted. These corrections are based on the formula:

Error in longitude = error in latitude × cot Az sec lat.

By means of this correction table, an observer who has worked out his morning Sun sight with a wrong latitude is able to find the longitude correction immediately the correct latitude by Sun's meridian altitude is determined, without having to re-calculate the hour angle from the morning sight. This is the original *longitude factor* or *longitude correction table* as it is now called, forming Table C of the A, B and C Tables.

Goodwin concludes his description of Laverty's pamphlet as follows:

> 'And, indeed, the author's style in dealing with the Sumner principle is so easy and lucid, and his mathematical artifices are so neat and concise, that one cannot but feel some sentiment of regret that his works have come down to us only in a fragmentary form.'

4. AZIMUTH TABLES IN POSITION-LINE NAVIGATION

The tangent method of position-line navigation is facilitated by *azimuth tables*. The earliest azimuth tables were designed for use with the Sun, and were used in particular for finding the Sun's true azimuth for measuring time (as was the case with Wakeley's tables), or for finding or checking the variation or compass error (as in the case of Lynn's, Burdwood's and Davis's tables).

(a) BURDWOOD AND DAVIS

Staff Commander John Burdwood is credited with the distinction of being the first to produce an azimuth table in a form suitable for facilitating position-line navigation. It was Burdwood's tables of 1852, enlarged by him in 1858, 1862, 1864 and 1866, that formed the basis of the comprehensive azimuth tables of Captain J. E. Davis, R.N., and his son Percy L. H. Davis, which are in common use at the present time.

(b) HEATH AND A, B AND C TABLES

Perhaps the most popular tables used in connection with position-line navigation are the A, B and C Tables to be found in collections of nautical tables such as those of Norie's and Burton's.

Tables of the A and B type were first published in the *Nautical Magazine* of 1846. These tables were sent to the Editor from H.M.S. *Iris* at Hongkong on June 27th 1845 by Lieutenant (later Admiral Sir) L. G. Heath, R.N.

William Allingham, a teacher of navigation well known to seamen during the early part of this century, informs us in an article that appeared in the *Nautical Magazine* for 1903, that at the time

POSITION-LINE NAVIGATION 285

of the first appearance of Heath's tables, tables of the A and B type had already passed their majority. The prototype of A and B correction tables appears to have been invented by M. Mazure Duhamel who published them in his *Mémoire sur l'Astronomie Nautique* in 1822.

Duhamel's table is based on the formulae:

$$\frac{dt}{dh} = \sec \phi \operatorname{cosec} Z \quad \ldots \quad (1)$$

and

$$\frac{dt}{d\phi} = -\sec \phi \cot Z \quad \ldots \quad (2)$$

in which t, h, ϕ and Z denote hour angle, altitude, latitude and azimuth respectively. These tables, a facsimile of which is here reproduced (see p. 286), appeared in the *Nautical Magazine* of 1832.

Heath's tables were accompanied with instructions for using the tables:

'First, for finding a line upon which the ship must be at the time of the morning sights; and second, for finding the ship's position immediately the true latitude at noon is known.'

In his communication to the Editor of the *Nautical Magazine*, Heath stated that he had

'... been informed that something of the same sort had previously appeared in the *Nautical Magazine* but not pointing out the uses to which it [the table] may be put.'

The table to which Heath referred is clearly the table of Duhamel's which was reprinted in the 1832 *Nautical Magazine*.

With the object of 'contributing all that he could' to make Sumner's method known, the Editor of the *Nautical Magazine* reprinted Heath's tables in the *Nautical Magazine* of 1860.

In 1873, Captain J. F. Trivett, the first headmaster of the training ship H.M.S. *Worcester*, and later the Board of Trade Principal Examiner of Masters and Mates, contributed an article to the *Nautical Magazine* in which he described Heath's tables, asking the Editor to reproduce them again.

A HISTORY OF NAUTICAL ASTRONOMY

Table as given in *Nautical Magazine* 1832

Partial Error in H.A. in secs of Time arising from a change of 1' in the alt. and lat.

Z	10		20		30		40		50		60		70		80		90	
Lat	h	φ	h	φ	h	φ	h	φ	h	φ	h	φ	h	φ	h	φ	h	φ
0	23.0	22.7	11.7	11.0	8.0	6.9	6.2	4.8	5.2	3.4	4.6	2.3	4.2	1.4	4.1	0.7	4.0	0.0
5	23.1	22.8	11.7	11.0	8.0	7.0	6.2	4.8	5.2	3.4	4.6	2.3	4.3	1.5	4.1	0.7	4.0	0.0
10	23.4	23.0	11.9	11.2	8.1	7.0	6.3	4.8	5.3	3.4	4.7	2.3	4.3	1.5	4.1	0.7	4.0	0.0
15	23.8	23.5	12.1	11.4	8.3	7.2	6.4	4.9	5.4	3.5	4.8	2.4	4.4	1.5	4.2	0.7	4.1	0.0
20	24.5	24.1	12.5	11.7	8.5	7.4	6.6	5.1	5.6	3.6	4.9	2.5	4.5	1.5	4.3	0.7	4.3	0.0
25	25.4	25.0	12.9	12.1	8.8	7.6	6.9	5.3	5.8	3.7	5.1	2.5	4.7	1.6	4.5	0.8	4.4	0.0
30	26.6	26.2	13.5	12.7	9.2	8.0	7.2	5.5	6.0	3.9	5.3	2.7	4.9	1.7	4.7	0.8	4.6	0.0
35	28.1	27.7	14.3	13.4	9.8	8.5	7.6	5.8	6.4	4.1	5.6	2.8	5.2	1.8	5.0	0.9	4.9	0.0
40	30.1	29.6	15.3	14.3	10.4	9.0	8.1	6.2	6.8	4.4	6.0	3.0	5.6	1.9	5.3	0.9	5.2	0.0
45	32.6	32.1	16.5	15.5	11.3	9.8	8.8	6.6	7.4	4.7	6.5	3.2	6.0	2.1	5.7	1.0	5.7	0.0
50	35.8	35.3	18.2	17.1	12.4	10.8	9.7	7.4	8.1	5.2	7.2	3.6	6.6	2.3	0.3	1.1	6.2	0.0
55	40.2	39.6	20.4	19.2	13.9	12.1	10.8	8.3	9.1	5.8	8.0	4.0	7.4	2.5	7.1	1.2	7.0	0.0
60	46.1	45.4	23.4	22.0	16.0	13.9	12.4	9.5	10.4	6.7	9.2	4.6	8.5	2.9	8.1	1.4	8.0	0.0
65	54.5	53.7	27.7	23.0	18.9	16.4	14.7	11.3	12.3	7.9	10.9	5.5	10.1	3.4	9.6	1.7	9.5	0.0
70	67.3	66.3	34.2	32.1	23.4	20.3	18.2	13.9	15.3	9.8	13.5	6.8	12.4	4.3	11.9	2.1	11.7	0.0

Enter the above table with the azimuth and the latitude the error in the H.A. in secs. of time will be found for a change of 1' in the altitude under h, and for a change of 1' in the latitude under φ. Thus in 50° N. when the Sun bore S. 60° E., and the latitude used in the morning sight was 20' in error, the error in the H.A. was, by above table, 3·6 secs. × 20 = 0·9' × 20 = 18'.

POSITION-LINE NAVIGATION 287

In 1878, Captain Trivett, in a letter to the Editor of the *Nautical Magazine*, stated that Heath's tables were used extensively by the officers of the P. and O. Company and other services. By this time the tables had become known as A and B Tables.

In his letter, Captain Trivett described how a former 'Worcester boy,' W. G. Hutchinson, had devised a method for finding azimuth using the A and B Tables. Hutchinson had written to Captain Trivett, his former teacher, asking him to check his method.

Hutchinson's rule is:

'With the correction for longitude (sum or difference of A and B) take away the decimal point (i.e. treat it as a whole number) and find the corresponding departure. With this departure as diff. lat. and 100 as departure, enter Traverse Table, and the degree corresponding to the course, as generally styled, is the azimuth.'

After giving an example, Hutchinson gave a very interesting demonstration of his method. It was the demonstration that prompted Captain Trivett to induce 'Old Worcesters' not to lay aside that mathematical knowledge which they once possessed. 'A knowledge of the theory of navigation,' wrote the Captain, 'confers an advantage on those who aspire to be called navigators.'

A well-known teacher of navigation, W. H. Bolt, a former Christ's Hospital boy, and a successor to Norie's Nautical Academy, wrote a paper on the equation of equal altitudes which appeared in 1874. Bolt drew attention to Trivett's article of 1873, in which Heath's tables were reproduced for the third time, and stated that the division of the error in longitude consequent upon an error in latitude into two parts—one part from Table A depending on latitude and hour angle, and the other from Table B depending on declination and hour angle—suggested the computing formulae, viz.:

$$\tan \phi . \cot t \text{ for Table A}$$
and
$$\tan \text{dec.} \cosec t \text{ for Table B}$$

In other words:

Error in longitude for 1' error in latitude
$$= \tan \phi . \cot t \pm \tan \text{dec.} \cosec t$$

Bolt's object in writing this paper was to describe a method in which the A and B Tables could be employed for finding the equation of equal altitudes.

The original tables of Heath's did not extend sufficiently for this purpose. Bolt, therefore, extended Table B from 24° to 60°, and altered the headings of both A and B Tables so that they could be used for the equal-altitude problem, as well as serving their original purpose for finding the longitude correction.

The rules given by Heath for using the tables were, according to Bolt, difficult to remember. To Bolt is due the credit of simplifying the rules.

'In my teaching,' wrote Bolt, 'I use:

latitude and declination SAME name—Diff. A and B
latitude and declination DIFF name—Sum A and B.'

These rules were certainly shorter and less complex than those of Heath's.

An interesting note appears at the foot of Bolt's paper, in which he points out that the original computer of Table B made a curious error in the column for six hours in which every result is zero, and that the column should be filled in with the numbers expressing the natural tangents of declinations.

In another paper published in the same year—1874—Bolt described a method in which A and B Tables could be used for correcting the middle time in the old double-altitude problem.

In 1878, R. W. Espinasse published a little-known but extended version of Heath's tables. Espinasse informs us that he was induced to enlarge Admiral Heath's tables and publish them for general information in consequence of the total loss of the *Loch Ard* on the Victorian coast of Australia. In publishing this table the author was particularly careful in choosing the title: *Admiral Heath's Tables—reprinted and enlarged by R. W. Espinasse.*

In 1883, being unaware of the work of Espinasse, H. S. Blackburne, then an officer in the P. and O. Company's service, had published by R. H. Laurie: *A and B Tables for Correcting the Longitude and Facilitating Sumner's Method on the Chart, etc.* As the work of a young man who evidently had the desire to be of service to his brother officers, Blackburne's tables are a credit to his industry.

Blackburne laboured unnecessarily hard in producing his A and B Tables. He informs us that he occupied, in calculating his tables, all the spare time he could afford during several years of busy sea-life as a junior officer. He computed the values for his A and B Tables using the methods of oblique trigonometry. His task would have been infinitely simpler and far less time-consuming had he used the less troublesome formulae of right-angled trigonometry. At a later time Blackburne admitted his earlier ignorance of the more concise methods for computing A and B values.

As far back as 1878, when Blackburne was a junior officer in the Venice line of the P. and O. Company, he had been in the habit of using that part of his manuscript tables suitable to the route.

W. H. Rosser, Principal of Norie's Nautical Academy and one-time editor of *Norie's Epitome and Nautical Tables*, had published in 1875 a book entitled *How to Find the Stars and their Use in determining Latitude, Longitude, and Error of Chronometer*. The second edition of this work appeared in 1883, some weeks after Blackburne's A and B Tables had been published. To the original title of his book Rosser added: *With A, B, and C Tables for finding Azimuth and Correction to Longitude*. In the same year, 1883, Rosser published: *Stellar Navigation with new A, B, and C Tables for finding, by easy methods, Latitude, Longitude, and Azimuth: Latitudes and Declinations ranging to 68° N. and S.* Rosser's A, B and C Tables were printed in the Norie collection for the first time in 1889.

Laverty's table of corrections for finding the error in longitude due to an error in latitude was extended by A. C. Johnson, the naval instructor; and it first appeared in Johnson's well-known pamphlet *On Finding Latitude and Longitude in Cloudy Weather*.

Captain Lecky obtained Johnson's permission to insert his table in the first edition of *Wrinkles in Practical Navigation*, published in 1881. Rosser also published Johnson's table in his *Stellar Navigation* of 1883.

The combination of Heath's table extended by Blackburne, Rosser and Lecky, and Laverty's table extended by Johnson and Lecky, now form the celebrated A, B and C Tables.

Captain Lecky's name is, perhaps, the most closely connected with A, B and C Tables, although credit for being first to combine them ought rightly to go to Rosser.

Lecky re-calculated and extended A, B and C Tables with the help of Captain Alfred Fry of Liverpool; and these tables form the greater part of Lecky's *General Utility Tables* first published in 1897. These tables were designed to replace the monumental azimuth tables of Burdwood's and Davis's. In this respect they were unsuccessful.

Extract's from Sumner's work were first published in France by M. Joseph Barthet, in the *Annales Maritime* of 1847. Sumner's method became popular among British seamen before it took hold on the Continent. However, the most significant advance of all in the history of modern astronomical navigation was due to the investigations made by French naval officers and astronomers during the period c. 1865–1875, when the so-called 'New Navigation' and the *méthode du point rapproché* were introduced.

Before discussing the methods of the *Nouvelle Navigation Astronomique* we shall deal with the practical aspect of determining position lines using the Sumner method.

In the extended title of his book, Sumner states:

'When the Latitude and Longitude and Apparent Time at the Ship are Uncertain, one altitude of the Sun with the True Greenwich Time, determines:

First

The True Bearing of the Land.

Second

The Errors of Longitude by Chronometer consequent to any error in the Latitude.

Third

The Sun's True Azimuth.'

A fault which often brought unwarranted severe criticism of Sumner's method is that the bearing of the land is not determined unless the land is near enough to assume that the rhumb line* and the circle of equal altitude between the ship and the land are coincident. The postion line on the chart was (and still is) invariably plotted as a rhumb line arc; and this may lead to material error in the solution. The distance over which a Sumner line and

* A rhumb line is a line of constant course and therefore makes a constant angle with the meridians it crosses.

rhumb line may be considered to be coincident, depends upon the radius of curvature of the projected circle of equal altitude on the Mercator chart. This, in turn, depends upon the latitude of the observer, and the azimuth and zenith distance of the observed body.

What surely must be one of the earliest of numerous criticisms of the Board of Trade examinations is that made in 1862 by James Gordon, a teacher of navigation. Gordon contributed an article entitled 'On Sumner's Method' which appeared in the *Nautical Magazine* of that year. Either through cussedness or false national pride, the author, after first proclaiming that Sumner's method was not original, stated that:

> 'American writers have boasted so much about this Sumner's method that we deemed it necessary to clear it of the halo of originality before proceeding to show its utter uselessness as a practical problem.'

Gordon's criticism of Sumner's method appears to have stemmed from the fact that it was 'patronized by the Board of Trade,' as he put it; because candidates for the Extra Certificate at the time were required to verify the double-altitude problem by Sumner's method. Gordon's criticism arose from the fact that an arc of a circle of equal altitude is not a rhumb line.

In view of his remarks on Sumner's method, it is interesting to note that James Gordon, A.M., published a little book in 1850 entitled *A New and Easy Method of Finding a Ship's Position at Sea*. In the preface to his book Gordon remarks:

> 'The method of finding a ship's position at sea by means of what is termed by Captain Sumner a parallel of equal altitude having attracted considerable notice in America, the author has thought it proper to present to British mariners what he considers to be an improvement on the method as originally proposed.'

The proposed improvement consisted in using a construction method on a plotting sheet instead of a Mercator chart. This permits a scale of any convenient size to be used; and, as Gordon puts it:

'... consequently the result may be obtained to the required degree of accuracy; whereas by using a common chart the scale may not be sufficiently large to give a correct result.'

Sumner's method became exceedingly popular in British ships; so much so that later improvements in position-line navigation were overlooked, and the older Sumner method retained its popularity, and is used extensively, in a modified form, even at the present time.

Several years after the great French contribution to nautical astronomy had been presented to the world we find W. H. Bolt, the well-known teacher of navigation, writing, in 1880, a long series of articles on the 'New Navigation and Sumner's Method.' Although Bolt described the French investigations, the greater part of his series of papers was devoted to Sumner's problem; and, according to M. le Comte du Boisy, who contributed to the *Nautical Magazine* of 1881 a valuable paper on the 'New Astronomical Navigation,' Bolt had made a poor attempt at describing the new methods from France. However, that part of Bolt's articles dealing with the Sumner method was well thought out and undoubtedly proved valuable to some seamen of the time.

After discussing the importance of carrying chronometers, and handling and managing them so that G.M.T. is always available, Bolt declared that simultaneous altitudes of two stars are unquestionably the best for determining a ship's position. With a good knowledge of the stars and planets, two objects may be selected at pleasure and in such relation to each other that the angle between their vertical circles shall be suitable; that is to say, between 60° and 120°. If there is any doubt, a third object will give, with the other two, a space or triangle of certainty within which the ship must be. Bolt quoted Raper, who stated that, in respect of star observations: 'The observer should make it a matter of special practice.'

Bolt dealt at length with the effects on position lines of errors in altitude and time. He recommended, as Sumner had done, calculating four hour angles using two latitudes embracing the ship's D.R. latitude, and suggested checking the directions of the position lines obtained by finding the bearings of the observed objects from time azimuth tables. He mentions the fact that although the circle of equal altitude is, at every point on it, perpendicular to the true bearing of the observed object at that point, this is not so for the line of position. The bearings at the ends of a position line are different; and, for this reason, according to Bolt, working two hour angles for each observation instead of an hour angle and an azimuth, is to be preferred.

The projection of a circle of equal altitude on a Mercator chart is a curve having a complex form. The smaller is the zenith distance of an observed object, the smaller is the projection of its circle of equal altitude which tends to be circular.

When the declination of a heavenly body is equivalent to the latitude of an observer, the body crosses the meridian of the observer at the zenith: so that at the time of meridian passage of such a body the observer is at the geographical position of the body. If the latitude of an observer and the declination of a body are within a few degrees of each other, then the observer readily may find his position from two observations of the body when it is near meridian passage, by plotting the curves of equal altitude on his Mercator chart as circles centred respectively at the projections of the geographical positions of the body at the times of observations. This method of fixing by means of Sun's altitudes was suggested by Captain Angus of the P. and O. Company in 1884, and was known as *Angus's method*. But the problem had been investigated a decade or more before Angus described it.

5. MARCQ ST HILAIRE AND THE NEW NAVIGATION

In 1875, a paper in the *Revue Maritime*, written by a French naval officer named Marcq St Hilaire contained in its title the term 'la Nouvelle Navigation.' The names of Marcq St Hilaire and Lieutenant Hilleret, also of the French Navy, were mentioned in a classic work entitled *Nouvelle Navigation Astronomique*, written by Yvon Villarceau and Aved de Magnac. Villarceau was an eminent astronomer of the Paris Observatory, and de Magnac a Commander in the French Navy.

Captain Marcq St Hilaire who, according to du Boisy, was endowed with a profoundly acute mathematical mind, was the inventor of the method known to the French as the *méthode du point rapproché*, and by British navigators as the *intercept method* of position-line navigation.

The work of the French astronomers and naval officers who introduced the methods known as the New Navigation, marked a collective, and very fruitful, attempt to study the problems of astronomical navigation in a systematic and scientific way.

Commander de Magnac, while serving as a navigating officer in about 1867, found himself in circumstances similar to those in which Captain Sumner found himself on that eventful day in 1837 when position-line navigation was born. Circumstances were such that it was very important for de Magnac to find his ship's position with great accuracy. In a clear spot near the zenith during twilight a star appeared and de Magnac succeeded in getting its altitude. Because the altitude was so great, none of the known methods of nautical astronomy afforded him the means of obtaining a reliable result from this observation; and thus the observation was useless. This observation appears to have led de Magnac to undertake the improvement of nautical astronomy.

De Magnac was aware that the theory of nautical astronomy was incomplete and that, in order to rectify the matter, the united efforts of astronomer and navigator were required. In particular, de Magnac considered that, above all, the scientific management of chronometers at sea must be settled.

On returning to France de Magnac consulted Villarceau with the object of finding out the most systematic and mathematical method for determining the algebraic formulae representing the rates of chronometers at sea. After lengthy investigations into the matter, de Magnac obtained formulae suitable for rating chronometers. He communicated his results to the Académie des Sciences in November 1868.

After further investigations de Magnac arrived at a rational scientific theory of employment of chronometers at sea. In 1874 he published a pamphlet entitled *Recherches sur l'emploi des Chronometres à la Mer*, in which he described his theories. Later, de Magnac collaborated with Villarceau to produce *Nouvelle Navigation Astronomique*, referred to above.

Having settled the scientific and practical management of chronometers at sea, Villarceau and de Magnac set about improving the methods of finding a ship's position at sea.

Du Boisy, commenting on the work of Villarceau and de Magnac, remarked that:

'... no books treating on nautical astronomy written prior to the publication of *Nouvelle Navigation Astronomique* exactly defined the problem of the determination of the place of a ship at sea, that is to say, [mathematically] the problem of the determination of a point: all those works give merely the problem of the determination of a line, either latitude or longitude. But what does the sailor want? A point, not merely a line. Now it is obvious that the problems relating to the determination of a point and of a line are quite different. Nautical Astronomy could not develop itself in all its excellence, so long as its most efficient and useful mathematical problem had not been defined and investigated.'

In investigating the problem, consideration was first given to the results of a single observation. In this case the zenith distance of the observed object in minutes of arc gives the radius of the circle of equal altitude in miles, the centre of the small circle being located at the geographical position of the object at the time of observation. The geographical position of the body is determined if the G.M.T. of the observation is known: and this, of course, depends upon knowledge of the chronometer error or, in other words, on the proper management of the chronometers on board.

The azimuth of the body at the time of the observation determines the direction of a great circle which passes through the geographical position of the object and the observer's position at the time of the observation. Consequently the ship's position lies at the intersection of a small circle of equal altitude and a great circle of azimuth.

If latitude is not known, azimuth cannot be found to the required degree of accuracy for fixing the ship to the nearest minute or so of longitude. The observation of a single body, therefore, is insufficient for determining a ship's position. Recourse, therefore, must be made to simultaneous observations of two bodies to give azimuthal distance between the two bodies as well as their altitudes.

Simultaneous altitudes and corresponding geographical positions of two celestial bodies furnish two circles of equal altitude which cross at two points on the Earth—one of which is the observer's position. The D.R. position of the ship or the bearing of either object may be used to determine which of the two intersections marks the ship's position.

The simplest method of determining a ship's position, using

this method, is to plot the geographical positions of the observed bodies on a model globe, and then to draw the two circles of equal altitude to intersect at the ship's position. This method is unsuitable because of the difficulty of obtaining a position to the necessary degree of accuracy.

Because the Mercator chart virtually replaces the model globe, it was natural to inquire into the nature of the circles of equal altitude when projected on a Mercator chart. This the French investigators did. They found that a circle of equal altitude, when projected on a Mercator chart, is a curve which takes one of three forms. Two of these are open and the third is closed. The closed curve resembles an ellipse; and, in this case, the curve may be considered to blend with a circle when the latitude of the observer, and the zenith distance of the observed object are both small. In this circumstance, it is an easy matter, as the French showed in 1874, and Captain Angus in 1884, to plot a circle of equal altitude on a Mercator chart.

This method of tracing the circle of equal altitude, as pointed out by de Boisy, is the only practical result afforded by the study of curves of equal altitude; but, he goes on to say:

> 'It is a precious one, for it permits of the very simple and rapid utilization of great altitudes in which the old methods of navigation were most generally deficient. This was so much the more to be regretted since a star very often appears near the zenith, while it is impossible to perceive others in a better and less elevated position.'

The only practical application of the projection of equal-altitude curves on a Mercator chart having been considered, it appeared to the French astronomers and navigators that computation is necessary in order to fix a ship by simultaneous observations.

Fig. 4 represents the Earth with the North Pole denoted by P. G_1 and G_2 represent the geographical positions of two stars whose altitudes are observed simultaneously. XY and X_1Y_1 are the resulting circles of equal altitude which intersect at F, the place of the ship.

Let the altitudes of the stars whose geographical positions are at G_1 and G_2 be h_1 and h_2 respectively, and let their respective declinations be d_1 and d_2. Let the latitude of F be ϕ.

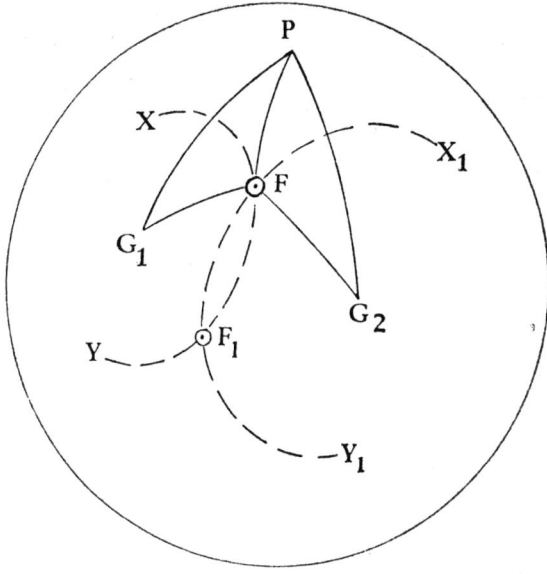

FIGURE 4

By applying the fundamental cosine formula to the spherical triangles PFG_1 and PFG_2, we have:

$$\cos G_1F = \cos PG_1 \cos PF + \sin PG_1 \sin PF \cos G_1PF$$
$$\cos G_2F = \cos PG_2 \cos PF + \sin PG_2 \sin PF \cos G_2PF$$

From which we have:

$$\sin h_1 = \sin \phi \sin d_1 + \cos \phi \cos d_1 \cos G_1PF \qquad . \quad (1)$$
$$\sin h_2 = \sin \phi \sin d_2 + \cos \phi \cos d_2 \cos G_2PF \qquad . \quad (2)$$

These formulae are complex and quite impracticable for use at sea. Investigation into the formulae, however, demonstrated that the most favourable mathematical conditions for finding position by astronomical observations are:

1. The difference between the azimuths of the two observed objects must be 90°.
3. The altitudes must not be too great.

The first condition follows from a consideration of the errors of altitude and their effects on the resulting fix. The second condition arises from the fact that if the altitudes are great, the circles of equal altitude are small, in which case the points of the intersection (F and F_1 in Fig. 4) are close to one another, and the mathematical solution may be indeterminate.

Du Boisy informs us that failure to employ the direct method of finding the ship's position from simultaneous altitudes using equations (1) and (2) led to the inquiry into the indirect method which he calls the 'Method of Equations of Condition.' This method gives the required result with fewer calculations than the direct method; and, moreover, it led to the so-called *méthode du point rapproché*.

The following description of the application of the method of equations of condition, and the principle of the *méthode du point rapproché*, to which the name of Marcq St Hilaire is closely associated, is that given by du Boisy in 1881.

'The method of equations of condition can only be employed when approximate values of the required quantities are known: then, instead of directly computing these quantities, the corrections to be made to those approximate values are calculated in order to obtain sufficiently accurate values of the required quantities. A ship at sea always possesses its estimated position by D.R. This position is always more or less approximate. Consequently the method of equations of condition can be applied to the determination of the place of a ship at sea; and, in this case, may be presented thus:

'Let h and h_1 be simultaneous altitudes at two celestial bodies observed at a known G.M.T. Let the latitude and longitude by D.R. be ϕ and λ respectively. Let the true latitude and longitude be ϕ_t and λ_t respectively.

'First consider ϕ and λ to be the true latitude and longitude of the ship. With these compute the respective altitudes and azimuths. Let these be H and H_1 and Z and Z_1.

'Let $\delta\phi$ and $\delta\lambda$ be the algebraic corrections to be applied to ϕ and λ to get ϕ_t and λ_t. That is, let us state the condition (hence the name of the method—equations of condition)

$$\phi_t = \phi + \delta\phi \quad \text{and} \quad \lambda_t = \lambda + \delta\lambda$$

'It is demonstrated that:

$$h - H = \cos Z \delta\phi + \sin Z \cos \phi \delta\lambda + a(\delta\phi)^2 \\ + b(\delta\lambda)^2 + c(\delta\phi)^3 + \cdots$$

POSITION-LINE NAVIGATION

$$h_1 - H_1 = \cos Z_1 \delta\phi + \sin Z_1 \cos \phi \delta\lambda + a_1(\delta\phi)^2 \\ + b_1(\delta\lambda)^2 + c_1(\delta\phi)^3 + \cdots$$

in which a, a_1, b, b_1, etc. represent certain algebraic combinations of trigonometrical functions of the known quantities.

'Neglecting terms of the second and higher degrees the above equations reduce to:

$$h - H = \cos Z \delta\phi + \sin Z \cos \phi \delta\lambda$$
$$h_1 - H_1 = \cos Z_1 \delta\phi + \sin Z_1 \cos \phi \delta\lambda$$

'If we put $p = (h - H)$ and $p_1 = (h_1 - H_1)$ we have

$$\delta\phi = \frac{p \sin Z_1 - p_1 \sin Z}{\sin (Z_1 - Z)} \qquad \cdot \quad \cdot \quad (3)$$

$$\delta\lambda = \frac{p \cos Z_1 - p_1 \cos Z}{\sin (Z_1 - Z) \cos \phi} \qquad \cdot \quad \cdot \quad (4)$$

'Applying these corrections, $\delta\phi$ to the latitude by D.R., and $\delta\lambda$ to the longitude by D.R., we obtain approximate values of the true latitude and longitudes which, except in rare cases, will be sufficiently correct.'

After the above discussion, du Boisy considered how equations (3) and (4) are represented geometrically on a Mercator chart. Fig. 5 serves to illustrate his discussion.

In Fig. 5 let E be the position by D.R. (lat. ϕ long. λ). Set off p in miles in the direction of (or opposite to as required) Z to obtain the point X, which is named *point rapproché*. At X draw a line P_1L_1 perpendicular to EX. This represents the position line obtained from the observation of the star whose altitude was h.

In a similar manner draw a line of length p_1 miles in the direction of (or opposite to as required) Z_1, to obtain the point Y which is the second *point rapproché*. Through Y draw P_2L_2 at right angles to EY to obtain the position line obtained from the observation of the star whose altitude was h_1. The two position lines are tangential to curves of equal altitude shown on the figure as dotted lines intersecting at C, which is the true place of the ship.

The method of equations of condition enables us to substitute for the curves of equal altitude two straight lines tangential to these curves at the *points rapprochés*. This substitution produces an error corresponding to the distance between C and F in Fig. 5.

The distance CF depends upon:
1. The distance EC.
2. The radii curvature of the curves of equal altitude.
3. The difference between the azimuths of the observed objects.
4. The latitude of the observer.

It increases with distance EC, the altitudes and the latitude: and also when the angle between the bearings of the observed bodies approaches 0° or 180°.

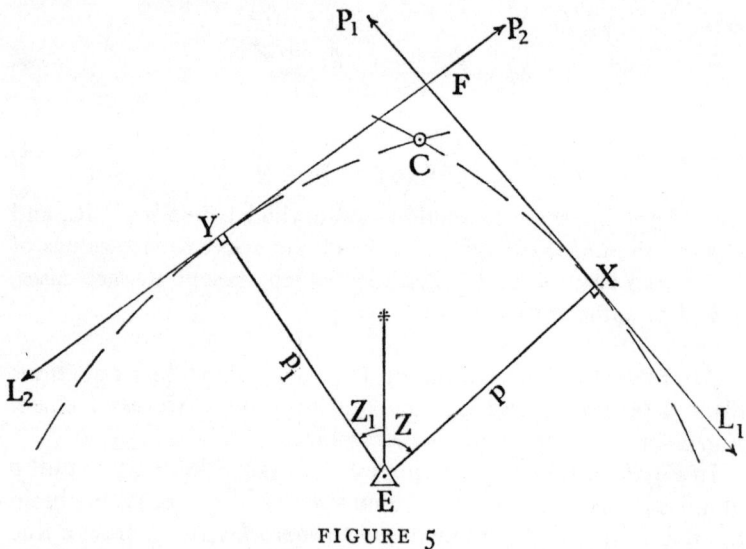

FIGURE 5

Determination of the limits of error CF shows that CF is nearly always inappreciable.

Du Boisy concludes his discussion on the New Navigation by pointing out that the calculation of the *point rapproché* is alone sufficient in any and every case. Thus, the method took the place of all the older methods of navigation apart from the lunar method.

'In short,' du Boisy states, 'the calculation of *point rapproché* is par excellence the Nautical Problem, and a Master who understands and uses it derives advantage from it which other astronomical observations do not give. According to opinions of eminent French officers who have practised the calculation of *point rapproché*, this problem and that of Latitude by Meridian Altitude or Pole Star will, before long, be the only ones employed at sea, and the last two will continue in use for the single reason that they are short, but they do not give better results.

'The results of New Astronomical Navigation have given rise to a remarkable coincidence ... circle and line are completely equivalent.

'Thus, results obtained from observations of heavenly bodies are the same as those afforded by lighthouses. Therefore we say that the New Astronomical Navigation proves that for the determination of the place of the ship, the heavenly bodies are, for the navigator's purpose, like so many lighthouses in the sky.'

Judging by its cool reception to the north of the English Channel, the brilliant work of the French in bringing nautical astronomy to a state of excellence was not fully appreciated by British seamen. And even today a century after its introduction, the intercept method for finding *point rapproché* does not receive due attention from seamen.

The first mention in print in England of the creditable French investigation into Nautical Astronomy appears to have been a review of the book by Villarceau and de Magnac, which was made for the *Nautical Magazine* as late as 1879. The reviewer (as is often the case with reviewers) clearly had not studied the contents of this remarkable book, and appears to have thought that it was nothing more than a restatement of the work of Dunn, Lynn and Sumner. The closing paragraph of the review is interesting:

'This is not the place to discuss merits and demerits of the problem in all its bearings ... suffice it to say that the problem has begun to take a considerable hold on nautical men, not only in England but abroad—Germans and Italians have written on the subject.'

It appears that the Editor of the *Nautical Magazine* invited Mr W. H. Bolt of Norie's Nautical Academy to write on the subject of the 'New Navigation and Sumner's Method'. We have had occasion to refer to the series of articles written by Bolt on Sumner's method. In describing the New Navigation, Bolt refers to the work of Villarceau and de Magnac, and also that of Fasci, Marcq St Hilaire, Hilleret, Boilard, Caspari, Mouchez and others.

'The partisans of the old methods,' wrote Bolt, 'are positively furious on the subject: while the propounders and promoters of the New Methods have determined upon nothing less than a radical change in the theory and practice of navigation.'

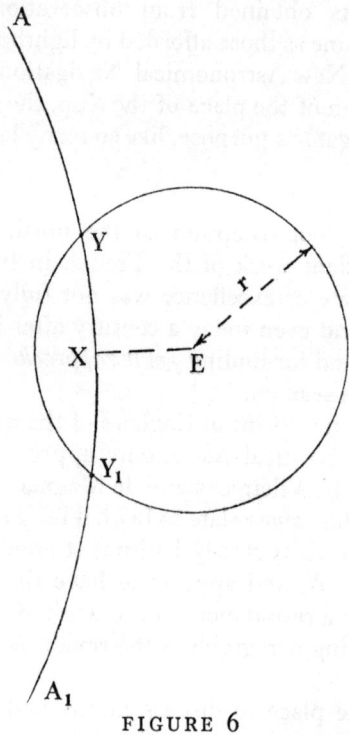

FIGURE 6

Had the promoters had their way, and the partisans of the old methods been more rational in their ideas, nautical astronomy, since 1875, would have been rationalized, and all methods would have been swept aside, having been replaced by the never-failing intercept method.

Bolt's treatment of the New Navigation is not without interest. Referring to Fig. 6:

Let E represent the ship's position by D.R. Let the circle of radius r be a circle of error within which it is assumed that the ship's actual position must lie. AA_1 is part of a circle of equal altitude which cuts the circle of error at Y and Y_1. The ship's actual position is certainly on arc YXY_1 of the circle of equal altitude, provided that the altitude of the body has been observed correctly. The exact position of the ship on arc YXY_1 is unknown; but the point X, which lies midway between Y and Y_1 and in a direction from E equivalent to that of the azimuth (or 180° therefrom) of the observed body, is the most likely position of the ship. Hence the name *point rapproché* given to X.

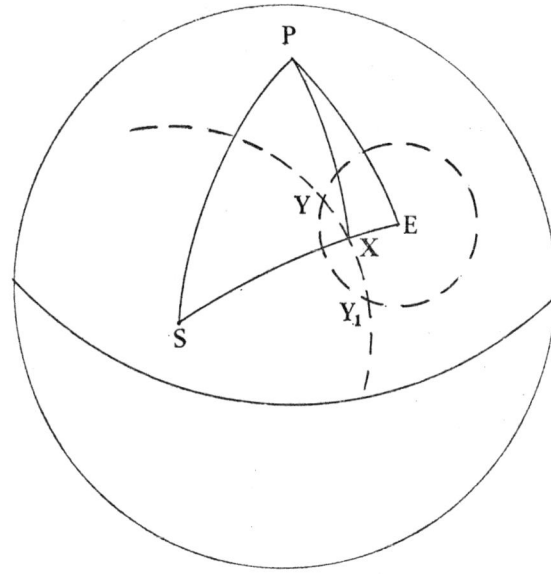

FIGURE 7

'Point X,' said Bolt, 'is a very important one in the New Navigation. It will be coincident with E when the D.R. position happens to be the true position, and in other circumstances it must always be nearer the true position than is the D.R.; for the arc EX is perpendicular to the arc YXY_1. It is also evident that the distance of E from any point on arc YXY, other than X is greater than EA. Consequently, in solving the problem, if we take the point X as the approximate position of the ship in preference to the position by D.R., we are making use of a point approximately nearer to the true one.'

Bolt then described the problem in relation to Fig. 7.

In Fig. 7, S is the geographical position of a heavenly body*. The problem is to find the position of the point X which lies on the great-circle arc SE.

For the data of the problem we have: an altitude α of a celestial body whose azimuth is Z, observed at a given G.M.T.

Let the latitude of E (the D.R. lat.) = ϕ
Let the west longitude of E (the D.R. long.) = λ
Let the declination of the body = d

Then:

$$\text{arc PE} = (90 - \phi)$$
$$\text{arc PS} = (90 \pm d)$$
$$\text{angle EPS} = (\text{G.H.A.}* - \lambda) = h$$

Using these two arcs and their included angle we may calculate ES and PES.

We have, moreover,

$$EX = ES - XS$$

Because E is never more than a few miles from X, the arc EX may be considered to be a rhumb line.

The computations involved are those of an altitude and a time-azimuth. The former comes through the latter, and this through an auxiliary arc θ, by the aid of three fundamental equations of spherical trigonometry.

$$\tan \theta = \cot d \cos h \quad \quad \ldots \quad \ldots \quad (1)$$

$$\tan Z = \frac{\sin \theta \tan h}{\cos (\phi + \theta)} \quad \quad \ldots \quad \ldots \quad (2)$$

$$\tan \alpha = \cos Z \tan (\phi \pm \theta) \quad \quad \ldots \quad \ldots \quad (3)$$

These formulae are absolutely general and may be used in all cases.

Following the above description of the New Navigation and the formulae required for finding the azimuth and the altitude of the observed body at the ship's estimated position, Bolt added:

'Our continental neighbours are great in projection and nice calculations. They have what may be called "squared paper", that is, paper so ruled as to be covered with an immense number of very small squares. On a sheet of this paper, having projected the point X, they rectify it for every conceivable error, including also the chronometric curve for error and rate.'

The articles on New Navigation by Bolt in 1880 and du Boisy in 1881 evidently interested many readers of the *Nautical Magazine*. In the 1882 *Nautical Magazine* we find the Editor writing as follows:

> 'At the request of many readers we now translate the formulae [given by Bolt] into a verbal rule, observing however that though rules and directions derived from formulae are generally long and often tedious, we have endeavoured to simplify the subject by the aid of a blank form.'

The rules given are no less complex than those given for numerous other navigational problems, but the blank form illustrates the relative ease with which the problem could be solved, and provides evidence of how the intercept problem was solved by perhaps a few British navigators of the last century.

BLANK FORM

```
                  o   '   "
     H.A.                          cos .... tan ....
     dec  _____                 cot ....
                                   _____
arc 1 (θ)                          tan      sin ....
est. lat                           _____
arc 2 (θ ± φ) _____             tan .... sec ....
                                   cos .... tan ....
                                   _____
arc 4 computed alt ..⟵......       tan      arc 3 Az = ....
     true altitude ..........
     _____
     Difference (+ or −)
```

The problem admits of the position being finally determined (that is position of *point rapproché*) by aid of Traverse Table, but this method is not preferable to that by projection on chart of large scale.

W. R. Martin, in his book on *Navigation and Nautical Astronomy* of 1888, devotes no more than half a page or so to the Marcq St Hilaire method of dealing with curves of equal altitude. In the second edition of his book, which appeared in 1891, the account of Marcq St Hilaire's method is identical with that given in the first edition, although in both editions Martin refers to the work of Brent, Walter and Williams, first published in 1886, and entitled *A Short and Accurate Method of Obtaining the Latitude at Sea from an Ex-meridian Altitude of a Heavenly Body, to which is added brief explanation, with examples, of the 'New Navigation.'*

This account of the *Nouvelle Navigation* appears to have been the first to have been given in a British navigation book. It seems that about eleven years after Marcq St Hilaire's method had been published in France, one of the three authors of the ex-meridian tables, was attracted to the method of the *point rapproché*, and it was included, as a sort of makeweight, to the ex-meridian tables. It is interesting to note that in the later editions of these tables, the 'additional' explanation of the New Navigation was omitted—a retrograde step to be sure.

In the ninth edition of his famous *Wrinkles in Practical Navigation* published in 1894, Captain Lecky, in describing the ex-meridian tables of Brent, Walter and Williams, mentions the fact that the book of tables includes: 'An explanation of that phase of the Sumner, or Double-Altitude problem, now known as the New Navigation.'

It is clear from Lecky's subsequent remarks that he, in company with numerous other British navigators, paid little attention to the New Navigation, and failed to appreciate its far-reaching consequences.

'This [description of New Navigation],' Lecky wrote, 'the reader can take or leave as he pleases. Whilst admitting its neatness, the writer is averse to the diagram part of the method, which somehow seems to him out of place in a ship . . . Like lunars, the New Navigation is of the fancy type—all very well with gentle zephyrs, but not suited to oilskin weather.'

The power of Lecky's pen, especially in making such statements as the last sentence above, may have turned (and probably did turn) many men's minds away from the marvellous method, the credit for the invention of which we owe to the French.

Soon after the turn of the century, it was ordained that the intercept method of Marcq St Hilaire should be the standard method to be taught to the junior navigating officers of the Royal Navy. Coinciding with this epoch a book by J. R. Walker entitled

POSITION-LINE NAVIGATION

An Explanation of the Method of Obtaining the Position at Sea known as the New Navigation was published in Portsmouth in 1901.

In 1907, E. B. Simpson-Baikie produced a work entitled *New Navigation Tables*. These tables were designed to substitute the simple method of plotting position lines by a tedious process of computation. Simpson-Baikie's method, with its long list of rules for naming corrections in latitude and longitude, is almost unintelligible. Its publication, however, indicates the strong tendency of many navigators to avoid graphical methods of solution.

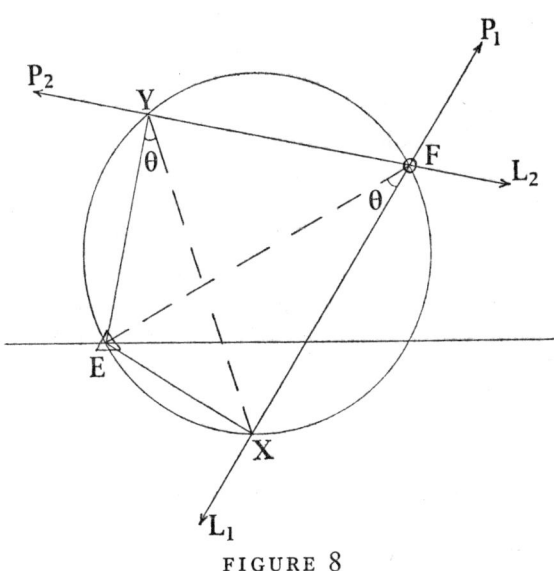

FIGURE 8

H. B. Goodwin proposed a tabular solution for finding the ship's position from two position lines obtained using the intercept method. His proposed method is very interesting but not nearly so practical as the solution by plotting.

Goodwin's method is described in relation to Fig. 8.

In Fig. 8 let E be the position used in calculating intercepts EX and EY. Let the lines P_1L_1 and P_2L_2 be the position lines resulting from simultaneous observations, and which intersect at the ship's position denoted by F.

Goodwin demonstrated that because EXF and EYF are each

90°, therefore, the points E, Y, F and X are concyclic; and that EF is a diameter of the circle through E, Y, X and F.

The angle EYX may be found using Traverse Tables, given intercepts EX and EY and the angle XEY which is the difference between the azimuths of the observed objects.

Now, EFX = EYX, because these angles both stand on the chord EX. Let each of these angles be θ.

By applying θ to the direction of P_1L_1, the direction of EF may be found.

The distance EF may be found from the Traverse Tables using θ and the intercept EX in the right-angled triangle EXF.

Knowing the direction and distance of EF and the position of E, the position of F may thus be found.

Goodwin proposed constructing a table from which, using as arguments angle between intercepts and a pair of intercepts, the distance EF and θ could be lifted.

The distinct advantage of the Marcq St Hilaire method over the older methods of position-line navigation rests in the fact that the method is applicable regardless of the bearing of the observed object. In the Sumner method the navigator has to discriminate: if the observed body is near the prime vertical circle, he may work his sight using the so-called 'longitude method'; but if the body is near the meridian, this method falls down, and the latitude by ex-meridian method may be used instead.

Now the least important of the uses of the Marcq St Hilaire method is its application to the construction of short-method navigation tables—a subject which we shall discuss in Chapter VIII.

CHAPTER VIII

Navigation tables

1. EARLY ASTRONOMICAL AND MATHEMATICAL TABLES

The ocean navigator, when practising the science of nautical astronomy, is dependent upon astronomical and mathematical tables designed to facilitate his computations. It is with these tables that this chapter is primarily concerned.

The astronomical tables designed specifically for navigational purposes provide astronomical data relating to the celestial positions and motions of the heavenly bodies used by navigators. The earliest astronomical tables used by the seaman were to be found in his navigation manuals which provided him with instructions for finding the position of his ship when out of sight of land, and for setting her course.

The earliest astronomical navigation was geared to finding latitude from an observation of the Pole Star—the Seaman's star—during hours of darkness; or from an observation of the Sun at noon when this luminary attains his greatest altitude for the day. The first astronomical tables designed for the use of the seaman were, therefore, Pole-Star tables and tables of Sun's declinations.

The earliest navigation manual appears to be the Portuguese work printed in 1509, entitled *Regimento do Astrolabio e do Quadrante*. Included in this work are a Regiment for the Pole Star and a Rule for the Sun. The Regiment for the Pole Star comprised a series of rules for finding latitude from an observation of the Pole Star, to the altitude of which a correction had to be applied, the correction depending upon the relative positions of the Pole Star and the two stars of the Lesser Bear known as the Guards. The determination of latitude from an observation of the Sun's meridian altitude was possible by means of the table of Sun's declination; which, in the work referred to, contained values of the Sun's declination in degrees and minutes of arc for noon each day of the year calculated for Lisbon.

To find the latitude from a meridian altitude observation of the Sun, the seaman was instructed to use one of nine rules: which one depended upon the relative names of the latitude and declination, and whether the shadow cast by the noon-day Sun was to the north or to the south.

The *Regimento do Astrolabio e do Quadrante* was designed for Prince Henry's navigators; and the table of Sun's declinations included in this work was based on tables published by Abraham ben Samuel Zacuto, the famous astronomer of Salamanca, in 1496.

That the science of ocean navigation had been enormously stimulated by the geographical discoveries of the 15th century is a factor of great importance. And when the need arose, the astronomical tables of the period were adapted for the seaman's use. These included the revised version of the 1272 Alphonsine tables which were published by Regiomontanus in 1474, and the *Almanach Perpetuum* of Zacuto which dates from 1496. George Sarton, in his *Six Wings*, reminds us that up until this time astronomical tables were used primarily by astrologers, and that this explains their relative abundance.

The first Spanish navigation manual appears to be the *Suma de Geographia*, by Fernandez de Enisco. This work was published in 1519, and was based on the earlier Portuguese manual. It was Fernandez's book that was translated into English, in about 1540, by Roger Barlow as his *Brief Sum of Geography*. Barlow's translation was not printed until recently. The manuscript, which is preserved in the British Museum, and which is referred to by Professor E. G. R. Taylor in her *Tudor Geography* as being the first English work if its kind, contains the first nautical tables of Sun's declinations with rules for their use in the English language.

In later manuals of navigation, a significant improvement consisted of the provision of a table of the Sun's declinations for noon each day in a four-year cycle. Such a table is included in Barlow's *Brief Sum of Geography*, although the original Spanish work contained a single table of the Sun's declinations for one year only. The improved tables of Sun's declinations meant four tables: one for use in leap years; a second for use in 'first years after leap years'; and a third and fourth for 'second' and 'third years after leap years.'

The famous Flemish astronomer Gemma the Frisian published his *De Principiis Astronomiae* in 1530. In this work a considerable amount of scientific information, including astronomical data as well as descriptions of instruments of importance to navigation, were included.

NAVIGATION TABLES

In 1545 the *Arte de Navegar* by the Spaniard Pedro de Medina appeared. This was followed by *Brevo de la Spera y de la Arte de Navegar* by Martin Cortes. Both these Spanish works on navigation were translated into several European languages. The favourite with the English navigators of the period was Cortes, whose work was translated by the Cambridge scholar Richard Eden, who had close contacts with many of the English explorers of the day, including Sebastian Cabot and Richard Chancellor.

John Tapp published nautical books as early as the beginning of the 17th century and had, as early as 1596, edited Eden's translation of Cortes's *Arte de Navegar*. Tapp's most significant publication was the *Seaman's Kalendar* containing astronomical tables, or ephemerides, of the Sun and Moon, and of the bright fixed stars. This work was first published in 1600, and it represents the first attempt to produce an almanac specifically for the use of the seaman. The *Seaman's Kalendar* and a rival *Mariner's New Calendar*, which first appeared in 1676, were published annually until well into the 18th century.

When lunar theory had been developed to a sufficiently accurate state, and it had become possible to find longitudes at sea from observations of lunar distances, the required astronomical data were given in an official nautical almanac. The first British *Nautical Almanac* was published in 1765 for the year 1767 under the hand of Nevil Maskelyne. This was exactly a century after the Royal Observatory at Greenwich had been established, with John Flamsteed as the first Astronomer Royal, for the purpose of improving knowledge of the motions of the Moon and planets and the positions of the stars, for the better prosecution of astronomical navigation.

At the same time as that of the publication of the first British *Nautical Almanac,* Maskelyne's famous work entitled *Tables Requisite to be used with the Nautical Ephemeris for finding Latitude and Longitude at Sea* was published. The second and third editions of this valuable work, also under Maskelyne's hand, appeared in 1781 and 1802 respectively.

For practising astronomical navigation the seaman needs, in addition to a nautical ephemeris, astronomical tables of refraction, dip, parallax and semi-diameter of the Sun and Moon, as well as mathematical tables of logarithms of numbers, and natural and artificial, or logarithmic, trigonometrical functions. These tables formed part of the equipment of navigators of the late 17th and early 18th centuries. A discussion on the application of early mathematical tables to practical navigation appears in Chapter III: a discussion on tables of dip, refraction, parallax and semi-diameter appears in Chapter IV, to which the reader is referred.

Maskelyne's *Tables Requisite* . . . contained all the tables necessary for navigational purposes, as well as instructions on how to use them for position finding at sea.

During the 18th and 19th centuries, in addition to the basic astronomical and mathematical tables mentioned above, navigators were provided with additional tables of great diversity, designed to assist them in their use of the many new methods of position finding introduced for their attention. These included tables for finding latitude by double altitudes, and latitude by ex-meridian observations, as well as tables for use with improved methods for finding longitude by lunar distance and by chronometer. Other tables of interest included those designed for finding the azimuth of a heavenly body given latitude, declination and hour angle, or altitude.

It is interesting to note that the original azimuth tables of Burdwood's and Davis's were designed primarily for finding the true azimuth of the Sun or other heavenly body to facilitate checking compasses—a need that grew with the development of the iron and steel ships of the period.

Tables designed to simplify the problem of finding latitude by double altitudes, and by ex-meridian observations, have been discussed in Chapter V; and those designed to simplify the lunar problem have been dealt with in Chapter VI. In the pages that follow immediately we shall discuss the seaman's nautical almanac and the major changes that have been made in the arrangement of the data used by seamen for navigational purposes. This will be followed by a discussion on some of the numerous short-method and inspection tables that have been devised to simplify the solution of the astronomical- or PZX-triangle.

The term *short method* is often used at the present time to denote a navigational method in which the PZX triangle is divided into two right-angled triangles each of which is solved by Napier's rules, the results being tabulated in a *short-method table*. Short-method tables are, at the present time, being superseded by inspection tables. Short-method tables are to be regarded as having provided an intermediate step between rigorous mathematical solutions, in which tables of trigonometrical functions and logarithms are used, and modern inspection tables, by means of which the solution of the PZX triangle may be obtained without recourse to the tables of trigonometrical functions and logarithms, and without computation or interpolation.

A significant feature of short-method tables is the inducement they offered to the navigator to work to a practical degree of accuracy. The old custom of using log trig. functions to six, or even more, places of decimals to solve sights in which the accuracy of the given information is relatively coarse, was quite unjustifiable; and the practice tended to create a false sense of security in those navigators who persisted in using the old so-called long methods of sight reduction.

2. THE *NAUTICAL ALMANAC*

The *Nautical Almanac*, founded by Nevil Maskelyne in 1765, was published annually, two or three years in advance, by Maskelyne until his death in 1811. During this long period of forty-six years, the British *Nautical Almanac* enjoyed a high reputation for its accuracy and presentation. Typical of Maskelyne's almanacs is the ephemeris for 1797. In this almanac twelve pages of data are provided for each month. Page *I* of the month contains a calendar of the month with Sundays, holidays and terms; the phases of the Moon; and times of astronomical phenomena such as eclipses of the Sun and Moon; position of the Sun in the ecliptic; and position of the Moon relative to the fixed stars. Page *II* of the month contains longitude, Right Ascension and declination of the Sun, and the equation of time for noon at Greenwich on each day of the month. Page *III* gives the Sun's semi-diameter; the hourly motion of the Sun; and times of eclipses of Jupiter's satellites. Page *IV* gives the longitudes and latitudes of the planets, together with their declinations and times of meridian passage. Page *V* gives the Moon's place using ecliptic coordinates (latitude and longitude), for noon and midnight at Greenwich each day of the month. Page *VI* gives the Moon's place using equinoctial coordinates (declination and Right Ascension) for noon and midnight at Greenwich each day of month; and also the times of the Moon's meridian passage at Greenwich. Page *VII* gives the Moon's semi-diameter and horizontal parallax for noon and midnight at

Greenwich for each day of the month. Pages *VIII* to *XI* contain lunar distances between the Moon and Sun and selected stars. Page *XII* gives the configurations of the satellites of Jupiter at 6 o'clock in the evening at Greenwich for each day of the month.

Following Maskelyne's death, the *Nautical Almanac* began to lose its character for accuracy; and, according to Francis Baily who, in 1822, published a pamphlet entitled *Remarks on the Present Defective State of the Nautical Almanac*, the official almanac

> 'eventually was reduced so low in the estimation of the world, that in the language of one of our senators who was fully competent to judge (Mr Croker in his speech in the House of Commons March 6th 1818, on introducing the Bill for new modelling the Board of Longitude), it was become a bye-word amongst the literati of Europe.'

Baily, who had published a small volume of *Astronomical Tables and Remarks for the Year 1822*, with the object of supplementing and improving upon the *Nautical Almanac*, had urged, in his pamphlet, that:

> '... the honour of the country, and the interests of science demand that Great Britain should not be eclipsed by any of the minor states of Europe. And if the *Nautical Almanac* be designed also for an astronomical ephemeris as well as for the quarter deck, it ought to contain more than it does now.'

In 1818, Dr Thomas Young was appointed Secretary to the Board of Longitude and Superintendent of the *Almanac*. Young did much to improve the *Almanac* but was adamant in his view that it was intended as an instrument essentially for seamen. Ten years after Young's appointment, the Board of Longitude was swept away and the *Almanac* was placed under the superintendence of the Admiralty, Dr Young being appointed scientific officer.

In 1829 a memorial to the House of Commons stated that the *Nautical Almanac* was designed for astronomers as well as for seamen, and that this was declared in the first *Nautical Almanac*. The memorial contained a list of errors and deficiencies in the *Almanac* and stated that it had not kept pace with navigation or astronomy. The memorial was supported by the celebrated astronomer John F. W. Herschel. These allegations were disputed by Young but pressure from many practical astronomers resulted in the *Nautical Almanac* being reformed.

The *Nautical Almanac* of 1834 contains the improvements suggested by a committee set up by the Royal Astronomical Society in 1831. From 1834 until 1931 the form of the *Nautical Almanac* was, by and large, retained.

It was recognized, towards the end of the last century, that many data contained in the *Nautical Almanac* were valueless to seamen. The first part, therefore, containing the ephemeris and lunar distances—data of importance for navigational purposes—was published separately.

From 1906, tables of lunar distances, comprising some seventy pages, were no longer given in the *Nautical Almanac*.

In 1914, on the recommendation of the Royal Astronomical Society, the *Nautical Almanac, Abridged for the use of Seamen* made its first appearance. This slender volume, the style of which remained standard until 1929, is in sharp contrast to the full *Nautical Almanac* for use in the observatory.

In 1925 a small slip pasted on the cover of the abridged *Nautical Almanac* contains the following remark:

'In both the abridged and complete *Nautical Almanacs* the times styled G.M.T. are NOW reckoned from midnight, as in civil use; but up to and including the year 1924, these times were reckoned from noon, in accordance with the then custom of astronomers.'

The abridged *Almanacs* of the period 1914–1929 gave the Right Ascension of the Mean Sun (R.A.M.S.), declination of the Sun and the equation of time for every second hour of G.M.T. starting at midnight at Greenwich for each day of the year. The Right Ascension and declination of the Moon, were, likewise, given. The Right Ascension and declination of the navigational planets were given for Greenwich mean noon for each day of the year, and a monthly table of R.A's and declinations of fixed stars was provided. In addition to these tables, sunrise and sunset and Pole Star tables were provided. So also was a table of proportional parts to facilitate finding the elements of the heavenly bodies at times different from those for which they were tabulated.

In 1929 a major change in the *Abridged Nautical Almanac* took place. Instead of tabulating R.A.M.S. and equation of time, as had hitherto been the custom, artificial quantities known as R and E were tabulated instead.

$$R = \text{R.A.M.S.} \pm 12 \text{ hours}$$
$$E = 12 \text{ hours} - (\pm \text{ equation of time})$$

The advantage of the artificial quantity R was that in reducing star (or Moon or planet) sights, the Ship's Mean Time (S.M.T.) was obtained directly from the Right Ascension of the Meridian (R.A.M.), because:

$$\text{R.A.M.} - \text{R.A.M.S.} \pm 12 \text{ hours} = \text{S.M.T.}$$
$$\therefore \text{R.A.M.} - R = \text{S.M.T.}$$

The advantage of the artificial quantity E was that not only was the trouble in dealing with the sign of the equation of time overcome, but S.M.T. could be obtained directly from the Local Hour Angle of the True Sun (L.H.A.T.S.), because:

$$\text{S.M.T.} = \text{L.H.A.T.S.} - 12 \text{ hours} + \text{equation of time}$$
$$\therefore \text{S.M.T.} = \text{L.H.A.T.S.} - E$$

The equation of time is defined as the excess of Mean time over Apparent (or true) time at any instant; so that when Mean time exceeds Apparent time, the equation of time is positive and E, therefore, is less than twelve hours. Similarly, when Mean time is less than Apparent time at any instant, the equation of time is negative and E is more than twelve hours.

In 1948, a year after its formation, the Institute of Navigation was asked by the Admiralty to advise on the re-design of the *Nautical Almanac* for surface navigation. Proposals had been put forward by Mr D. H. Sadler, the Superintendent of H.M. Nautical Almanac Office, for the revision of the *Abridged Nautical Almanac*. The Institute was invited by the Hydrographer of the Navy to comment upon these proposals. Mr Sadler described his proposals to a special meeting of the Institute on May 7th 1948 under the chairmanship of its President, Sir Harold Spencer Jones. The outcome of Mr Sadler's proposals was the introduction, in 1952, of a remodelled *Nautical Almanac*.

The 1952 and subsequent *Nautical Almanacs* are designed along the lines of the *Air Almanac* of 1937, in which Greenwich

Hour Angles of the Sun, Moon, navigational planets and the First Point of *Aries* are tabulated against G.M.T.

In the present-day *Nautical Almanac*, G.H.A's, of the Sun, Moon, planets and the First Point of *Aries*, are tabulated for every integral hour of G.M.T. for the whole year. Interpolation tables are provided to facilitate finding the G.H.A. of a celestial body at any time. To provide for stellar navigation the quantity (360°— R.A. in degrees, etc.) known as Sidereal Hour Angle (S.H.A.) is tabulated for a large number of fixed stars. The G.H.A. of a star may be found from the relationship:

$$G.H.A.* = G.H.A.\Upsilon + S.H.A.*$$

The great advantage of the tabulation of G.H.A's is that longitude is found direct from L.H.A., because:

$$Long. = G.H.A.* \sim L.H.A.*$$

where * is any celestial body.

3. SHORT-METHOD AND INSPECTION TABLES

The term *sight reduction* is used to describe the process by which a navigator solves a PZX triangle in order to find his latitude or longitude or, more generally, to determine a position line, somewhere on which his ship's position lies.

In the astronomical- or PZX-triangle, the angles P and Z may be computed if the three sides are known. From the angle P the L.H.A. of the observed body at the time of the observation is known. This, compared with the G.H.A. of the object obtained from the *Nautical Almanac*, gives a longitude. This longitude, together with the latitude from which the side PZ is deduced, defines a point on a position line, the direction of the position line being found from the angle Z of the PZX triangle. This is the basis of Sumner's solution, and it is usually known as the *longitude by chronometer* method.

An alternative method of determining a position line is the *intercept method* of Marcq St Hilaire. In the intercept method, the zenith distance of the observed object at a chosen position is calculated. This calculated zenith distance is compared with the observed zenith distance corrected for dip, refraction, etc. The difference between these, in minutes of arc, is known as the intercept. The position line is drawn at the end of the intercept (which is drawn from the chosen position), in a direction at right angles to the azimuth of the object at the time of the observation.

To find the longitude of the ship at the time of the sight from a single observation using the longitude method, the observer requires knowledge of the latitude of his ship. If the latitude is not known, the side PZ—which is the complement of a latitude—is also not known. In this case the angle P (from which the L.H.A. of the observed object is obtained) cannot be found, unless the azimuth of the object at the time of the observation is exactly 090° or 270°.

The common practice in early times was to make the observation for longitude when the observed object (invariably the Sun during the morning) bore due east or west: that is to say, when it was on the prime vertical circle. In this event, any error in the latitude used to compute angle P in the PZX triangle did not produce error in the calculated hour angle.

In cases when the morning Sun was observed off the prime vertical circle, it was not uncommon for the observer to find the latitude from an observation of the Sun at meridian passage, and then, by working back to the time of the morning sight, to find the latitude at sights. By so doing, the computed longitude at sights was reliable. With the introduction of the *A and B Tables* (see Chapter VII) this process became redundant, the morning sight being worked out with a D.R. latitude, the longitude at sights being then worked up to noon, and corrected if necessary by using the combined A and B corrections, after the latitude by meridian altitude observation of the Sun had been found.

As long ago as 1770, Cassini calculated and published tables of solutions of PZX triangles against arguments: latitude, declination and zenith distance. The object of these tables was to eliminate the necessity of the labour of computation. In 1793, Lalande published similar tables, extended to cover all latitudes up to 60° N. and S. Cassini's tables must stand as the first of the inspection tables of nautical astronomy.

Credit for being the first Englishman to publish a set of inspection tables for sight reduction belongs to Commander Thomas Lynn. Lynn, a merchant seaman, published his *Horary Tables* in 1827, dedicating them to the Honourable East India Company. Lynn's horary tables were designed to give the solutions of hour angle against latitude, declination, and altitude for each whole degree, as arguments, requiring triple and, therefore, troublesome interpolation.

In 1863, Monsieur Louis Hommey published a set of horary tables (*Table d'angles Horaires*) based on the same principle as those of Cassini's and Lalande's. Solutions of angle P of the PZX triangle were given for all latitudes.

In 1871, Sir William Thomson (later Lord Kelvin) read a paper before the Royal Society of London entitled 'On the Determination of a Ship's Place from Observations of Latitude.' The author described Sumner's method of plotting a position line, and added:

'Were it not for the additional trouble of calculating a second triangle, this method ought to be universally used instead of the ordinary practice of calculating a single position with the most probable latitude taken as if it were the true latitude.'

It may well be wondered what were the circumstances which led Sir William Thomson to interest himself in navigational affairs. He was a keen yachtsman; he was a brilliant mathematician and natural philosopher; and he was gifted in having a great natural ability not only to discover scientific principles, but also to put principles into practical use. His work in connection with the sounding machine and the magnetic compass is proof enough of this.

Thomson, in his role as scientific adviser to the Atlantic Telegraph Company, gained first-hand experience at discovering and learning about some of the problems which faced navigators of the time. He relates, in the paper mentioned above, how he first learnt Sumner's method of navigation in 1858 from Captain Moriarty on board H.M.S. *Agamemnon*. He informs us that Moriarty used the method regularly in the Atlantic Telegraph expeditions of 1858 and 1865-1866.

Thomson pointed out that very little experience at sea suggested the desirability of abolishing calculations as far as possible in the ordinary day's work. His own words are interesting:

'... When we consider the thousands of triangles daily calculated among all the ships at sea, we might be led for a moment to imagine that every one had been already solved, and that every new calculation is merely a repetition of one already made. But this would be a prodigious error; for nothing short of accuracy to the nearest minute in the use of the data would thoroughly suffice for practical purposes. Now there are 5,400' in 90° and therefore there are $5,400^3$ or 157,464,000,000 triangles to be solved for a single angle. This at 1,000 a day would take 400,000 years. Even with an artifice such as that to be described below, for utilizing solutions of triangles with their sides integral numbers of degrees, the number to be solved (being 90^3 or 729,000) would be too great, and the tabulations of the solutions would be too complicated (on account of the trouble of entering for the three sides) to be convenient for practice; and tables of this kind which have been actually calculated and published (as for example Lynn's *Horary Tables* of 1827) have not come into general use.'

It occurred to Thomson that, by dividing the problem into the solution of two right-angled triangles, the ship's position could be found without recourse to calculation. Thomson's method, whereby the PZX triangle is split into two right-angled triangles, is the first of numerous methods in which the same technique is employed.

John Thomas Towson, a figure famous in the last century in British mercantile marine circles, is credited with being the first to divide the spherical triangle used in great-circle sailing into two right-angled triangles, and to produce a tabular method for solving great-circle sailing problems. Towson's *Tables to Facilitate the Practice of Great-Circle Sailing* were first published in 1847.

Course angles in great-circle sailing correspond to hour angles and azimuths in nautical astronomy, so that Towson's tables could be (and were in later years) adapted for astronomical problems.

Thomson's method is described with reference to Fig. 1.

In Fig. 1 let O be a point on PZ (or PZ produced) where XO is an arc of a great circle perpendicular to PZ (or PZ produced).

Using PX and angle P in the right-angled triangle XPO, PO and XO may be found. Arc ZO may be found by taking the difference between PZ and PO.

With ZO and OX, in the right-angled triangle XOZ, ZX, the zenith distance, and XZO, the azimuth, may be found.

NAVIGATION TABLES 321

Suppose now that the solution of the right-angled triangle XPO for PO and XO, to the nearest integral number of degrees, would suffice. Further, suppose PZ to be the integral number of degrees nearest to the estimated co-latitude; then ZO will also be an integral number of degrees. Thus, the two right-angled spherical triangles XPO and XZO each have arcs of integral numbers of degrees for legs.

Thomson found that the two steps indicated above can be so managed as to give, with all attainable accuracy, the whole infor-

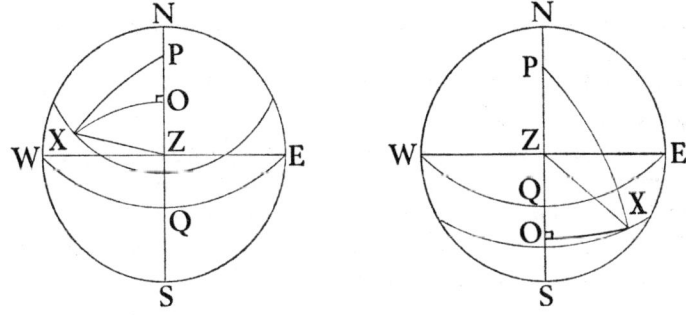

FIGURE I

mation deducible from them regarding the ship's position. It followed, therefore, that the necessity for calculating the solutions of spherical triangles in the ordinary day's work at sea is altogether obviated provided that a convenient table of the solutions of no more than 90^2, i.e. 8,100 triangles is available.

Thomson, together with Mr E. Roberts of the Nautical Almanac office, put the necessary calculations in hand with a view to publishing a table in order to facilitate the solution to the problem of finding a ship's position from astronomical observations. These tables, entitled, *Tables for Facilitating Sumner's Method at Sea*, appeared in 1876.

Thomson's tables are arranged with all the values (to integral degrees from 0° to 90°) for the side ZO denoted by b, in the right-angled spherical triangle ZOX, in a vertical column at the head of which is written the value of XO denoted by a, the perpendicular dropped from the observed body on to the observer's celestial meridian. On the same level as the value of b, in the column corresponding to a, the table shows the value of the co-hypotenuse and angle A opposite to side a.

By clever design the tables are amongst the briefest of those produced for the same purpose. The rules for using them are, however, complex; and it is little wonder that the tables were not popular amongst seamen. The remarks of a reviewer, made in the *Nautical Magazine* of 1876, are interesting:

> 'We cannot but admire the ingenuity and mathematical talent in the method now proposed by Sir William Thomson, and also in the construction of the tables. But truth compels us to state that our admiration of the work must cease there. Indeed, we are surprised that so much talent and ingenuity should have been expended in a problem which admits generally of such an easy solution, and by the means of any of the usual epitomes in daily use by navigators.
>
> 'It appears that the author has endeavoured by the method now proposed to disguise the problem as much as possible; and, instead of it appearing to the navigator as a plain straightforward question, easy of solution, it would appear to the plain sailor as something very mysterious. . . .'

The rules for using the tables are indeed complex; and this is probably the reason why seamen, in general, found little or no use for Thomson's tables. The reviewer mentioned above regarded the work, rather contemptuously, as '. . . a very pretty puzzle for the amusement of the amateur navigator.'

It was little realized at the time of the appearance of Thomson's tables that the author had initiated a new epoch in the practice of astronomical navigation. In the paths that the brilliant Thomson had trodden, others followed. Not many decades were to pass before a veritable spate of short-method tables appeared for the approval of the seaman.

A few years after the first publication of Thomson's tables, a Russian mathematician and astronomer named Kortazzi produced a modified version of Thomson's tables; but these, like those of the originator, found no favour with seamen. Kortazzi's work, entitled *Modification des Tables d'Azimuth de Thomson*, was published in Paris in 1880.

In an interesting paper published in Vol. 4 (1951) of the *Journal of the Institute of Navigation*, by Captain F. Radler de Aquino of the Brazilian Navy, a pioneer in short-method tables, Sir William

Thomson is described as being the first to use Greenwich Hour Angle (G.H.A.) and assumed positions. Aquino points out that Thomson's table, although more naturally arranged for finding altitude (or zenith distance) and azimuth, is based on Sumner's method, in which Local Hour Angle (L.H.A.) and azimuth are required. In this connection it is not without interest to note that Thomson's tables, although not published until 1876, were devised and described at least four years before Marcq St. Hilaire described his remarkable and far-reaching intercept method in an article in *Revue Maritime et Coloniale*.

An interesting navigational method was published in 1873 by the Italian G. F. Martelli.

Martelli's original work, entitled *Tables of Logarithms*, contained no more than forty-nine small pages comprising five logarithmic tables. It appears that Martelli never divulged the principles upon which his tables are based, thus earning for them the appellation 'Martelli's Mysteries', by which the tables generally became known by British seamen.

Martelli's tables became very popular, and numerous editions have been published. Their popularity remains, and many present-day navigators employ them for solving their Sun- and star-sights.

Martelli's method was designed for finding the angle P of the PZX triangle. The formula used was arranged to simplify tabulation and arithmetic. Although accuracy is not high, the tables are handy and nicely arranged, and the solution of the angle P is speedy. The rules for using the tables are simple and the solution is obtained from six inspections of the tables. The principal disadvantage of the early editions of Martelli's tables was that no means were provided for finding the azimuth of the observed object. This deficiency was overcome on the later editions of the tables.

Volume three of the *Admiralty Manual of Navigation* describes the quantities tabulated in each of Martelli's five tables. A mere glance at these descriptions is sufficient to indicate the complexity and apparent mystery of Martelli's methods.

We have mentioned in Chapter II, Wakeley's *The Mariner's Compass Rectified*, first published in 1665. The principal table in this work, which ran into numerous editions, is to be regarded as the first table of Sun's true azimuth.

Thomas Lynn, whose *Horary Tables* we have described, published a comprehensive azimuth table, for use with the Sun, in the year 1829.

The most famous of all azimuth tables, are those computed by Staff Commander John Burdwood, R.N., first published in 1852. Burdwood's *Tables of Sun's True Bearing or Azimuth* were enlarged in 1858, 1862, 1864 and 1866.

Captain J. E. Davis, R.N. (d. 1877) who distinguished himself in various nautical subjects, is best known through the famous azimuth tables that bear his name. For many years Captain Davis served in the Hydrographic Office of the Admiralty. His valuable azimuth tables, giving azimuths of the Sun for values of latitude and declination in one-degree steps, and values of local hour angle in four-minute intervals, were computed by Davis and his son Percy L. H. Davis. Davis's azimuth tables were first published in 1871.

In 1890, Captain Patrick Weir's azimuth diagram was published. This diagram appears to have been a favourite among Royal Naval officers, and its principles and use are still described in the current *Admiralty Navigation Manual*.

Perhaps the most popular azimuth tables, especially in the British Merchant Navy, are the *A, B and C Tables* which we have described in Chapter VII. These tables, like other azimuth tables, are used not only for finding azimuths for the purpose of finding the error of the magnetic compass, but also for finding the direction of a position line obtained from an altitude observation.

Percy L. H. Davis published his *Altitude-Azimuth Tables* in 1917 and 1921. In this work, azimuths and corresponding altitudes are tabulated together. These tables are especially useful for star identification.

Whilst Chief Assistant at the Nautical Almanac Office, Percy L. H. Davis published his *Chronometer Tables* in 1897. These tables were designed for use with the Sun for latitudes up to 50° N. and S. Later, the tables were extended for use with all celestial bodies regardless of their declination. The chronometer tables are arranged so that each whole degree of latitude within the limits of the tables appears on an opening, with all declinations up to 24°. There are thirty selected altitudes suitable for a.m. sights of the Sun. The table is arranged as follows:

	Lat. 37°			
	same	Dec.	19°	name
Alt.	hrs. mins. secs.	A	L	D
40°	3 34 52	−5·1	−0·6	+2·7
41°	3 29 50	−5·1	−0·6	+2·7

The columns headed A, L and D, are the variations in hour angle for 1′ in altitude, latitude and declination respectively. These values are used to facilitate interpolation. A ready reckoner is provided with the tables in order to save the trouble of multiplication when interpolating.

The longitude of the point through which to draw the position line is found by inspection right away. Separate azimuth tables are required to find the direction of the position line. An alternative method, often recommended, of laying down a position line using Davis's *Chronometer Tables*, is to work out two longitudes corresponding to two integral latitudes differing by 1°. This method obviates the need for azimuth tables.

In 1905, Davis published his *Requisite Tables*. These tables contained all the mathematical tables required for the practice of astronomical navigation using the cosine-haversine method of sight reduction. They included tables of logarithms of numbers, and logarithms of sines, tangents, etc. The unique feature of Davis's tables is the table of logarithmic and natural haversines, placed in juxtaposition in a combined table. This innovation, which proved immensely valuable for solving sights by means of the long methods, was introduced by Davis, who, in the preface of his *Requisite Tables* made the modest observation:

'The table of logarithmic and natural haversines will be found to save entering the tables in cases where the log. and nat. haversines of the same angle are needed.'

The comments on the number of decimal places in mathematical tables for navigational purposes made by Davis in the

Preface to his *Requisite Tables*, are of great interest. Davis informs us that:

'... in deciding this [the number of decimal places] the author has been guided by the following considerations.

'If all logarithmic work were done without interpolation, it would be as easy to deal with six as with five places of decimals. ...

'Signor Boccardi, writing exhaustively on the subject of computation, quotes Gauss, the German mathematician, who says that the time occupied in a calculation when using five, six, and seven places of decimals is in proportion to one, two, and three, adding as to speed that his own experience indicates one, two, and four as possibly a more correct ratio. On the point of accuracy it must be remembered that the chronometer is the ultimate datum of all nautical calculations, and that the seconds are not trustworthy to a unit. Now, in the log haversines, the fifth figure of the log decides the second of time up to 8 hours. ... There is nothing new in the suggestion now put forward that five-figure logarithms are sufficient for seamen's use; so long ago as 1781 it was stated by Nevil Maskelyne that five-figure logarithms were abundantly sufficient for the general purposes of navigation, and that sines alone were needed to six places; and it is probable that had it not been for the importance at that time attached to the lunar problem, six-figure logarithms would never have come into general use. ...'

Davis's *Chronometer Tables*, like those of Lynn's and Hommey's, met with little success. The utility of the table was discounted by the necessity of making separate adjustments for the odd minutes of the latitude, declination and the altitude, so that a computer using the tables was sometimes left in some doubt whether the measure of the saving of time and trouble enjoyed was not a somewhat problematical quantity. In other words, Davis's inspection method is little, if at all, better than the normal method in which logarithmic trigonometrical functions tables are used. Interpolation in tables was, and still is, regarded as being an undesirable process in the minds of most seamen.

It is in the general absence of interpolation that the advantage of the intercept method of sight reduction consists from the point of view of a short-method table maker. The important feature of the intercept method, in this connection, is that one may work from any position in the vicinity of the D.R. position, without loss of accuracy in the final result.

An inspection table constructed for each integral degree of latitude and longitude (or every four minutes of hour angle) renders it possible to eliminate the need for interpolation in latitude and hour angle, leaving only interpolation for declination. In other words, by choosing a position, the latitude and longitude of which give integral numbers of degrees to work from, only a third of the interpolation normally required is necessary.

The Rev. Frederick Ball, an instructor in the Royal Navy, is credited with being the first to produce a set of inspection tables based on the above principles. Ball perceived that since in the Marcq method the observer may take any position near the D.R. position of the ship, it is possible so to choose the position for solving the sight as to have always an integral number of degrees of latitude and hour angle. Ball's work, entitled *Altitude or Position-Line Tables*, first published in 1907, gave altitudes for each integral degree of latitude and declination. Interpolation was necessary only for odd minutes of declination, the correction for these being effected by means of a small supplementary table. The longitude used is chosen so that the angle P is a round four minutes in time.

A second edition of Ball's tables appeared in 1910. A reviewer of the tables, writing in the *Nautical Magazine* of 1911, expressed the opinion that:

'Mr Ball is probably the first compiler of tables of this nature who has succeeded in gaining for his work the distinction of passing into a second edition.'

A third edition of the tables was published in 1915.

H. B. Goodwin's comments on Ball's tables are interesting. The following remarks are contained in an article which appeared in the *Nautical Magazine* of 1912:

'But probably the simplest and most complete work of this nature is that published by Mr F. Ball. Mr Ball boldly taking the bull by the horns, has worked out the altitude for each degree of latitude and declination and for 4 minutes of H.A., within the limits 0° to 60°, both in latitude and declination. Mr Ball's work might perhaps be accepted as representing the very latest word in table construction, so far as the majority of people are concerned, but for one rather serious objection, which arises from the very completeness with which the task has been carried out. To place on record the result of the solution of so vast a number of triangles involves a very large amount of space, and three bulky volumes with a total of nearly 800 pages in all are therefore required. Each volume being priced at 15 shillings the cost of the whole work is upwards of £2—a sum which must be considered prohibitive in the case of a majority of those in whose interests the tables have been calculated.'

Another opinion of Ball's tables made by Captain H. S. Blackburne, which appeared in the *Nautical Magazine* of 1911, is interesting. Captain Blackburne wrote:

'... In the navy where all the officers have been imbued with what is termed the "new navigation," Mr Ball's book will, of course, be much appreciated and very generally used, as I presume that his work will be supplied to each of H.M. ships; but in the Merchant Service (except perhaps in the case of specially liberal companies or ship-owners) it can hardly be expected that ship-owners will supply the work to their ships, and not many officers and masters even, will feel that they can afford to buy a set at 45 shillings when they can get the same results by the aid of other tables with equal rapidity and at less than quarter the cost.'

Blackburne himself published in 1914, his *Excelsior Azimuth and Position-Finding Tables*. These tables provided a solution to the longitude by chronometer problem, the method employed generally by seamen in the merchant service.

A French professor of hydrography named Souillagouet, is credited with being the first to devise navigation tables based on the division of the PZX triangle by dropping a perpendicular from the zenith on to the hour circle of the observed body. Souillagouet's tables, entitled *Tables du Point Auxilaire*, were published in 1891.

Bertin, another Frenchman, devised tables similar to those of Souillagouet. Bertin's tables entitled *Tablette de Point Sphérique* were published in 1919.

Captain Guyou of the French Navy produced an original work in 1914 entitled *Nouvelles Tables de Navigation* which was published in Paris. The central notion in Guyou's tables is 'reduction to the equator.' Instead of calculating the altitude directly, Guyou finds from Table 1 such values for a new declination and altitude as at a place on the equator would give the actual values of hour angle and azimuth at the time and place of the observation. Table 2 is entered with arguments declination obtained from Table 1 and same hour angle as before. At a single opening the calculated altitude is obtained.

Guyou's tables contained 700 pages in two bulky volumes.

Lieutenant (later Vice Admiral) Radler de Aquino of the Brazilian Navy ranked as a great authority on short-method tables. He first published his famous *Altitude and Azimuth Tables*, which he described as the 'Simplest and Readiest in Solution,' in 1907.

Aquino's tables were designed for use with the intercept method. In this process the zenith distance of the observed object is found using the sides PZ and PX and the included angle of the astronomical triangle. Aquino's aim was to provide a means of dispensing with logarithms and with a limited amount of interpolation to determine the zenith distance and the azimuth of the observed object at an assumed position.

Aquino's tables are based on the splitting of the PZX triangle into two right-angled triangles by dropping a perpendicular great circle from the observed object on to the side PZ or PZ produced. It follows that while Ball's tables involve the three arguments—latitude, declination and hour angle—Aquino's tables involve only two; and the tables, therefore, are much less voluminous than those of Ball's.

Aquino's original tables have much in common with Towson's tables for facilitating great-circle sailing; but Towson's tables proceed for each whole degree of latitude (of the vertex) whereas Aquino's proceed by half degrees.

The Rev. William Hall, in an article entitled 'The Newest Navigation' which appeared in the *Nautical Magazine* for 1910, described Aquino's tables in the following terms:

'De Aquino's tables are beautiful in theory and excellent for a skilled man. I have rarely been so pleased as when I got a copy. But I dare not affirm that everyone will take the trouble to learn them, and they need clear understanding. ...

'... The direct method in Davis's [Chronometer Tables] as well as Ball's [Altitude and Position-Line Tables] is satisfactory, because you know exactly what you have got. On the other hand, solutions depending on splitting up the PZX triangle into two right-angled triangles must always have a variety of cases, and you cannot avoid this, because sines and cosines are what they are.'

Aquino's tables form, in effect, a spherical traverse table for the right-angled triangles PMX and ZMX, M being the foot of the perpendicular from X on to the observer's celestial meridian. The side MX, being common to both triangles, acts as a link.

The Rev. Hall informs us that it took him 'a whole afternoon' to understand the tables thoroughly, and it was not until he had altered all the headings of the tables that he felt he could use them properly. His advice to prospective users of Aquino's tables was to alter the headings as he had done.

Aquino brought out several editions of his tables (there were three British editions published in 1910, 1912 and 1924) and they have been popular amongst navigators who employ the intercept method of sight reduction. They have never been popular in the merchant service, the British merchant naval officer having a predilection for the longitude by chronometer method worked out using logarithms.

A compatriot of Aquino, Lieutenant Raul Romeo Braga of the Brazilian Naval College proposed, in 1912, a method for finding zenith distance which impressed H. B. Goodwin. Goodwin was granted permission by Braga, and proceeded to prepare the table. Braga proposed to proceed by way of sines, but Goodwin adopted the method to the haversine.

From the well-known formula:

$$\text{Hav } ZX = \text{hav } (PX \sim PZ) + \sin PX \sin PZ \text{ hav } P$$

Goodwin deduced the following formula:

$$\text{hav } ZX = \underbrace{\text{hav } (PX \sim PZ) \cos^2 P/2}_{\phi} + \underbrace{\text{hav } (PX + PZ) \text{ hav } P}_{\theta}$$

The two expressions for ϕ and θ are tabulated in the one case for $(PX \sim PZ)$ and P, and in the other for $(PX + PZ)$ and P, the natural haversine of the zenith distance being given by the sum of the quantities extracted from the table. To keep the tables within reasonable limits in point of size, Goodwin proposed calculating for even degrees, and employing artifices to render interpolation unnecessary.

Braga himself had tables entitled *Taboas de Alturas* published in Paris in 1924. Braga's tables are based on the relationship:

$$\text{hav alt.} = A + B$$

where

$$A = \text{hav } P - \{\text{hav } (L + D) \text{ hav } P\}$$

and

$$B = \text{hav } (L \sim D) - \{\text{hav } (L \sim D) \text{ hav } P\}$$

Goodwin published *The Alpha Beta Gamma Navigation Tables* in 1921. This work, comprising two tables in no more than thirty-four pages, is adapted for finding the hour angle; but instructions are given for finding the intercept as well. Goodwin included in his *Alpha Beta Gamma Tables*, a Rust Azimuth Diagram.

Captain A. Rust of the United States Navy is the author of *Practical Tables for Navigation and Aviation* published in Philadelphia in 1918. Rust's tables for finding hour angle are based on the formula:

$$\log \text{hav } P = \log \sec L + \log \sec D + \log \tfrac{1}{2} \cos (L \sim D) - \sin P$$

He included in his tables an azimuth diagram, referred to above, based on the formula:

$$\sin Z = \sin P \cos \text{dec} \sec \text{alt}$$

which had first been published in a set of ex-meridian tables in 1908.

An interesting navigational method based on the Marcq St Hilaire principle of position-line navigation was published in 1920 by the Hydrographic Department of the Japanese Navy. This publication, the title of which is *New Altitude and Azimuth Tables between Latitude 65° N. and 65° S. For the Determination of the Position Line at Sea*, was the work of S. Ogura of the Japanese Navy.

In using the tables of Ball's or Aquino's the navigator makes use of a chosen position having an integral number of degrees of latitude and a longitude which yields an hour angle of an integral number of degrees. The same idea is used in the Ogura method which is now to be described.

If H, ϕ, d and h represent altitude, latitude, declination and hour angle respectively, we have, from the fundamental formula of spherical trigonometry:

$$\sin H = \sin \phi \sin d \pm \cos \phi \cos d \cos h$$

Assume that:

$$A \sin K = \sin \phi \quad \quad \quad \quad (1)$$

$$A \cos K = \cos \phi \cos h \quad \quad \quad \quad (2)$$

so that

$$\sin H = A \sin K \sin d \pm A \cos K \cos d$$

i.e.

$$\sin H = A \cos (K \pm d) \quad \quad \quad \quad (3)$$

From (1) and (2) we may obtain by division:

$$\tan K = \tan \phi \sec h \quad \quad \quad \quad (4)$$

If K is thus found, A may be found from (1) thus:

$$A = \sin \phi \operatorname{cosec} K$$

If A and K are known, H may be found from (3):

$$\sin H = A \cos (K \pm d)$$

For the purpose of his table, Ogura expressed equation (3) as:

$$\operatorname{cosec} H = \frac{1}{A} \sec (K \pm d)$$

From which:

$$\log \operatorname{cosec} H = \log \frac{1}{A} + \log \sec (K \pm d)$$

Ogura's table gives for each whole degree of latitude and hour angle corresponding values of log $1/A$ and K, the values of log $1/A$ being magnified by 100,000.

A fragment of Ogura's original table is reproduced:

Lat.	H.A. 45°		H.A. 46°	
	$\log \frac{1}{A}$	K	$\log \frac{1}{A}$	K
34°	9143	43° 39'	9544	44° 09'
35°	8875	44° 43'	9262	45° 14'
36°	8607	45° 47'	8970	46° 17'

Ogura's method was a distinct advance in navigation tables. To find the required altitude at a chosen position, all that was necessary was to lift log $1/A$ and K from the table: combine K with the declination (sum when latitude and declination are of different names, and difference when they have the same names) and add to log $1/A$ the log sec ($K \pm d$) to give the log cosec of the altitude.

The principal feature of Ogura's method is its conciseness, and the table involves complete freedom from interpolation. The table embraces all latitudes up to 65°, and is contained in small compass.

H. B. Goodwin, in describing Ogura's table in the *Nautical Magazine* of 1921, suggested that a table on the Ogura model might well be included in Inman's or Norie's tables.

In 1924 a table entitled *Short Method for Zenith Distance*, was first published in Norie's *Nautical Tables*. This table, popularly known as *A and K Tables* was based on Ogura's method. The most recent edition of Norie's *Nautical Tables* (1963) includes an improved table based on Ogura's method, but incorporating an azimuth table as well as an altitude (zenith distance) table.

Concurrently with the publication of Ogura's tables, a new method of sight reduction was introduced by A. Yonemura of the Japanese Navy.

Yonemura's tables, published to facilitate his method, contain logarithms of haversines and secants, arranged for the easy solution of the PZX triangle for altitude and azimuth.

Yonemura's method is similar to the cosine-haversine method of P. L. H. Davis which, since it was introduced in 1905 in Davis's *Requisite Tables*, has formed the standard method of sight reduction in the Royal Navy. Davis's method is based on haversines and cosines, in the formulae:

$$\text{hav } z = \text{hav } (L+D) + \text{hav } \theta$$

where
$$\text{hav } \theta = \cos L \cos D \text{ hav } P$$

Yonemura's formulae, in contrast, are:
$$\frac{1}{\text{hav } \theta} = \sec L \sec D \left(\frac{1}{\text{hav } P}\right)$$
where
$$\text{hav } \theta = \cos L \cos D \text{ hav } P$$

Whereas in Davis's method additional azimuth tables are necessary to find the azimuth of the observed body, Yonemura included in his tables a table for finding azimuth based on the formula:
$$\sin Z = \sin P \cos D \sec \text{alt}$$

Since 1920, numerous navigational tables have been designed to facilitate the solution of the PZX triangle; and there seems to be no limit to the ingenious mathematical and tabular artifices that may be employed to produce navigation tables different from any that have been produced before.

Notable tables are *Hughes' Tables for Sea and Air Navigation* first published in 1938, and produced by Dr L. J. Comrie while Superintendent of the Nautical Almanac Office.

Comrie's method is similar to that of Ogura's. The tabulated quantities were mechanically computed and the tables are skilfully designed. In a tribute to Comrie following his death in 1950, D. H. Sadler considers these tables to be 'probably the finest book of navigational tables of their type.'

4. GRAPHICAL SOLUTIONS OF THE PZX TRIANGLE

In addition to tabular methods, several graphical and mechanical solutions of the PZX triangle have been devised.

The use of the stereographic and gnomonic projections of the sphere have been suggested and used in many cases to facilitate the solution of spherical triangles. The principal property of the gnomonic projection is that all great circles are projected as straight lines. That of the stereographic is that all circles—great and small—are projected as circles or straight lines. The earliest use to the mariner of these projections was for solving great-circle sailing problems, but they were later suggested for solving PZX triangles.

Hugh Godfray is usually credited for inventing the method of solving great-circle sailing problems by means of a gnomonic chart, although the method was described by Samuel Sturmy in his *Mariner's Magazine* as early as 1679. Sturmy wrote as follows:

'You may set down (on slate) therein the two places you are to sail between according to their latitudes and longitudes, and then only by your ruler draw a straight line which will represent the great circle which passeth between those two places, and will cross those degrees of latitude and longitude which you must sail by.'

Sturmy gave in his book a blank chart on the gnomonic projection in the quadrant form having a radius of $6\frac{1}{2}$ inches. For illustration he laid down the great circle between Lundy Island in the Bristol Channel to Barbados, and from it he tabulated the latitudes for every 5° difference of longitude, and finally compared the tabulated with the computed values to show their correctness.

Hugh Godfray of St. John's College, Cambridge, designed a 'Time-Azimuth Diagram' based on the gnomonic projection which was published in 1858.

We have already mentioned the time azimuth diagram of Captain Patrick Weir which was published by the British Admiralty in 1890. Weir's diagram, which is still obtainable and still used at sea, is one on which hour angles and latitude parallels are represented by confocal hyperbolae and ellipses respectively.

In 1896, F. A. L. Kitchin, a Naval Instructor of the Royal Navy, designed an original altitude-azimuth diagram which the author claimed to be simpler to use than Weir's diagram. Kitchin described his diagram in the *Nautical Magazine* of 1896. In the same year T. Wood Robinson, also a Royal Naval instructor, devised a 'New and Simple Chart for Time Azimuth' which he described in the *Nautical Magazine*.

Molfino and Alessio of Italy designed azimuth diagrams in 1901 and 1908 respectively.

Lieutenant A. Rust of the United States Navy published in 1908 his *Ex-meridian Altitude, Azimuth and Star-Finding Tables* in which two diagrams, one an altitude azimuth and the other a time-azimuth diagram, were included. Many other azimuth diagrams have been produced in addition to those mentioned above.

As well as graphical solutions for azimuths, several interesting graphical solutions for hour angle and zenith distance (or altitude) have been devised.

An early graphical solution for hour angles and altitudes, as well as for azimuths, was suggested by Margetts in his *Horary Tables* of 1790 which were designed for:

'shewing by inspection the Apparent Diurnal motion of the Sun, Moon and Stars the Latitude of a ship and the Azimuth, Time, or Altitude corresponding with any celestial Object.'

Graphical methods for solving the PZX triangle may be divided into two classes:
1. Those in which geometrical constructions are made in order to find the required quantities.
2. Those in which diagrams are used from which the required quantities may be obtained by inspection.

The latter methods have one obvious advantage over the former; and, during the latter part of the 19th century, several suggestions were made for diagrams from which the solution of the PZX triangle may be made by inspection. Notable in this connection was M. Maurice d'Ocagne whose diagram, which was published during the closing decade of the last century, was based on the all-haversine formula.

The d'Ocagne diagram consists of a square the sides of which are graduated in haversine θ from $\theta = 0°$ to $\theta = 180°$. The basis of the principle whereby altitudes and azimuths may be found by its use is expressing the all-haversine formula as an equation of a straight line.

G. W. Littlehales of the United States Hydrographic Department, whose name was well known in connection with original work on nautical astronomy and navigation, employed similar principles to those used by d'Ocagne in devising his *Altitude-Azimuth and Hour Angle Diagram* which was published by the U.S. Navy Hydrographic Office in 1917. Littlehales, like d'Ocagne, adapted the all-haversine formula for the purpose.

The denominator in the ordinary haversine formula for calculating an angle given three sides, is a product of two sines. This may be expressed in terms of a difference of two haversines, as follows:

$$\begin{aligned}
\sin A \sin B &= \tfrac{1}{2}(2 \sin A \sin B) \\
&= \tfrac{1}{2}\{\cos(A-B)\cos(A+B)\} \\
&= \tfrac{1}{2}[\{1-\cos(A+B)\}-\{1-\cos(A-B)\}] \\
&= \operatorname{hav}(A+B) - \operatorname{hav}(A-B)
\end{aligned}$$

so that for the PZX triangle:

$$\operatorname{hav} P = \frac{\operatorname{hav} z - \operatorname{hav}(p \sim c)}{\operatorname{hav}(p+c) - \operatorname{hav}(p \sim c)}$$

or

$$\operatorname{hav} z - \operatorname{hav}(p \sim c) = \{\operatorname{hav}(p+c) - \operatorname{hav}(p \sim c)\} \operatorname{hav} P \quad . \quad (1)$$

in which:

$$\begin{aligned}
z &= \text{ZX} = (90° - \text{altitude}) \\
c &= \text{PZ} = (90° - \text{latitude}) \\
p &= \text{PX} = (90° \pm \text{declination})
\end{aligned}$$

Now, if p and c are regarded as constants, formula (1) takes the form:

$$y = mx + c$$

which is the general form of the equation of a straight line.

It follows, therefore, that on the d'Ocagne diagram a straight line may be drawn which, for given values of p and c, will form a link connecting values of P and z for all values of P between 0° and 180°.

The straight line may be determined by putting P = 0° and P = 180° in formula (1), as follows:

If P = 0°, hav P = 0. Putting this value for hav P in (1) we have:

$$\operatorname{hav} z = \operatorname{hav}(p \sim c) \quad \text{and} \quad z = (p \sim c)$$

If P = 180°, hav P = 1. Putting this value for hav P in (1), we have:

$$\operatorname{hav} z = \operatorname{hav}(p+c) \quad \text{and} \quad z = (p+c)$$

When P is 0° and 180°, the body is at upper and lower transit respectively, so that by marking on the margins of the diagram, and then joining, points corresponding to the meridian zenith distances at upper and lower transits, a straight line is obtained which connects hour angle and zenith distance for every instant between P = 0° and P = 180°. If, therefore, one of these is given the other may be found by inspection.

In a similar manner, using the zenith distance and co-latitude, a straight line may be drawn on the diagram which links p and z for all values of Z between 0° and 180°.

For finding zenith distances, the scale of the diagram would have to be very large for the diagram to give results to the required degree of accuracy; but, for azimuths, a relatively small scale would be sufficient.

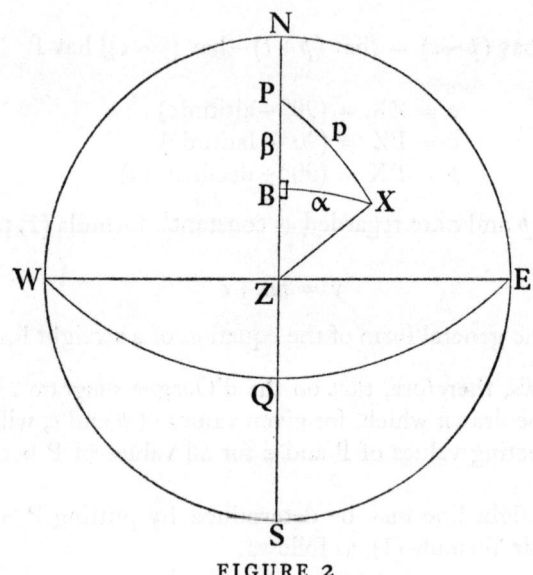

FIGURE 2

An ingeniously contrived graphical method for solving spherical triangles was produced in 1892 by Favé and Rollet de l'Lisle. An account of their diagram appeared in the *Annales Hydrographiques*, a translation of the account being made for the *Nautical Magazine* of 1892.

The Favé and Rollet de l'Lisle diagram is based on the splitting of the PZX triangle by a perpendicular from the observed object on to the observer's celestial meridian.

Fig. 2 represents the celestial sphere drawn on the plane of the celestial horizon of an observer whose latitude is equivalent to the arc NP. P is the projection of the celestial pole and N, E, S and W are the projected positions of the cardinal points of the celestial horizon of the observer whose zenith is projected at Z. WQE is the projection of the equinoctial and X is that of an observed celestial body. Arc XB is a great circle through X which crosses the observer's celestial meridian at right angles at B.

The formulae connecting P, PB (β), BX (α) and PX (p) in the triangle PBX are:

$$\cos p = \cos \alpha \cos \beta \quad \ldots \quad (1)$$
$$\cot P = \cot \alpha \sin \beta \quad \ldots \quad (2)$$

By giving p selected values (say every 10° from 0° to 90°) a series of curves representing equation (1) may be plotted.

Similarly, by giving P selected values (say every 10° from 0° to 90°), a series of curves representing equation (2) may be plotted.

The two series of curves form a lattice of curvilinear quadrilaterals, the curves, in effect, being projections of hour circles and parallels of declination on the plane of the diagram.

In a similar manner, the right-angled triangle BZX is dealt with. Solution for zenith distance and azimuth is made in two steps. A Favé diagram covers an eighth of a sphere, so that, for full coverage, four diagrams are required.

Another interesting graphical method, published by George Littlehales of the United States, is based on the stereographic projection of a model globe of radius 6 feet. The projection is divided into several sheets which are bound together; and, although the work is bulky, accuracy to minute of arc is possible.

Commander Baker of the British Royal Navy invented a navigation machine employing prepared altitude curves which are traced on a transparent tape wound on two rollers. The tape, which moves across a Mercator plotting chart, is marked with altitude curves for a series of suitable stars. Several tapes are provided, each occupying a 4° range of declination from 0° to 24° N. and S.

The fact that latitude and Local Sidereal Time are defined by simultaneous altitudes of two celestial bodies of given declinations and Sidereal Hour Angles led to the invention in 1924 by Beij of the U.S.A., of the *Two Star Diagram*, on which latitude and L.S.T. are abscissae and ordinates respectively. Curves on the diagram correspond to altitudes of selected stars. Simultaneous observations of the altitudes of the two stars enables the observer to find, by inspection, the latitude and L.S.T. The longitude is found by taking the difference between the Local and the Greenwich Sidereal Times.

Captain P. V. H. Weems of the United States Navy devised his well-known *Star Altitude Curves* on similar lines to those of Biej. Weems's *Star Altitude Curves*, which first appeared in 1928, are plotted on a chart on which the latitude scale conforms with that on a Mercator projection, so that the azimuth of the body is at every point at right angles to the tangent to the curve at the point.

Star altitude curves placed on a film and projected on to a plotting sheet form the basis of the 'Astrograph' designed for air navigation in the early part of the Second World War by Pritchard and Lamplough of the Royal Aircraft Establishment.

5. MECHANICAL AIDS TO CALCULATION

The slide rule has been put to good service for the purpose of solving navigational problems. This marvellous device was invented in England and, according to de Morgan:

> 'for a few shillings most persons might put into their pockets some hundred times as much power of calculation as they have in their heads: and the use of the instrument is attainable without any knowledge of the properties of logarithms on which the principle depends.'

Edmund Gunter is credited with the invention of the straight logarithmic scale, by means of which calculations were made with the aid of a pair of compasses.

The principle of the slide rule is that the logarithm of a product of two numbers is equal to the sum of the logarithms of the numbers. Thus, if two successive segments are set off along a straight line, of lengths equal to $\log A$ and $\log B$ on a given scale, their sum is the log of the product of A and B on the same scale. Edmund Gunter's 'line of numbers,' on the scale which bears his name, is based on this principle. The 'line of numbers' is a scale from one end of which are set off lengths proportional to the logarithms of numbers between 1 and 10. The addition and subtraction of lengths in the scale is equivalent to multiplication and division of the corresponding numbers, and was effected by the aid of a pair of compasses. Gunter's scale has, in addition to the 'line of numbers,' other scales showing the logarithms of trigonometrical functions used in navigation.

The straight logarithmic slide rule is the invention of William Oughtred (1575–1660), and dates from about 1620. Oughtred's slide rule superseded Gunter's compasses by making two logarithmically-divided Gunter's scales slide one along the other. Oughtred is also said to have: '... cast Mr Gunter's scale of numbers into a ring with another moveable circle upon it.'

The famous Newton is said to have used three parallel logarithmic scales for the purpose of solving cubic equations. He also made the suggestion that a 'runner' would facilitate the use of the logarithmic scales.

In 1903 the Rev. William Hall, the well-known naval instructor, devised a slide rule for nautical astronomical purposes which was referred to somewhat irreverently as 'Mr Hall's devil stick.'

Professor Charles Poor of the U.S.A. devised a circular slide rule, which he called the 'Poor line of Position Indicator,' in about 1920.

In 1922 a position-line slide rule was invented by Captain L. C. Bygrave. This slide rule was designed at the Air Ministry Laboratory in London, and was placed on the market by Messrs. Henry Hughes & Son, the well-known nautical instrument makers.

The Bygrave slide rule is 9 inches long, $2\frac{1}{2}$ inches in diameter, and weighs about 2 lbs. Two scales are engraved, one on each of two cylinders which slide relative to one another. The inner cylinder, from which the results are read, is graduated with logarithmic tangents, and the spiral scale which is engraved on it is 24 feet long, being divided to minutes of arc. The outer scale is graduated with logarithmic cosines. Two pointers are provided on a third sliding cylinder, one pointer for each scale. After a little practice the determination of altitude and azimuth may be performed to an accuracy of a minute of arc, in about two minutes.

Numerous attempts have been made to provide an instrumental solution to the nautical astronomical problem. The globe of the early navigators marked the first navigational instrument that could be adapted for this purpose. The fault with early globes was the coarse degree of accuracy of the solutions obtained by their use. However, of the many methods of getting an instrumental solution of the PZX triangle, the most obvious is that by means of an accurately made globe on which the geographical positions of observed objects may be plotted and arcs of circles of equal altitude drawn, to intersect at the ship's position. It is interesting to note that as late as 1943 an American engineer, D. McMillen, designed a globe of radius 7 inches, for this purpose.

The 'modern' machines which come into this category consist of several graduated arcs, each representing one of the principal circles of the celestial sphere, such as the equinoctial, celestial meridian and vertical circle. The precision navigation machine, designed in 1932 by the American Edward J. Willis, consists of five such arcs. By setting them for any given three of the arcs or angles of the PZX triangle, the required unknown parts may be read off the appropriate scales. Willis designed several navigation machines, in the latest of which three arcs only are used.

In 1936, Hagner of the U.S.A. designed a machine called the Hagner Position Finder. This was designed on a principle used in 1895 by Lieutenant Beehler of the United States Navy. Beehler invented an instrument called the 'Solarometer' which, as its name suggests, was used to find position from Sun observations. The Solarometer consisted of a bowl set in gymbals in which a sighting and measuring device, forming in effect a miniature celestial sphere, floated in mercury. By sighting the Sun, the model sphere is automatically oriented to the celestial concave and the Earth, this providing the means of finding terrestrial position.

In an interesting paper entitled *Diagrammatic Solutions for Astronomical Navigation*, which appeared in Vol. 4 of the *Journal of the Institute of Navigation*, Dr H. C. Freiesleben, of the German Hydrographic Institute, described an astronomical computing instrument made by Messrs. Zeiss and known as ARG. The basis of the ARG is a stereographic projection of a hemisphere on the plane of a celestial meridian. The declination and hour angle of an observed body are set on the instrument by means of a microscope and a cross-shaped micrometer. The projection is then rotated so that the point initially representing the elevated pole now represents the observer's zenith. The body's altitude and azimuth are then read off on lines which initially represented parallels of declination and hour circles but which now represent parallels of altitude and vertical circles.

APPENDIX 1
Spherical astronomy

For purposes of nautical astronomy, the Sun, Moon, planets and stars are imagined to lie on the surface of a sphere of infinite radius known as the celestial sphere.

Because the Earth spins, the celestial sphere appears to revolve around the Earth; so that, in general, the celestial objects continually change their directions relative to both the observer's meridian and his horizon. That is to say, all celestial objects change their altitudes and azimuths because of the Earth's rotation. When the apparent diurnal motion of the celestial bodies is considered, the Earth is regarded as occupying the centre of the celestial sphere.

The Earth not only rotates about her polar axis; she also revolves around the Sun in a nearly-circular orbit taking a year to complete one revolution. Because of the Earth's revolution around the Sun, the Sun appears to revolve around the celestial sphere, describing, in the course of a year, a great circle which lies in the plane of the Earth's orbit around the Sun. This great circle is called the *ecliptic*. When the Sun's apparent annual motion across the background of fixed stars is considered, the Sun is regarded as occupying the central position of the celestial sphere.

The great circle on the celestial sphere which is co-planar with the Earth's equator, which is the great circle lying in the plane of the Earth's spin, is called the *equinoctial*. This great circle divides the celestial sphere into the northern and southern celestial hemispheres.

The equinoctial and ecliptic intersect at two diametrically opposed points on the celestial sphere which are known as the First Points of *Aries* and *Libra* respectively. When the Sun occupies either of these positions he moves from one celestial hemisphere into the other. The days on which these events take place are called the *equinoxes*.

For about three months following the spring equinox, at which time the Sun crosses from the southern into the northern celestial hemisphere, the Sun increases his angular distance north of the equinoctial. The day on which he changes his motion relative to the equinoctial is called the summer solstice. After the day of the summer solstice, the Sun moves towards the equinoctial, arriving at the First Point of *Libra* about six months after leaving the First Point of *Aries* on the day of the spring equinox. For about three months after the day of the autumnal equinox, the Sun moves southwards until the day of the winter solstice, after which he moves towards the equinoctial, arriving at the First Point of *Aries* about three months later.

The twelve constellations through which the Sun passes during his annual apparent orbit are called the *signs of the zodiac*. These constellations lie in the zodiacal belt; and the Sun occupies each sign for a twelfth of a year.

The ancient astronomers of Babylon described the Sun's position in the celestial sphere relative to the First Point of *Aries*, the arc of the ecliptic contained between the First Point of *Aries* and the Sun being a measure of the Sun's celestial longitude. The positions of celestial bodies other than the Sun were denoted by giving their *celestial latitudes* and *celestial positions*. The celestial latitude of a celestial position is the arc of a secondary to the ecliptic contained between the position and the ecliptic.

A more useful method of defining celestial positions is based on the equinoctial instead of the ecliptic. The coordinates of a celestial position using the equinoctial system are related to those used to describe terrestrial positions. These coordinates are called *Right Ascension* (R.A.) and *declination*. The Right Ascension of a celestial position is a measure of the arc of the equinoctial contained between the First Point of *Aries* and the celestial meridian through the position. Celestial meridians are perpendicular to the equinoctial just as terrestrial meridians are perpendicular to the equator. Celestial meridians extend from the north to the south celestial poles. The great circle on the celestial sphere which lies in the same plane as the observer's terrestrial meridian is called the *observer's celestial meridian*. That part of the observer's celestial meridian which extends from the elevated celestial pole through the observer's zenith is called the observer's upper celestial meridian. The other half, which extends from the elevated celestial pole and passes through the point antipodal to the observer's zenith, is called the observer's lower celestial meridian.

The declination of a heavenly body is a measure of the arc of a celestial meridian between the body and the equinoctial. If the body lies in the northern (southern) celestial hemisphere it is said to have north (south) declination.

The Sun's declination is 0° on the days of the equinoxes. He reaches his maximum north declination of $23\frac{1}{2}°$ on the day of the summer solstice, and his maximum south declination on the day of the winter solstice.

Because the Moon revolves around the Earth and the planets revolve around the Sun, the declinations and R.A's of these bodies continually change. The process of predicting the positions of the Sun, Moon, and planets involves accurate knowledge of the orbital movements of the Moon and planets (including the Earth). This knowledge, since the days of the famous Newton, has been based on dynamical principles which enable astronomers to compute *ephemerides* of the celestial bodies of the solar system.

Fixed points on the celestial sphere maintain their declinations and R.A's. Although stars are tremendously far from the solar system, they have real movements. Moreover, the stars have apparent motions due to the *precession* and *nutation* of the Earth's polar axis, to *abberration*, and to annual *parallax*.* The real motion, and the apparent motions of stars due to precession and nutation, abberration and annual parallax, result in the celestial positions of the stars changing, albeit very slowly. For practical purposes, stars are regarded as having constant declinations and R.A's. over relatively short periods of time.

Because the Earth rotates she has gyroscopic inertia which results in her tending to maintain her plane of spin. Because of this, the celestial bodies appear to sweep out diurnal circles in the opposite direction to that of the Earth's spin.

Every point on the Earth, except the extremities of her spin axis, continually moves in a direction called east. The opposite direction to east is west; and the directions which are 90° to the left and right of east are called north and south respectively. The directions E., W., N. and S. are called the *cardinal points* of the compass.

The natural compass of an observer is his celestial horizon. This is a great circle which divides the celestial sphere into the visible and invisible hemispheres. The pole of the visible hemisphere is called the *observer's zenith*. Semi-great circles which extend from

* For precession, see p. 123. For these other terms see pp. 105-6.

the zenith, and which cut the celestial horizon at right angles are called *vertical circles*.

The vertical circles through the N. and S. points of an observer's celestial horizon lie in the plane of the observer's celestial meridian. All celestial objects that lie east of the observer's celestial meridian are rising objects, that is to say, their altitudes are increasing. All celestial objects that lie west of the observer's celestial meridian are setting objects, that is to say, their altitudes are decreasing.

The altitude of a celestial body is a measure of the *arc* of a vertical circle contained between the object and the celestial horizon vertically beneath the body. The *azimuth* of a celestial body is a measure of the arc of the observer's celestial horizon contained between the vertical circle through the elevated pole, and the vertical circle through the body. In other words, the azimuth of a body is the angle at the observer's zenith contained between his upper celestial meridian and the vertical circle through the body.

The altitude and azimuth of a celestial body are coordinates of a system used for defining celestial positions. This system is called the *horizon system*.

When a celestial body crosses an observer's upper celestial meridian, the object ceases to rise and commences to set. In other words, the body reaches its greatest daily altitude. This altitude is referred to as the body's *meridian altitude*.

When an object attains its meridian altitude it is said to *culminate*. At the instant of culmination, the object has a Local Hour Angle (L.H.A.) of 0° or 0 hrs.

The *Local Hour Angle* of a celestial body is a measure of the angle at the celestial pole, or the arc of the equinoctial, contained between the observer's upper celestial meridian and the celestial meridian of the object. When considering an object's diurnal motion, its celestial meridian is usually known as an *hour circle*.

The hour circle of the Sun may be imagined to sweep out an angle of 360°, or 24 hours, in a day. This forms the basis of the clock dial. When the Sun culminates, the time by a clock keeping correct solar time is 12 o'clock. Twelve hours before the time at which the Sun culminates he is at *lower meridian passage*, and the solar day commences. Twelve hours after the Sun culminates the solar day ends and the succeeding day commences.

Both R.A. and hour angle of a celestial body are measured as arcs of the equinoctial, so that, when defining the celestial position of a celestial body using the equinoctial system, hour angle may be used instead of R.A.

The hour angle of a celestial body measured from the Greenwich celestial meridian is called the *Greenwich Hour Angle* (G.H.A.) of the celestial body.

The difference between the G.H.A. and the L.H.A. of a celestial body at any instant is a measure of the longitude of the local, or observer's, meridian.

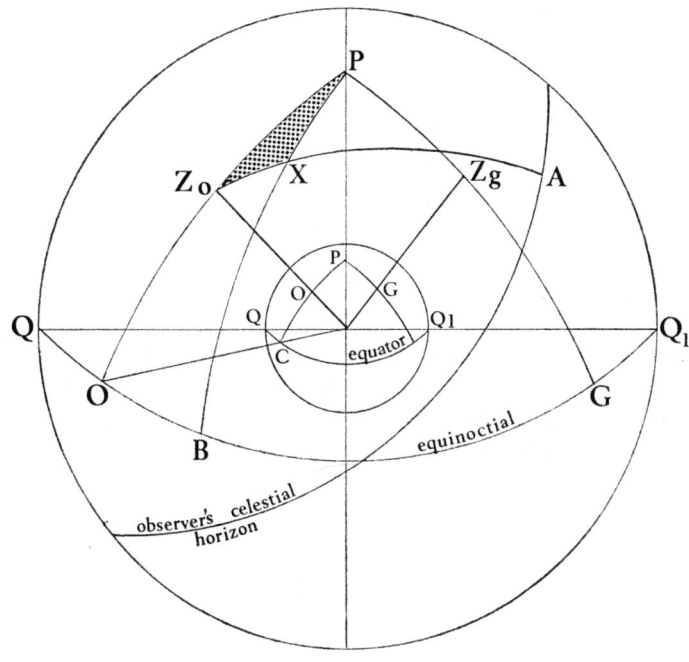

FIGURE 1

The basis of finding terrestrial position by astronomical means is the relating of the position of a celestial body for a particular instant of time, using the equinoctial system of coordinates, and its position for the same instant using the horizon system of coordinates. This may readily be seen from Fig. 1.

Fig. 1 illustrates the celestial sphere with the Earth lying at its centre; p is the Earth's North Pole and qq_1 is the equator; P and QQ_1 are the North Celestial Pole and the equinoctial respectively;

o represents an observer whose zenith is at Z_o; Z_g represents the zenith at Greenwich.

PG represents the Greenwich celestial meridian
PO represents the observer's celestial meridian
PB represents the celestial meridian of body X
arc BX = declination of X
arc AX = altitude of X
arc co = arc OZ_o = latitude of observer
angle Z_oPX = L.H.A. of X
angle Z_gPX = G.H.A. of X
angle gpo = angle Z_gPZ_o = longitude of observer.

In the spherical triangle PZ_oX:

arc PX = co-declination of X
arc PZ = co-latitude of observer
arc ZX = co-altitude of X
angle Z_oPX = L.H.A. of X
angle PZ_oX = azimuth of X

Given latitude of observer, and altitude and declination of X, the angle Z_oPX may be computed. This gives the L.H.A. of X. When this is compared with the G.H.A. of X (obtained from an ephemeris), the longitude of the observer is found.

APPENDIX 2

Spherical trigonometry

Spherical trigonometry is concerned with the several methods of solving spherical triangles.

A *spherical triangle* is formed on a sphere by the intersection of three great circles: a *great circle* is a circle on the sphere's surface on the plane of which the centre of the sphere lies.

Two great-circle arcs intersect to form a spherical angle, the magnitude of which is equivalent to the plane angle between the tangents to the great-circle arcs at the point of intersection.

The measure of a spherical arc or side of a spherical triangle is equivalent to the angle at the centre of the sphere contained between the radii which terminate at the ends of the arc.

Every spherical triangle has six parts, three of which are angles, and the other three sides. It is conventional to denote an angle of a spherical triangle by a capital letter, and a side by a small letter corresponding to the letter used for the opposite angle. Thus, if an angle is denoted by X, the side opposite is denoted by x.

If three parts of any spherical triangle are known it is possible to compute any of the other parts directly by means of one of three fundamental formulae. These are the *spherical sine, cosine* and *four-parts formulae*.

1. THE SPHERICAL SINE FORMULA

In any spherical triangle XYZ:

$$\frac{\sin x}{\sin X} = \frac{\sin y}{\sin Y} = \frac{\sin z}{\sin Z}$$

Proof: Referring to Fig. 1, in which the spherical triangle XYZ is depicted on a sphere whose centre is at O.

Drop a perpendicular from X on to plane OYZ at P.

Drop perpendiculars from P on to radii OY and OZ at A and B respectively.

Because XA and XB lie in the planes of the arcs XY and XZ respectively, it follows that:

plane angle XAP = spherical angle XYZ
plane angle XBP = spherical angle XZY

now

$$\frac{\sin z}{\sin Z} = \frac{AX/OX}{XP/BX} = \frac{AX \cdot BX}{OX \cdot XP}$$

and

$$\frac{\sin y}{\sin Y} = \frac{BX/OX}{XP/AX} = \frac{BX \cdot AX}{OX \cdot XP}$$

therefore:

$$\frac{\sin z}{\sin Z} = \frac{\sin y}{\sin Y}$$

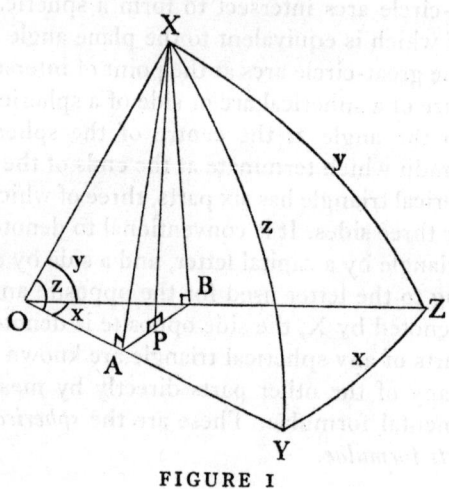

FIGURE I

By dropping a perpendicular from Y on to the opposite plane OXZ, and proceeding as above, it may be proved that:

$$\frac{\sin z}{\sin Z} = \frac{\sin x}{\sin X}$$

Therefore:

$$\frac{\sin x}{\sin X} = \frac{\sin y}{\sin Y} = \frac{\sin z}{\sin Z}$$

The spherical sine formula may be used to find an angle given the opposite side and another angle and its opposite side; or to find a side given the opposite angle and another side with its opposite angle.

Because $\sin \theta = \sin (180 - \theta)$, the spherical sine formula is ambiguous.

2. THE SPHERICAL COSINE FORMULA

In any spherical triangle XYZ:

$$\cos X = \frac{\cos x - \cos y \cos z}{\sin y \sin z}$$

or

$$\cos x = \cos X \sin y \sin z + \cos y \cos z$$

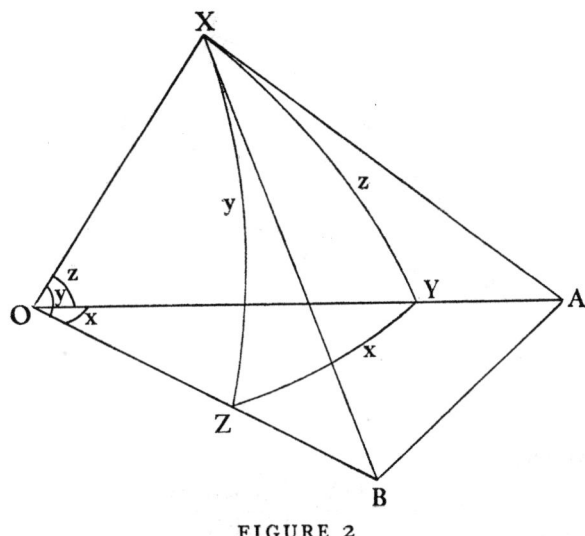

FIGURE 2

Proof: Referring to Fig. 2: Let XYZ be a spherical triangle on the sphere whose centre is at O. Tangents at X drawn in the planes of the sides XY and XZ meet the plane OYZ at A and B respectively.

Because XA and XB are tangents in the planes of the arcs XY and XZ, the plane angle BXA is equal to the spherical angle X. Also OXB and OXA are right angles.

By the plane cosine formula applied to triangles OAB and AXB:

$$AB^2 = OA^2 + OB^2 - 2 \cdot OA \cdot OB \cdot \cos x \quad . \quad . \quad (1)$$
$$AB^2 = AX^2 + BX^2 - 2 \cdot AX \cdot BX \cdot \cos X \quad . \quad . \quad (2)$$

Subtract (2) from (1):

$$O = OA^2 + OB^2 - 2.OA.OB.\cos x - (AX^2 + BX^2 - 2.AX.BX.\cos X)$$
$$= OA^2 + OB^2 - 2.OA.OB.\cos x - AX^2 - BX^2 + 2.AX.BX.\cos X$$
$$= (OA^2 - AX^2) + (OB^2 - BX^2) - 2.OA.OB.\cos x + 2.AX.BX \cos X$$
$$= 2.OX^2 - 2.OA.OB.\cos x + 2.AX.BX.\cos X$$

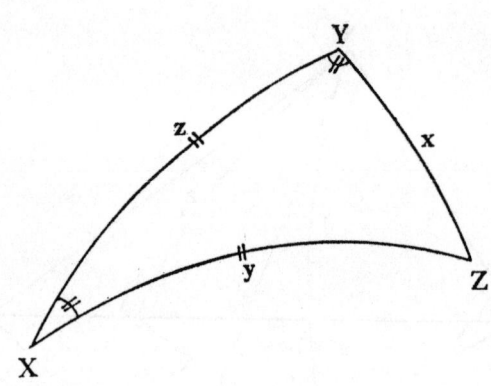

FIGURE 3

From which:

$$\cos X = \frac{OA.OB.\cos x - OX^2}{AX.BX} \qquad . \quad . \quad (3)$$

Divide (3) by OA.OB:

$$\cos X = \frac{\cos x - \cos y . \cos z}{\sin y . \sin z}$$

The spherical cosine formula suffers the disadvantage in that it is not suitable for logarithmic computation.

3. THE FOUR-PARTS FORMULA

In any spherical triangle XYZ, if three of any four adjacent parts are known, the fourth may be found directly by means of the four-parts formula.

In the triangle XYZ depicted in Fig. 3, the four-parts formula

SPHERICAL TRIGONOMETRY

connecting angles X and Y and the sides y and z, is:
$$\cos z \cos X = \sin z \cot y - \sin X \cot Y$$

Proof: By the spherical cosine formula:

$$\cos y = \cos Y \sin z \sin x + \cos z \cos x \quad . \quad . \quad (1)$$
$$\cos x = \cos X \sin y \sin z + \cos y \cos z \quad . \quad . \quad (2)$$

By the spherical sine formula:

$$\sin x = \frac{\sin X \sin y}{\sin Y} \quad . \quad . \quad . \quad . \quad (3)$$

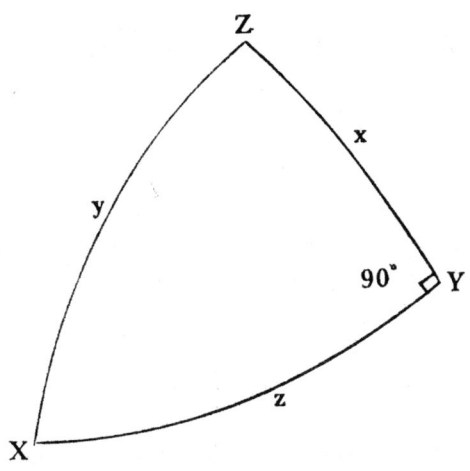

FIGURE 4

Substitute (2) for cos x in (1), and (3) for sin x in (1): Thus:

$$\cos y = \cos Y \sin z \, \frac{\sin X \sin y}{\sin Y}$$
$$+ \cos z \, (\cos X \sin y \sin z + \cos y \cos z)$$
$$= \cot Y \sin X \sin y \sin z$$
$$+ \cos z \cos X \sin y \sin z + \cos y \cos^2 z$$
$$\cos y - \cos y \cos^2 z = \sin y \sin z (\cot Y \sin X + \cos z \cos X)$$
$$\cos y \, (1 - \cos^2 z) = \sin y \sin z \, (\cot Y \sin X + \cos z \cos X)$$
$$\frac{\cos y \sin^2 z}{\sin y \sin z} = \cot Y \sin X + \cos z \cos X$$

or
$$\cos z \cos X = \sin z \cot y - \sin X \cot Y$$

4. NAPIER'S RULE OF CIRCULAR PARTS

If one of the angles in a spherical triangle XYZ is 90°, the fundamental formulae reduce to simple expressions, each involving three terms only. This is so because $\sin 90° = 1$, and $\cos 90° = 0$.

In the spherical triangle XYZ depicted in Fig. 4:

Because $Y = 90°$:

$$\sin Z = \sin z \operatorname{cosec} y$$
$$\cos x = \cos y \cos z$$
$$\cot X = \cot x \sin z$$

It is possible to derive ten such formulae which, collectively, provide the means of solving every case of right-angled triangles.

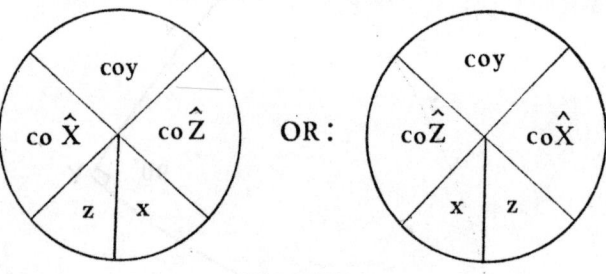

FIGURE 5

Instead of deducing from these formulae ten distinct rules for the solution of the various cases, the whole, by means of the assistance of an ingenious contrivance invented by the illustrious Baron Napier, may be comprehended in two simple rules known as *Napier's Rules*.

The parts of the right-angled triangle illustrated in Fig. 4 (not including the 90° angle) are written in order in the five sectors of the cartwheel illustrated in Fig. 5:

The two angles X and Z and the side opposite to the right angle Y are prefixed with 'co' meaning complement.

Of any three of the five parts in the cartwheel, one is a 'middle' part, and the other two are either 'opposite' or 'adjacent' parts.

Napier's mnemonic rules are:

sine middle part = product cosines opposites
sine middle part = produce tangents adjacents

Any oblique spherical triangle may be divided into two right-angled triangles by dropping a perpendicular great circle from any apex on to the opposite side or side produced. It follows, therefore, that Napier's simple formulae may be used to solve any oblique triangle indirectly and without resort to the fundamental formulae of spherical trigonometry. They are, therefore, powerful artifices in the practice of navigation; and they are particularly important in the construction of short-method navigation tables.

In astronomical navigation the more important spherical trigonometrical problems are those in which it is required to find an angle given three sides; or those in which it is required to find a side given the other two sides and the included angle. The spherical cosine formula is, therefore, the basis of the solutions of most nautical astronomical problems.

Because the spherical cosine formula is not suitable for logarithmic computation, other formulae derived from the cosine formula, and which are suitable for use with logarithms, are invariably used by navigators.

The trigonometrical functions versine and haversine are functions used almost exclusively by navigators.

$$\text{vers } \theta = 1 - \cos \theta$$
$$\text{hav } \theta = \tfrac{1}{2}(1 - \cos \theta)$$

The great advantage of the versine is that its sign is positive for all angles, so that the various forms of the versine and haversine formulae help to eliminate or reduce the seaman's traditional difficulty of dealing with trigonometrical functions of angles over 90°.

Bibliography

Abbreviations

P.T.R.S   Philosophical Transactions of the Royal Society
N.M.   Nautical Magazine
J.I.N.   Journal of the Institute of Navigation
R.A.S.   Royal Astronomical Society

1. Allingham, W. *A Graphic Method of Determining a Ship's Geographical Position.* N.M. 1892.
2. ——— *Short Cuts in Navigation.* N.M. 1898.
3. ——— *Self Tuition for Future Extra Masters.* N.M. 1903.
4. Anderson, E. W. The Principles of Navigation. 1966.
5. Andrew, J. Astronomical and Nautical Tables. 1895.
6. Angus, T. S. *Sumner's Method in Low Latitudes.* N.M. 1884.
7. Anon. *Directions for Seamen bound for far Voyages.* P.T.R.S. No. 8. 1665.
8. ——— *An Appendix to the Directions for Seamen bound for far Voyages.* P.T.R.S. No. 9. 1665.
9. ——— *Instructions concerning the use of Pendulum Watches for finding Longitude at Sea.* P.T.R.S. No. 47. 1669.
10. ——— *Letter from a Spanish Professor of the Mathematicks proposing a New Place for the First Meridian.* P.T.R.S. No. 118. 1675.
11. ——— *An Account of an Observation of an Eclipse of the Moon observed at Moscua, in Russia.* P.T.R.S. No. 192. 1690.
12. ——— *A true Copy of a Paper found in the Hand-Writing of Sir Isaac Newton, among the Papers of the late Doctor Halley, containing a Description of an Instrument for observing the Moon's Distance from the Fixt Stars at Sea.* P.T.R.S. No. 192. 1742.
13. ——— *Longitude by Chronometer at Sunrise or Sunset. Weston's Method.* N.M. 1848.
14. ——— *Description of Pagel's Method from 'La Latitude par les Hauteurs hors du Méridien etc.' par Louis Pagel.* N.M. 1861.
15. ——— *The New Navigation. Practical Rules.* N.M. 1882.
16. ——— *Errors of the Sextant.* N.M. 1883.
17. ——— *Sumner Lines.* N.M. 1891.
18. ——— *The Solarometer.* N.M. 1893.
19. ——— *To identify the Stars in Cloudy Weather.* N.M. 1894.
20. ——— Nautical Almanacs. 1897, 1920, 1925, 1931, 1947, 1952.
21. ——— Admiralty Manual of Navigation. 1914.
22. ——— Admiralty Manual of Navigation. 1922.

23. Anon. Admiralty Manual of Navigation. 1928.
24. ——— Admiralty Manual of Navigation. 1938.
25. ——— Admiralty Manual of Navigation. 1954.
26. ——— *A Modern View of Lunar Distances* (H.M. Nautical Almanac Office) J.I.N. Vol. 19. 1966.
27. Aquino, R. F. de. The 'Newest' Navigation and Aviation, Altitude and Azimuth Tables. 1924.
28. ——— *A New Wrinkle in Astronomical Navigation.* J.I.N. Vol. 4. 1951.
29. Atkinson, J. Epitome of the Art of Navigation. Revised edition by W. Mountaine. 1765.
30. Baily, F. Remarks on the Present Defective State of the Nautical Almanac. 1822.
31. Ball, F. *On Determining the Position Line.* N.M. 1907.
32. ——— Altitude or Position Line Tables. 1st ed. 1907.
33. ——— *Stellar Observations and their Practical Applications.* N.M. 1911.
34. ——— Altitude or Position Line Tables. 3rd ed. 1915.
35. Barlow, R. A Brief Summe of Geography. Hakluyt Society Edition by E. G. R. Taylor. 1932.
36. Bayley, W. H. *Time from the Sun's Altitude.* N.M. 1861.
37. Becher, A. B. *Lieut. Becher's Horizon for Astronomical Observations at Sea or on Shore.*
38. ——— *The Pendulum Marine Artificial Horizon invented by A. B. Becher, Commander R.N.* N.M. 1844.
39. Bethune, C. R. D. *Note on a Method of Reducing the Apparent Distance of the Moon from the Sun.* N.M. 1861.
40. ——— *The Determination of a Ship's Place.* N.M. 1871.
41. Bevis, J. *An Observation of an Eclipse of the Sun at the Island of Newfoundland, August 5, 1766, by Mr. James Cook with the Longitude of the Place of Observation deduced from it.* P.T.R.S. Vol. 57. 1767.
42. Blackburne, H. S. *The Value of Stellar Observations.* N.M. 1896.
43. ——— *A B Tables.* N.M. 1903.
44. ——— *Ex-Meridian Position Lines.* N.M. 1906.
45. ——— Modern Up-to-date Navigation. 1914.
46. ——— Tables for Azimuth, Great Circle Sailing and Reduction to the Meridian with a New and Improved Sumner's Method. 1918.
47. ——— The 'Excelsior' Ex-Meridian and Position Finding Tables. 4th ed. 1923.
48. Blundeville, T. His Exercises ... verie necessarie to be read and learned of all young Gentlemen etc.' 1594.
49. ——— The Theoriques of the Seven Planets. 1602.

BIBLIOGRAPHY

50. Blythe, J. H. et al. *Sight Reduction Tables for Marine Navigation.* J.I.N. Vol. 19. 1966.
51. Boisy, M. le Comte du. *The New Astronomical Navigation.* N.M. 1881.
52. Bolt, W. H. *Correction for the Middle Time in the Double Altitude Problem.* N.M. 1874.
53. ———— *Raper's Navigation.* N.M. 1876.
54. ———— *Raper's Navigation.* N.M. 1877.
55. ———— The New Navigation. 1880.
56. ———— *On the Connection between the Equation of Equal Altitudes and the Angle of Position.* N.M. 1889.
57. Bouguer, M. Nouveau Traité de Navigation. New edition by de la Caille. 1781.
58. Bourne, W. Inventions or Devises. 1578.
59. ———— A Regiment for the Sea. 1574. Hakluyt Society issue, edited by E. G. R. Taylor, 1961.
60. Bowditch, N. American Practical Navigator. 1914.
61. ———— American Practical Navigator. 1943.
62. ———— American Practical Navigator. 1958.
63. Brent, Walter and Williams. Ex-Meridian Altitude Tables. 4th edition. 1901.
64. Brown, J. (Philomath). The Description and Use of the Triangular Quadrant. 1671.
65. Brown, Ll. A. The Story of Maps. 1949.
66. Bunbury, E. H. A History of Ancient Geography. Dover edition. 1959.
67. Burdwood, J. *To find the Latitude and also their Azimuths by the Altitudes of Two Stars observed at the same instant.* N.M. 1865.
68. Card, S. F. Navigation. 1917.
69. Cassini, J. D. *Doctrine of Refraction.* P.T.R.S. No. 84. 1672.
70. ———— *Configurations of the Satellites of Jupiter for the years 1676 and 1677.* P.T.R.S. No. 128. 1676.
71. Chapin, S. L. *A Survey of the Efforts to Determine Longitude at Sea. 1660–1760* (in three parts). Journal of the U.S. Institute of Navigation. Vol. 3. 1953.
72. Chaucer, G. The Complete Works, edited by F. N. Robinson N.D.
73. Chauvenet, W. A Manual of Spherical and Practical Astronomy. 1906.
74. Clagett, M. Greek Science in Antiquity. 1957.
75. Clissold, P. C. H. *An Eighteenth Century Voyage.* J.I.N. Vol. 9. 1956.
76. Coleman, G. Lunar and Nautical Tables. 2nd edition. 1857.
77. Collier, J. Compendium Artis Nauticae. 1729.

78. Collins, J. Navigation by the Mariner's Plain Scale new-plained. 1679.
79. ——— *Extracts of Two Letters by Mr. Flamsteed to Mr. Collins.* P.T.R.S. No. 110. 1674.
80. Colson, J. *The Construction and Use of Spherical Maps.* P.T.R.S. No. 440. 1736.
81. ——— *The Mariner's New Calender.* 1754.
82. Comrie, L. J. Hughes' Sea and Air Navigation Tables. 1938.
83. Costard, G. The History of Astronomy. 1767.
84. Cotter, C. H. The Elements of Navigation. 1953.
85. ——— *Edward Wright.* N.M. Vol. 170. 1953.
86. ——— *Maritime Geography.* N.M. Vol. 181. 1959.
87. ——— *The Mariner and his Map.* N.M. Vol. 182. 1959.
88. ——— *The Sextant and Precision Celestial Navigation.* J.I.N. Vol. 16. 1963.
89. ——— *Reduction to the Prime Vertical Circle.* J.I.N. Vol. 17. 1964.
90. ——— *The Navigational Revolution of the Eighteenth Century.* N.M. Vol. 195. 1966.
91. ——— *An Historical Review of the Ex-Meridian Problem.* J.I.N. Vol. 17. 1964.
92. Craven, H. L. D. Short Tables for use at Sea. 1910.
93. Crone, G. R. *Early Books and Charts in the Royal Geographical Society's Collection.* J.I.N. Vol. 6. 1953.
94. Croudace, W. S. Star Formulary for finding Latitude and Longitude by Sumner's Method. 1873.
95. ——— *Sun Formulary.* 1873.
96. ——— *Formulary for finding Latitude by Meridian Altitude and the Longitude from an Altitude of the Sun.* 1873.
97. Dampier, C. W. A History of Science. 4th ed. 1948.
98. Davis, P. L. H. Requisite Tables. 1905.
99. Dijksterhuis, E. J. The Mechanization of the World Picture. (translated from the Dutch by C. Dikshoorn). 1961.
100. Dobson, T. *A New and Simple Demonstration of the Rules for computing the Amplitude and the Time of Rising and Setting of a Heavenly Body.* N.M. 1879.
101. ——— *Another New and Simple Method for computing the Amplitude and Time of Rising and Setting.* N.M. 1879.
102. Dolland, P. *Letter to N. Maskelyne describing some Additions and Alterations made to Hadley's Quadrant to render it more serviceable at Sea.* P.T.R.S. Vol. 62. 1772.
103. Dreyer J. L. E. Tycho Brahe. (Dover edition) 1963.
104. Dundas-White, J. *Longitude by Eclipses.* N.M. 1900.

BIBLIOGRAPHY

105. Dunn, S. A New and General Introduction to Practical Astronomy. 1774.
106. ——— The Theory and Practice of the Longitude at Sea. 1776.
107. Dutton, B. Navigation and Nautical Astronomy. 1943.
108. Duval, R. Ch. *Admiral Marcq de Blond de St. Hilaire.* J.I.N. Vol. 19. 1966.
109. Edmonds, H. H. E. *The Calculation of Altitude for a Heavenly Body.* N.M. 1911.
110. Elton, John. *The Description of a New Quadrant for taking Altitude without an Horizon, either at sea or on Land.* P.T.R.S. No. 423. 1732.
111. Fitzmaurice, L. T. *On Finding the Position by Double Altitude with only One Latitude.* N.M. 1854.
112. Flamsteed, J. *An Account of the Eclipses or Ingresses of Jupiter's Satellites into his Shadow, etc.* P.T.R.S. No. 151. 1683.
113. ——— *Concerning the Eclipses of Jupiter's Satellites for the Year following 1684, etc.* P.T.R.S. No. 154. 1683.
114. ——— *Concerning the Eclipses of Jupiter's Satellites for the Year 1685.* P.T.R.S. No. 165. 1684.
115. ——— *Description and Uses of an Instrument for finding the Distance of Jupiter's Satellites from its Axis, etc.* P.T.R.S. No. 178. 1685.
116. ——— *Calculations of the Eclipses of Jupiter's Satellites for the Year 1687.* P.T.R.S. No. 184. 1686.
117. ———*A Remark concerning the Longitude of the Cape of Good Hope.* P.T.R.S. No. 185. 1686.
118. Forbes, E. G. *The Foundations and Early Development of the Nautical Almanac.* J.I.N. Vol. 18. 1962.
119. Forbes, G. History of Astronomy. 1909.
120. Frankel, J. P. *Polynesian Navigation.* Navigation. (Journal of the U.S. Institute of Navigation.) Vol. 9. 1962.
121. Freiesleben, H. C. *Early Pole Star Tables.* J.I.N. Vol. 8. 1955.
122. Fryer, D. H. *Cartography and Aids to Navigation*, in 'A History of Technology,' edited by Singer et al. Vol. 5. 1958.
123. Gatty, H. The Raft Book. 1943.
124. Gellibrand, H. An Institution Trigonometrical. 2nd ed. 1652.
125. ——— Trigonometria Brittanica. 1633.
126. George, C. *On Finding Time and Latitude.* N.M. 1843.
127. Gill, J. Textbook on Navigation. New ed. by W. V. Merrifield. 1918.
128. Godfray, H. A Treatise on Astronomy. 2nd ed. 1874.
129. Goodwin, H. B. *A Method of finding exact Greenwich Time from an observed Lunar Distance.* N.M. 1888.

BIBLIOGRAPHY

130. Goodwin, H. B. *New Applications of the Burdwood, Davis, and Johnson Azimuth Tables.* N.M. 1894.
131. —— *The Ex-Meridian treated as a Problem in Dynamics.* 1894.
132. —— *Position of the Moon in Nautical Astronomy.* N.M. 1896.
133. —— *Ex-Meridians: Ancient and Modern.* N.M. 1896.
134. —— *A Sumner Pamphlet of Fifty Years ago.* N.M. 1897.
135. —— *The Azimuth Tables and their Work.* N.M. 1897.
136. —— *The New Nautical Astronomy on the Continent.* N.M. 1898.
137. —— *The Problem of the Latitude treated as a Quadratic Equation.* N.M. 1898.
138. —— *The Simplification of Formulae in Nautical Astronomy.* N.M. 1899.
139. —— *An Italian Contribution to Modern Navigation.* N.M. 1899.
140. —— *Raper's Approximate Solution of the Double Altitude.* N.M. 1899.
141. —— *Spherical Traverse Tables and Their Use.* N.M. 1900.
142. —— *Napier's Analogies and the Double Chronometer.* N.M. 1901.
143. —— *Longitude by Equal Altitudes.* N.M. 1902.
144. —— *The Lunar Problem in Extremis.* N.M. 1902.
145. —— *A Kinematical Ex-Meridian Table.* N.M. 1903.
146. —— *On Reduction to the Prime Vertical.* N.M. 1904.
147. —— *The Passing of the Lunar.* N.M. 1905.
148. —— *An Improved Ex-Meridian Table.* N.M. 1905.
149. —— *On Finding Position Lines by Star Altitudes.* N.M. 1905.
150. —— *Prime Vertical Reduction Tables.* N.M. 1906.
151. —— *The Nautical Almanac and Defunct Lunar.* N.M. 1907.
152. —— *A New Application for the Table for finding Latitude By Polaris.* N.M. 1907.
153. —— *The New Lunar and its Limits of Accuracy.* N.M. 1908.
154. —— *Graphic Method of Reduction to Meridian and Prime Vertical.* N.M. 1908.
155. —— *A New kind of Sumner Line.* N.M. 1909.
156. —— *The Haversine in Nautical Astronomy.* N.M. 1910.
157. —— *A New Form of Table for calculating Altitude.* N.M. 1912.
158. —— *A New Form of Table for use with the Marcq Position Lines.* N.M. 1913.
159. —— *Up-to-date Position Lines.* N.M. 1913.
160. Gordon, J. *A New and Easy Method of Finding a Ship's Position at Sea.* 1850.

161. Gordon, J. *On Sumner's Method*. N.M. 1862.
162. ———— *On the Double Altitude Problem*. N.M. 1879.
163. Gordon, R. B. *The Attainment of Precision in Celestial Navigation*. J.I.N. Vol. 17. 1964.
164. Gould, R. T. *The Marine Chronometer: Its History and Development*. 1923.
165. Graham, R. *The Description and Use of an Instrument for taking the Latitude of a Place at any time of the Day*. P.T.R.S. No. 435. 1734.
166. Grant, Charles. *The Means for Finding the Longitude at Sea*. 1810.
167. Gunter, E. *The Description and Use of the Sector. The Crosse Staffe and other Instruments*. 1624.
168. Gunther, R. T. *Early Science in Oxford. Volumes 1 and 2*. 1923.
169. ———— *The Mariner's Astrolabe*. Geographical Journal. Vol. 72. 1928.
170. ———— *Early Science in Cambridge*. 1937.
171. Hadley, J. *The Description of a New Instrument for taking Angles*. P.T.R.S. No. 420. 1731.
172. ———— *An Account of Observations made on board the Chatham Yacht, August 30th and 31st and September 1st 1732, in pursuance of an Order made by the Right Honourable the Lords Commissioners of the Admiralty, for the Trial of an Instrument for taking Angles described in the Philosophical Transactions No. 420*. P.T.R.S. No. 425. 1732.
173. ———— *A Spirit Level to be fixed to a Quadrant for taking a Meridianal Altitude at Sea when a Horizon is not visible*. P.T.R.S. No. 430. 1733.
174. Hall, B. *Equal Altitudes*. N.M. 1842.
175. Hall, W. *Navigation a Century ago*. N.M. 1902.
176. ———— *The Lunar: One more Method*. N.M. 1903.
177. ———— *The Slide Rule in Navigation, or 'Gunter redivivus.'* N.M. 1903.
178. ———— *Modern Navigation*. 1904.
179. ———— *Appendix to Raper's Practice of Navigation*. 1914.
180. Halley, E. *Catalogus Eclipsium omnium Satellitum Joviatium Anno 1688*. P.T.R.S. No. 191. 1687.
181. ———— *Cassini's Tables for the Eclipses of the First Satellite of Jupiter reduced to the Julian Stile and Meridian of London*. P.T.R.S. No. 214. 1694.
182. Handson, R. *Trigonometry, or the Doctrine of Triangles, First written in Latine by Bartholemus Pitiscus, and now translated*

into English by R. Handson where-unto is added certaine nautical Questions. 1614.
183. Hansen, L. F. Trigonometry and Navigation. 1919.
184. Harboard, J. B. Lectures on Elementary Navigation. 1895.
185. ——— Glossary of Navigation. 4th edition by C. W. T. Layton. 1938.
186. Harris, J. The Description and Use of the Globes. 8th edition. 1757.
187. Heath, L. G. *The Ship's Position by the Sun's Altitude—Sumner's Method.* (Reprint from N.M. 1845.) N.M. 1860.
188. Hebden, E. H. *Clearing the Lunar Distance.* N.M. 1858.
189. ——— *Method of deducing Time from the Sun's Altitude.* N.M. 1860.
190. ——— *A New Method of finding the Latitude by Double Altitudes of the Sun by means of Logarithmic Differences.* N.M. 1861.
191. Hewson, J. B. A History of the Practice of Navigation. 1951.
192. Hind, J. The Elements of Plane and Spherical Trigonometry. 2nd ed. 1828.
193. Holmes, Major. *A Narrative concerning the success of Pendulum Watches at Sea for the Longitudes; and the grant of a Patent thereupon.* P.T.R.S. No. 1. 1665.
194. Horsley, J. *Extract of a Letter to N. Maskelyne from John Horsley Fourth Mate on board the Glatton, East India Ship, giving an Account of his Observations at Sea for finding out the longitude by the Moon.* P.T.R.S. Vol. 54. 1766.
195. Hosier, J. The Mariner's Friend. 1809.
196. Huddart, Joseph. *Observations on Horizontal Refraction which affect the Appearances of Terrestrial Objects, and the Dip or Depression of the Horizon at Sea.* P.T.R.S. Vol. 87. 1797.
197. Hues, R. Tractatis de Globis. Hakluyt Society issue. Edited by C. R. Markham. 1889.
198. Inman, J. Navigation and Nautical Astronomy. 10th ed. 1855.
199. ——— Nautical Tables, edited by W. Hall and H. B. Goodwin. 1940.
200. Jeans, J. H. The Growth of Physical Sciences. 1950.
201. Johnson, A. C. *On Certain Short Methods in Navigation.* N.M. 1879.
202. ——— Time Altitudes for Expediting the Calculation of Apparent Time. 1894.
203. ——— Brief and Simple Methods of finding Latitude and Longitude. 3rd ed. 1895.
204. ——— A Handbook for Star Double Altitudes. 1898.
205. ——— Short Tables and Rules for finding the Latitude and Longitude. 2nd ed. 1900.

206. Johnson, A. C. On Finding Latitude in Cloudy Weather. 26th ed. 1903.
207. Jones, H. Spencer-. *The Development of Navigation.* J.I.N. Vol. 1. 1948.
208. —— *The Calendar.* In 'A History of Technology.' edited by Singer et al. Vol. 3. 1957.
209. Kelly, P. A Practical Introduction to Spherics and Nautical Astronomy. 1796.
210. King-Hele, D. G. *The Shape of the Earth.* J.I.N. Vol. 17. 1964.
211. Kitchin, F. A. L. *An Altitude Azimuth Diagram.* N.M. 1896.
212. —— *A Rising and Setting Diagram.* N.M. 1896.
213. —— *Maximum and Minimum Altitudes.* N.M. 1897.
214. —— *Practical Solution of a Spherical Triangle.* N.M. 1898.
215. —— *Kitchin's Ex-Meridian Diagram.* N.M. 1901.
216. Lande, M. de la. *Extract of Letter from N. Maskelyne concerning Lunar Method for Longitude.* P.T.R.S. Vol. 52. 1762.
217. Lax, W. *A Method of Finding the Latitude of a Place by Means of Two Altitudes of the Sun and the Time elapsed betwixt the Observations.* P.T.R.S. Vol. 89. 1799.
218. Leach, E. R. *Primitive Time Reckoning.* In 'A History of Technology,' edited by Singer et al. Vol. 1. 1954.
219. Lecky, S. T. L. Wrinkles in Practical Navigation. 9th ed. 1894.
220. —— General Utility Tables. 1st ed. 1897.
221. —— General Utility Tables. 4th ed. 1923.
222. —— Wrinkles in Practical Navigation. 22nd ed. by G. P. Bowen. 1937.
223. Leigh, Charles. *A Description of a Water Level to be fixed to Davis' Quadrant, whereby an Observation may be taken at Sea in thick and hazy Weather without seeing the Horizon.* P.T.R.S. No. 451. 1738.
224. —— *The Description and Use of an Apparatus added as an Improvement to Davis' Quadrant consisting of a mercurial Level.* P.T.R.S. No. 451. 1738.
225. Logan, J. *An Account of Mr. Thomas Godfrey's Improvement of Davis' Quadrant transferred to the Mariner's Bow.* P.T.R.S. No. 435. 1734.
226. Lord, W. B. *The Nautical Almanac Office.* N.M. 1897.
227. Mackay, A. The Theory and Practice of the Longitude. 1793.
228. —— The Complete Practical Navigator. 1802.
229. Mansilla, G. *The Altazimetro.* N.M. 1911.
230. Marguet, F. Histoire générale de la Navigation du XVe au XXe Siècle. 1931.
231. Markham, C. R. A Life of John Davis. 1889.
232. Martin, W. R. Navigation and Nautical Astronomy. 1888.

233. Martin, W. R. Navigation and Nautical Astronomy. 2nd ed. 1891.
234. Maskelyne, N. *Results of Observations of the Distance of the Moon from the Sun and fixed Stars made on a Voyage from England to the Island of St. Helena in order to determine the Longitude of the Ship from Time to Time, together with the whole Process of Computation used on this Occasion.* P.T.R.S. Vol. 52. 1761.
235. ——— *Concise Rules for computing the Effects of Refraction and Parallax in varying the Apparent Distance of the Moon from the Sun or Stars, etc.* P.T.R.S. Vol. 54. 1764.
236. ——— *Remarks on Hadley's Quadrant tending principally to remove the Difficulties which have hitherto attended the Use of the Back Observations, and to obviate the Errors that might arise from a Want of Parallelism of the true Surface of the Index Glass.* P.T.R.S. Vol. 62. 1772.
237. ——— *Concerning the Latitude and Longitude of the Royal Observatory at Greenwich.* P.T.R.S. Vol. 77. 1787.
238. ——— Tables Requisite to be Used with the Nautical Almanac. 2nd ed. 1781.
239. ——— The Nautical Almanac. 1797.
240. ——— Tables Requisite to be used with the Nautical Almanac. 3rd ed. 1802.
241. Maurice, J. *Longitude by the R.A. of the Moon.* N.M. 1900.
242. May, W. E. *The Double Altitude Problem.* J.I.N. Vol. 3. 1950.
243. Merrifield, J. A Treatise on Navigation. 1883.
244. ——— A Treatise on Nautical Astronomy. 1886.
245. ——— A Treatise on Navigation. New ed. 1905.
246. Moody, A. B. *Early Units of Measurement and the Nautical Mile.* J.I.N. Vol. 5. 1952.
247. Moore, J. H. The New Practical Navigator. 5th ed. 1788.
248. ——— The New Practical Navigator. 12th ed. 1796.
249. ——— The New Practical Navigator. 20th ed. by J. F. Dessiou. 1828.
250. Moriarty, H. A. Article on Navigation in Encyclopaedia Britannica. 9th ed. 1884.
251. Morris, C. Handbook to the B.O.T. Examinations. 3rd ed. 1913.
252. Moskowitz, S. *From Simple Quadrant to Space Sextant.* Navigation. (Journal of the U.S. Institute of Navigation.) Vol. 12. 1965.
253. Moxon, J. A Tutor to Astronomie and Geography. 1686.
254. Neugebauer, O. Ancient Mathematics and Astronomy. In 'A History of Technology,' edited by Singer et al. Vol. 1. 1954.
255. Nevins, A. E. *Star Observations at Sea.* N.M. 1884.
256. Newhouse, D. The Whole Art of Navigation. 1701.
257. Newton, J. *The Equal Altitude Problem.* N.M. 1861.

258. Norie, J. H. Complete Epitome of Navigation. 4th ed. 1814.
259. —— Complete Epitome of Navigation. 8th ed. 1825.
260. —— Complete Epitome of Navigation. 9th ed. 1828.
261. —— Complete Epitome of Navigation. 21st ed. by A. B. Martin. 1877.
262. —— Complete Epitome of Navigation. New ed. by W. H. Bolt. 1900.
263. —— Complete Epitome of Navigation. New ed. by J. W. Saul. 1917.
264. —— Nautical Tables. New ed. by G. P. Burris. 1950.
265. —— Nautical Tables. New ed. by F. N. Hopkins. 1963.
266. Norwood, R. The Seaman's Practice containing a Fundamental Problem in Navigation experimentally verified. 1644.
267. —— Trigonometrie: or the Doctrine of Triangles with exact Tables of the Sun's Declination etc. 2nd ed. 1657.
268. Parker, P. H. R. *A Deduction from the New Navigation.* N.M. 1890.
269. —— *Notes on Sumner Lines.* N.M. 1891.
270. —— *Navigation: A New Version of an Old Problem.* N.M. 1892.
271. —— *Double Altitude Problem.* N.M. 1893.
272. Parry, J. H. The Age of Reconnaissance. 1963.
273. Pemberton, H. *Some Considerations on a late Treatise intituled a New Set of Logarithmic Solar Tables etc., Intended for a more commodious Method of finding the Latitude at Sea by Two Observations of the Sun.* P.T.R.S. Vol. 51. 1760.
274. Phillippes, H. The Geometrical Seaman, or the Art of Navigation performed by Geometry. 2nd ed. 1657.
275. (Philonauticus). The Seaman's Calendar. 1667.
276. Plumstead, E. *Notes on Lunars, Sumners, and Double Altitudes.* N.M. 1899.
277. Price, D. J. *Precision Instruments to 1500,* in 'A History of Technology,' edited by Singer et al. Vol. 3. 1957.
278. —— *The Manufacture of Scientific Instruments from c. 1500 to c. 1700,* in 'A History of Technology,' edited by Singer et al. Vol. 3. 1957.
279. —— *Two Mariners' Astrolabes.* J.I.N. Vol. 9. 1957.
280. Purey-Cust, H. E. *Sumner: or the Whole Art of Navigation.* N.M. 1897.
281. Quill, H. John Harrison: the man who found Longitude. 1966.
282. Raper, H. *On the Longitudes of the Principal Points on the Globe.* N.M. 1839.
283. —— *Remarks on the Modes of Determining Longitude.* N.M. 1839.

284. Raper, H. *On the Necessity of Adopting Secondary Meridians.* N.M. 1839.
285. ——— *On the Chronometric Determination of Longitude.* N.M. 1839.
286. ——— *Discussion of the Longitude of the Principal Maritime Points of the Globe.* N.M. 1839.
287. ——— Practice of Navigation. 1st ed. 1840.
288. ——— *On Captain Sumner's Method of Determining the Position of a Ship.* N.M. 1844.
289. ——— Practice of Navigation. 6th ed. 1857.
290. ——— Practice of Navigation. 20th ed. by W. Hall and H. B. Goodwin. 1914.
291. Reeves, G. M. *On Captain Weir's Asimuth Diagram.* N.M. 1892.
292. Richey, M. W. *The Haven-Finding Art.* J.I.N. Vol. 10. 1957.
293. Riddle, E. Treatise on Navigation. 1824.
294. Riddle, J. *Captain Sumner's Method of Finding Latitude Longitude at Sea.* N.M. 1854.
295. ——— Treatise on Navigation. 6th ed. 1855.
296. Rios, J. (de Mendoza y) *Recherches sur les principaux Problèmes de l'Astronomie Nautique.* P.T.R.S. Vol. 87. 1797.
297. ——— Nautical Tables. 2nd ed. 1802.
298. Robertson, J. The Elements of Navigation in 2 volumes. 5th ed. by W. Wales. 1786.
299. Rosser, W. H. Self Instructor in Navigation. 1885.
300. Sadler, D. H. *The Navigational Work of the Nautical Almanac Office.* J.I.N. Vol. 1. 1948.
301. ——— *The Provision for Astronomical Navigation at Sea.* J.I.N. Vol. 2. 1949.
302. ——— Astronomy and Navigation. (Occasional Notes R.A.S. No. 13. Vol. 2. 1949.)
303. ——— *L. J. Comrie's Contribution to Navigation.* J.I.N. Vol. 4. 1951.
304. ——— *Margett's Horary Tables.* J.I.N. Vol. 6. 1953.
305. ——— *The Doctrine of Nautical Triangles Compendious.* J.I.N. Vol. 6. 1953.
306. ——— *Vice Admiral Radler de Aquino.* J.I.N. Vol. 7. 1954.
307. Sambursky, S. The Physical World of the Greeks. 1956.
308. ——— The Physical World in Late Antiquity. 1962.
309. Sarton, G. Six Wings. 1957.
310. Saul, J. W. Newton's Guide. 1924.
311. Seller, J. Practical Navigation. 1711.
312. Shadwell, F. A. Notes on the Management of Chronometers. New ed. 1861.

313. Short, J. *An Account of an Horizontal Top invented by Mr. Serson.* P.T.R.S. Vol. 47. 1752.
314. Simmons, R. E. G. *Observations of the Green Flash.* J.I.N. Vol. 4. 1951.
315. Sinclair, G. The Principles of Astronomy and Navigation. 1688.
316. Singer, C. and Holmyard, Hall and Williams. Editors of 'A History of Technology,' in 5 volumes. 1954–1958.
317. Smeaton, J. *Observations on the Graduation of Astronomical Instruments: etc.* P.T.R.S. Vol. 76. 1786.
318. —— *Description of an Improvement in the Application of the Quadrant of Altitude to a Celestial Globe for the Resolution of Problems dependant on Azimuth and Altitude.* P.T.R.S. Vol. 79. 1789.
319. Sommerville, W. E. The St. Hilaire Method in Practice. 1929.
320. Speidel, J. A Brief Treatise of Spherical Triangles. 1627.
321. Stebbing, F. C. Navigation and Nautical Astronomy. 1896.
322. Sturmy, Samuel. The Mariners Magazine. 3rd ed. by J. Colson. 1684.
323. Sumner, T. A New and Accurate Method of Finding a Ship's Position at Sea. 1843.
324. Tate, W. G. Theory and Practice of Navigation and Nautical Astronomy. 1900.
325. Taylor, E. G. R. *Jean Rotz: his neglected Treatise on Nautical Science.* Geographical Journal. Vol. 73. 1929.
326. —— Tudor Geography. 1930.
327. —— Late Tudor and Early Stuart Geography. 1934.
328. —— *Ideas on the Shape, Size, and Movements of the Earth.* Historical Association Pamphlet No. 126. 1943.
329. —— *The Navigator in Antiquity.* J.I.N. Vol. 1. 1948.
330. —— *The Sailor in the Middle Ages.* J.I.N. Vol. 1. 1948.
331. —— *The Dawn of Modern Navigation.* J.I.N. Vol. 2. 1949.
332. —— *Five Centuries of Dead Reckoning.* J.I.N. Vol. 3. 1950.
333. —— *The Navigating Manual of Columbus.* J.I.N. Vol. 5. 1952.
334. —— *Hariot's Instructions for Raleigh's Voyage to Guiana, 1595.* J.I.N. Vol. 5. 1952.
335. —— The Mathematical Practitioners of Tudor and Stuart England. 1954.
336. —— *An American Collection of Source Books for the History of Navigation.* J.I.N. Vol. 8. 1955.
337. —— *John Dee and the Nautical Triangle 1575.* J.I.N. Vol. 8. 1955.
338. —— The Haven-Finding Art. 1956.

339. Taylor, E. G. R. *Cartography, Survey and Navigation to 1400*, in 'A History of Technology,' edited by Singer et al. Vol. 3. 1957.
340. ——— *Cartography, Survey and Navigation, 1400–1750*, in 'A History of Technology,' edited by Singer et al. Vol. 3. 1957.
341. ——— *Mathematics and the Navigator in the Thirteenth Century*. J.I.N. Vol. 13. 1960.
342. ——— *Four Steps to the Longitude*. J.I.N. Vol. 15. 1962.
343. ——— (and M. W. Richey). The Geometrical Seaman. 1962.
344. ——— The Mathematical Practitioners of Hanoverian England. 1966.
345. Taylor, Janet. The Principles of Navigation Simplified. 3rd ed. 1837.
346. Thomson, W. *On the Determination of a Ship's Place from Observations of Altitude (Reprint from P.T.R.S.)*. N.M. 1871.
347. ——— Navigation: A Lecture. 1875.
348. Tonkin, J. *Rule for Finding the Time at Ship by Equal Altitudes, thence the Longitude*. N.M. 1838.
349. Toynbee, H. *A Few Words on Lunars*. N.M. 1859.
350. Trivett, J. F. *Latitude and Hour Angle by Simultaneous Altitudes of Two Stars*. N.M. 1850.
351. ——— *Finding the Azimuth at Sea*. N.M. 1878.
352. Turnbull, W. The Chronometer's Companion. 1856.
353. Vince, S. A Complete System of Astronomy. 2nd ed. 1823.
354. Wakeley, A. The Mariners Compass Rectified. Edited by W. Mountaine. 1766.
355. Walker, J. R. An Explanation of the Method of Obtaining the Position at Sea known as the 'New Navigation'. 1901.
356. Wargentin, P. *Letter to N. Maskelyne containing an essay of a New Method of determining the Longitude of Places from Observations of the Eclipses of Jupiters Satellites*. P.T.R.S. Vol. 56. 1766.
357. Waters, D. W. *Early Pole Star Tables*. J.I.N. Vol. 8. 1955.
358. ——— *Early Pole Star Tables*. J.I.N. Vol. 9. 1956.
359. ——— *A Tenth Mariner's Astrolabe*. J.I.N. Vol. 10. 1957.
360. ——— The Art of Navigation in England in Elizabethan and Early Stuart Times. 1958.
361. Waterton, J. J. *On Finding the Latitude, Time and Azimuth*. N.M. 1842.
362. Weems, P. V. H. Marine Navigation. 1940.
363. Whiston, W. The Longitude discovered by Eclipses, Occultations and Conjunctions of Jupiters Satellites. 1738.
364. White, J. *A Note on the Ex-meridian Problem*. N.M. 1896.
365. ——— *On Finding Longitude*. N.M. 1898.

BIBLIOGRAPHY

366. White, G. *Shortened and Simplified Method of Finding Latitude and Longitude by Double Altitudes.* N.D.
367. Wilson-Barker, D. *Star Observing.* N.M. 1884.
368. Wilson, H. Navigation new-modelled. 6th ed. 1750.
369. Wilson, J. *A Dissertation on the Rise and Progress of the Modern Art of Navigation,* in Elements of Navigation, by J. Robertson. 1st ed. 1772.
370. Wolf, A. A History of Science, Technology and Philosophy in the 16th and 17th centuries. 2 Vols. 1935.
371. Wood-Robinson, T. *Time Azimuth: A New and Simple Chart.* N.M. 1896.
372. Wright, E. The Haven-Finding Art. 1599.
373. —— Certaine Errors in Navigation, etc. Edition 1610.
374. —— A Description of the Admirable Table of Logarithmes. Published posthumously by Samuel Wright in 1616.
375. Young, J. R. Practical Astronomy, Navigation and Nautical Astronomy. 1856.

INDEX

INDEX

Index

Aberration, 106, 345
Adams, G., 89, 92
Ainsley, T., 154–5, 240
Airy, Sir G. B., 208, 218–19
Al Battani, 19, 58
Alessio, 335
Al Farisi, 99
Al Hazen, 19–20, 99
Al Kindi, 19, 99
Al Mamun, 19
Altitude, 268, 346
  absolute, 257–9
  apparent, 114–15
  circles of equal, 268–308
  corrections, 97–122
  equal, 257–65, 288
  maximum, 162–5
  measuring instruments, 57–96
  meridian, 162–5, 346
    of star, 129, 139–40
    of sun, 137–9
  See also Double-altitude problem and Ex-meridian problem
Anaximander, 180
Andrew, J., 151–2, 155
Angle of position, 51
Angus, Capt. T. S., 293, 296
  method of fixing, 293
Aphelion, 34
Apian, P., 197, 198
Apparent
  altitude, 114–15
  place, 119
  time, 316
Aquino, Capt. F. R. de, 322, 329
Arabian astronomy, 19–20, 61
  navigation, 129
  views on light, 99

Arctic Circle, 11, 128
ARG, 342
Aristotle, 8–9, 97–8
Aristyllus, 11
Arithmetical complement (Ar Comp), 213–14
Armillary sphere, 17, 60
Artificial horizon
  Becher's, 94–5
  Beechey's, 95
  Booth bubble, 96
  bubble, 91
  gyroscopic, 94
  mercury, 92–3
  Paget, 95–6
Aspley, J., 46
Astrograph, 340
Astrolabe
  plane, 17, 60–1
  Ptolemy's, 17
  seaman's, 60–3, 65–6, 117–18, 139, 140
Astrology, 6, 310
Astronomical Ephemeris, 28–9
  See also Nautical Almanac
Astronomical ring, 63
  See also Astrolabe
Astronomical (PZX) triangle, 51–2, 97, 243–54
  graphical solution, 334–40
  mechanical solution, 341–2
  tables, 247, 312–13, 317–34
Augmentation, 118–19
Azimuth, 51, 271, 346
  compass, 47
  diagrams, 324, 331, 335
  tables, 284–90, 321, 323–4, 328

INDEX

Babylonian, *see* Phoenician
Back-staff, 70–2
Bacon, R., 21, 99
Baffin, W., 109–10, 197
Baily, F., 314
Baker, Comdr., 339
Ball, Rev. F., 327–8
Barlow, R., 310
Barthet, M. J., 290
Bategnius, 155
Becher, Lt. A. B., 94
Beechey, Adm. F. W., 95
Beij, 339
Bertin, 329
Bessel, F., 106
Bethune, Adm. C., 274
Biot, 116
Bird, J., 89, 183, 204
Blackburne, Capt. H. S., 178–9, 288–9, 328
Blish, Comdr., 91
   prism, 91
Blundeville, T., 45, 62, 144, 198
Board of Longitude, 29, 149, 188, 193–4, 202, 203, 205, 218, 225, 226, 314
Boilard, 302
Boisy, Comte du, 292, 294, 296, 298–300, 305
Bolt, W. H., 262, 287–8, 292, 301–5
Borda, Chevalier de, 83, 87, 208, 209–12, 247, 250
Bouguer, M., 236
Bourne, W., 65, 66, 134, 140, 142
Bowditch, N., 161, 279
Bradley, Dr. J., 81, 104, 105, 106, 108, 202
Braga, Lt. R. R., 330–1
Brahe, Tycho, 24–5, 58, 60, 99, 142
Brent, 176, 306
Briggs, H., 46
Brinkley, J., 156, 157
Burdwood, J., 162, 284, 312, 324
Bygrave, Capt. L. C., 341

Caille, Abbé de la, 103–4, 108, 177, 200, 204, 208, 236
Caillet, M., 155
Calendar, 5, 10, 35–6
Callippus, 9–10
Campbell, Capt., 81, 202
Cardinal points of compass, 345
Carpenter, 198
Cary, 90, 94
Caspari, 302
Cassini, J. D., 100–3, 185–6, 187, 318
Celestial
   latitude and longitude, 268, 344
   meridians, 13, 33
   poles, 33, 123, 125
   sphere, 32, 124–8, 137–8, 180, 343, 347–8
Chaldean astronomers, 34–6
Chancellor, R., 83
Chaucer, G., 61
Chaulnes, Duc de, 89
Chauvenet, W., 84, 163
Chilmead, J., 143
Chronometer, 52–6, 194–5
   checking, 254–67
   longitude by, 29–30, 243–54
   rate, 53–4, 264, 294
   temperature effect on, 63–4
   winding of, 54–5
Circumpolarity, 125–7
Clairaut, 201
Clavius, 82
Clearing the distance, 29, 208
   See also Lunar distances
Climata, 128–9
Climatic zones, 129
Clocks
   mechanical, 22, 27, 42–3
   sand, 37
   See also Chronometer
Coignet, M., 39, 67
Coleman, G., 237
Collier, J., 40–1

INDEX

Collins, J., 156
Columbus, C., 66, 182
Commissioners of Longitude, *see* Board of Longitude
Comrie, Dr. L. J., 334
Connoissance des Temps, 48, 103, 203–4, 236
Constellations, 6, 11–12
  Orion, 6
  Pleiades, 6
  Southern Cross, 140–1
  See also Great Bear *and* Lesser Bear
Cook, Capt. J., 183, 255
Copernicus, 22–3, 58
Cortes, M., 24, 43, 62, 66, 132, 142, 311
Cosine formula, 245–6, 248, 250–3
Cosine-haversine method, 253, 325, 333
Cross bearings of the Sun, 270
Cross-staff, 23–4, 64–9, 139, 140, 197
Culminations, 32, 129, 346

Davis, J., 45, 68, 70, 141–2, 143
  quadrant, 72–3
Davis, Capt. J. E., 284, 324
Davis, P. L. H., 253, 284, 312, 324–6, 333–4
Day
  apparent solar, 43
  astronomical, 49
  civil, 49–50
  mean solar, 43, 49, 264
  sidereal, 32, 33, 36, 264
  solar, 32, 33–4, 36, 43, 49, 264
Dead reckoning, 127
Declination, 123, 125–6, 268, 344
  of Pole Star, 134
  of stars, 13, 141
  of Sun, 33, 141, 142
Dee, Dr. J., 43–4

Dekades, 35, 36
Delambre, 171, 208, 212–14, 260
Democritus, 180
Descartes, R., 100
Diagonal scale, 83
Dicearchus, 127
Differences, method of, 240–2
Digges, T., 53
Dip of horizon, 91, 111–18
Ditton, H., 192–3
Diurnal circles, 125, 345
Doctrine of the Spheres, 1, 46
Dollond, J., 89, 90, 204
Doppler navigation systems, 31
Double-altitude problem, 143–62, 271–2, 279, 288, 291–2
  Ainsley's method, 154–5
  Andrew's method, 151–2
  Burdwood's method, 162
  direct method, 161–6
  Douwes's method, 148–51, 158
  Duillier's method, 146–7
  Dunn's method, 151, 272
  Graham's method, 147–8
  graphical methods, 156
  IOU method, 161
  Ivory's method, 152–4
  Lynn's method, 156–60
  Riddle's method, 156
Double-chronometer problem, 272–4
Douwes, C., 148–9, 168
Ducom, 177
Duhamel, M., 285
Duiller, N. F., 146–7
Dunn, S., 151, 271–2
Dunthorne, 208, 225–6, 231–2
Dyson, Sir F. W., 195

Earth
  circumference of, 12
  figure of, 28, 111–12, 114, 120
  nutation of, 106

INDEX

Earth (contd)
  orbit of, 5, 34, 343
  precession of, 15, 28, 123, 345
  rotation of, 32, 33, 123, 345
Eclipses, 18, 181
  of Jupiter's satellites, 26, 27, 103, 184–9, 193, 313, 314
  of Moon, 182–3
  of Sun, 183–4, 256
  Saros cycle, 6, 200
Ecliptic, 11, 33, 35, 123, 343
  obliquity of, 13, 15, 33
  system, 268
Eden, R., 311
Elton, J., 91
Empedocles, 97–8
Enisco, F. de., 310
Epact, 40
Ephemeris, 28–9, 48
  See also *Connoissance des Temps* and *Nautical Almanac*
Equation of time, 15, 34, 316
Equations of condition, method of, 298–9
Equinoctial, 13, 33, 343
  system, 268, 347
Equinoxes, 14, 33, 34, 41, 343
  precession of, 15, 28, 123, 345
Eratosthenes, 12–13, 60, 127, 128
Espinasse, R. W., 288
Establishment of a port (H. W. F. & C. constant), 40
Euclid, 98
Eudoxus, 9
Euler, L., 201, 205
Ex-meridian problem, 163, 165–79
  double, 176–7
  graphical solution of, 176, 178
  triple, 177–8

Fasci, 302
Favé, 338–9
Ferguson, J., 236

Fermat, 100
First Point of Aries, 14, 33, 317, 343, 344
  of Libra, 343, 344
Fitzmaurice, L. T., 161–2, 279–80
Flamsteed, J., 28, 186–7, 193, 199, 311
Fleurais, Adm., 96
Fludd, R., 68
Fomalhaut, 237
Fortunate Isles, meridian through, 181
Foscolo, Prof., 176
  parabola, 176
Fusee, 43

Galileo, 27, 184–5
Gama, V. da, 66, 70
Gemma Frisius, 24, 29, 42, 194, 197, 310
Geocentric distance, 237
Geographical position, 268
George, Capt., 93
Gilbert, W., 23
Globe, 45, 143–5, 341–2
Gnomon, 7, 36–7, 128
Gnomonic projection, 334–5
Godfray, H., 178, 335
Godfrey, T., 81
Golden Number, 40
Goodwin, H. B., 177, 252–4, 261–3, 283–4, 307–8, 327–31, 333
Gordon, J., 291
Gould, Cmdr. R. T., 195
Graham, G., 89
  R., 147–8
Gravitation, 26–7
Great Bear, 3, 11–12, 39–40
Great circle, 13, 349
  sailing, 320, 334–5
Greek astronomy, 2, 6, 7–19, 180, 182
  optics, 97–8

INDEX

Green flash, 267
Greenwich
  hour angle, 268, 317, 323, 347
  mean time, 50-1, 237-42
  meridian, 49, 50, 180-2
  Royal Observatory, 28, 199, 257, 311
Grimaldi, 103
Grosseteste, R., 99
Guards, 38
  See also Lesser Bear
Gunter, E., 44, 46, 68, 340
  Scale, 46, 340
Guyon, Capt. M. E., 198, 204, 329
Gyroscopic
  inertia, 123
  platform, 188

Hadley, J., 74, 75, 77-82, 88, 91, 204
Hagner position finder, 342
Half-angle formulae, 250-1
Hall, Rev. W., 218, 232-5, 265-6, 329-30, 341
Halley, E., 75, 183, 193, 199-201
Hansen, L. F., 218
Hariot, R., 66
  T., 44, 109, 115, 136
Harrison, J., 29, 195, 243, 255
Hauksbee, F., 108
Haversines, 152, 355
  all-haversine formula, 336-7
Heath, Lt. L. G., 284-8
Height of the pole, 137
Henry the Navigator, Prince, 4, 130
Herne, 191
Hero of Alexandria, 98
Herschel, J. F. W., 315
Hilleret, Lt., 293, 302
Hipparchus, 13-16, 17, 18, 44, 60, 61, 123, 128-9, 182
Hire, A. de la, 100

Holywood, J., 21
Hommey, L., 319
Hood, T., 45, 69
Hooke, R., 74-5, 103, 105-6, 186
Horizon
  dip of, 91, 111-18
  irradiation, 122
  observer's celestial, 345
  system, 268, 346
Hornby, Rev., 184
Hour angle, 244, 268, 331, 346-7
  Greenwich, 268-9, 317, 323, 347
  Local, 51, 323, 346
  Sidereal, 317
Hour circle, 346
Hues, R., 44, 45, 143
Hutchinson, W. G., 287
Huyghens, 27, 43, 103
Hyperbolic navigation systems, 30

Inertial navigation systems, 31
Inman, Rev. J., 110, 160, 161, 168, 175, 217, 242, 252, 279
Intercept method, 279
IOU method (double altitude), 161
Irradiation, 121-2
Irwin, 188
Ivory, J., 110, 152, 155

Jacob ben Makir, 64
John, King of Portugal, 136
Johnson, A. C., 176-7, 274, 289
Jupiter, satellites of, 26, 27, 184-9, 193, 313, 314
  and speed of light, 103

Kamal, 69-70, 129
Kelly, P., 156
Kepler, J., 25-7, 99, 142, 185, 198
Kitchin, F. A. L., 178, 335
Kochab, 38, 130-2

Kortazzi, 322
Krafft, 208, 215–17

Lagrange, 189
Lalande, 146, 271, 272, 318
Lamplough, 340
Laplace, 189
Latitude, 123, 127–8, 180
  determination of, 123–79
    by double-altitude, 143–62
    by ex-meridian altitude, 163, 165–79
    by maximum altitude, 162–5
    by meridian altitude, 137–40
    by Pole Star, 130–7
    by Southern Cross, 140–1
  parallels of, 11, 12, 124, 128–9
  running down the, 131
Laverty, J. N., 282–4
Leadbetter, 147
Leap years, 142–3
Lecky, Capt. S. T. L., 90, 232, 263, 281, 289–90, 306
Lemonnier, 192, 200
Lenticular lens, 90–1
Lesser Bear, 8, 11–12, 37–9, 130–6, 309
Levi ben Gerson, 64
Light
  aberration of, 106
  nature of, 97–8, 103
  speed of, 106, 186
Lippershey, 27
Lisle, R. de l', 338
Littlehales, G. W., 336, 339
Littrow, 135
Local
  apparent time, 245
  hour angle, 51, 323, 346
  mean time, 50–1, 245
  sidereal time, 245
Logarithms, 27, 46
  proportional, 239

Longitude, 127–8, 180
  Board of, *see* Board
  determination of, 180–267
    by green flash, 267
    Jupiter's satellites, 27, 184–9, 193
    lunar distance, 23–4, 28–9, 194–242, 255–6
    lunar eclipses, 181, 182–3
    lunar occultations, 189–91
    lunar transit, 191–2
    solar eclipses, 183–4
    sunset and sunrise, 366–7
    timepiece, 24, 29–30, 181, 194, 243–66, 317, 328
    trial and error method, 151
  rewards for discovering, 27, 28, 149, 185, 192–5, 202–3, 205, 226, 255
Longomontanus, 198
Lull, Ramond, 62
Lunar distances, 23–4, 28–9, 194–242, 255–6
  approximate methods, 222–37
    Dunthorne, 225, 231–2
    Hall, 232–6
    Maskelyne, 226
    Merrifield, 227–30
  exact methods, 208, 222
    Airy, 218–22
    Borda, 209–12
    Delambre, 212–14
    Krafft, 215–17
    Young, 214–15
  graphical methods, 236–7
  G.M.T. found from, 237–42
  measuring instruments, 81, 83–7
  principles and practice, 205–8
  tables, 199–205
Lunation, 35
Lyons, I., 208, 225, 226, 237
Lynn, Comdr. T., 156–7, 159–60, 272, 318, 320, 324

INDEX

Mackay, A., 108, 110, 114, 116, 136, 174, 184, 198, 231, 236, 247, 249
Mackenzie, 90
Magnac, A. de, 293, 294, 301, 302
Magnitude of star, 38
Mannevillette, d'A. de, 201
Maps
  ancient, 17, 18, 127, 180
  projections, 127, 334–5, 339
Margetts, 236, 336
Mariner's New Calendar, 311
Mars, 25, 26
Martelli, G. F., 323
Martin, A. B., 237
  Comdr. W. R., 136, 148, 160, 170, 263, 267, 305–6
Martin Behaim (of Bohemia), 22, 62
Maskelyne, Dr. N., 28–9, 80, 88, 110, 114, 115, 148, 150, 188, 203–5, 208, 226, 239, 248–9, 311, 313, 326
Maurice, J., 191
Mayer, J. T., 28, 83, 108, 201–3
Mazaroth, 7
McMillen, D., 342
Mean
  Sun, 34, 49
  solar day, 49, 264
  time, 316
  time hours, 37
Medina, P. de, 24, 311
Mercator chart, 296, 299
Meridian
  altitude, 129, 137–40, 162–5, 346
  distance, 244
  lines
    Eratosthenes's, 12, 127
    prime, 180, 182–2
  passage, 97, 129, 346
  of moon, 40–1
Merrifield, Dr. J., 93, 174, 227–8
Méthode du point rapproché, 290, 293–308

Metonic cycle, 10, 40
Molfino, 335
Molyneux, S., 106
Moon
  apparent motion, 4, 190
  declination, 142, 163, 164
  eclipses, 14, 181–3
  epact, 40
  horizontal parallax, 120, 121
  maximum altitude, 164
  meridian passage, 40–1
  nodes, orbit, 15
  occultations, 189–91
  parallax-in-altitude, 120
  phases, 5, 6, 35, 40
  retardation, 191
  right ascension, 142, 191
  semi-diameter, 118–19, 121
  tables, 24, 28, 29, 142
  theory of motion, 199, 201
  transits, 191–2
  See also Lunar distances
Moore, J. H., 110, 116, 172–3, 251
Moriarty, Capt., 319
Morin, J. B., 198
Mouchez, 302
Müller, *see* Regiomontanus

Napier, Baron, 27, 46, 354
  mnemonic rules, 47, 354–5
Nautical Almanac, 28–9, 48–9, 120, 134–5, 156, 157, 169, 188–91, 195, 203–5, 225–6, 232, 237–242, 270, 311, 313–17
New Navigation, 30, 290, 292, 293–308
Newton, Sir I., 27–8, 75–6, 103, 142, 199, 341
Nicetas, 10
Nicol prism, 90
Nocturnal, 39–40
Norie, J. W., 110, 116, 118, 136, 173, 231, 236–7, 242, 289, 333

INDEX

Nuñez, P., 65, 82–3, 135, 143
Nuremberg eggs, 43
Nutation of Earth's axis, 106, 345

Observer's
  celestial horizon, 345
  celestial meridian, 344
  zenith, 345
Ocagne, M. d', 336, 337
Occultation, 189–91
Octant
  Newton's, 76–7
  Hadley's, 78–82
Ogura, S., 331–3
Oughtred, W., 341
Owen, Comdr., 272

Pagel, L., 274
Paget, 95–6
Parallactic
  angle, 51
  rotula, 236
Parallax
  annual, 25, 106, 345
  equatorial, 120
  horizontal, 119
  -in-altitude, 119–20
  ocular, 66, 68, 121
Parker, Capt. P., 274
Parmenides, 8
Pemberton, Dr. H., 148
Perihelion, 34
Personal error, 122
Philip III of Spain, 27, 28, 185
Philolaus, 10
Phoenician (and Babylonian)
  astronomy, 4–7, 34–6
  navigation, 2–4, 11–12
Picard, J., 103, 203, 260
'Plain' scale, 46
Planetary motion, Laws of, 26–7, 28

Planispheric astrolabe, 17, 60–1
Plato, 8–9
Pliny, 13
Polar distance, 126–7
Pole Star (*Polaris*), 37–8, 39, 59–60, 69–70, 130–6, 140, 168–9
Pol Fernohr, 136
Polynesian navigators, 130
Poor, Prof. C., 341
Portuguese navigators, 4, 21–2, 62, 130, 131, 136, 139, 140
Position-line navigation, 30, 164, 268–308
  Goodwin's method, 307–8
  méthode du point rapproché, 290, 293–308
  New Navigation, 30, 290, 292, 293–308
  St. Hilaire's (intercept) method, 30, 253, 293–308, 317, 323
  sight reduction, 317–40
  Simpson-Baikie's method, 307
  Sumner's (chord) method, 30, 160, 270, 275–84, 290–3, 308, 319, 321, 323
  tangent method, 282, 284
Precession of the equinoxes, 15, 28, 123, 345
Precision navigation machine, 342
Primum mobile, 9
Pritchard, 340
Proportional logarithms, 239
Ptolemy, 16–18, 19, 60, 98, 182
  theorem, 16
Purbach, 21, 44
Pythagoras, 8, 111
Pytheas, 10–12, 128
PZX triangle, *see* Astronomical triangle

Quadrant, 57–60, 66, 131, 139, 140, 183
  Davis, 72–4, 83, 117

INDEX

Quadrant (*contd.*)
 Hadley's, 74, 75, 77–82, 88, 91, 118
 reflecting, 74–83, 89–90
Quill, Col. H., 195

Radio
 time-signals, 30, 55
 direction-finding, 30
Ramsden, J., 89
 dividing engine, 89, 90
Raper, Lt. H., 94, 110, 116, 145, 155, 160, 173, 177, 242, 247, 254–6, 275, 278–9
Reduction to the meridian, 170–2
Reflecting circle
 simple, 83–4
 prismatic, 84
 repeating, 84–7
Refraction
 atmospheric, 17, 20, 25, 98–104, 107–11, 197, 207
  Bradley's formula, 107, 110
  Cassini's formula, 103
  dip of horizon, 91, 113–14
 horizontal, 234
 Snell's law, 100–1
Regiment, *see* Rules
Regiomontanus, 21, 22, 44, 65, 310
Rhumb lines, 290
Riddle, E., 110, 135, 155, 156, 247, 280
 J., 110, 155, 156, 161–2, 279–80
Right Ascension, 14, 46, 48, 123, 268, 344
 of Moon, 142, 191
 of Pole Star, 134–5
 of Sun, 48, 49
Ring dial, 42
Rios, M. del, 120, 149, 208, 217
Rising (2 sin²h/2), 173, 175
 See also Versine

Risings, achronychal, cosmical and heliacal, 5
Roberts, E., 321
Robertson, J., 91, 92, 94, 116, 136, 145, 147, 148, 169, 177
Roemer, 103, 186
Rosser, W. H., 173, 289
Roy, General, 114
Rules
 of the North Star, 131–6, 309
 of the Sun, 138–9
Running down the latitude, 131
Rust, Capt. A., 331, 335

Sadler, D. H., 316
St. Hilaire, Marcq, 30, 164, 253, 293–308, 317, 323
Saros cycle, 6, 200
Seaman's Kalendar, 311
Seller, J., 1, 42, 134, 139, 141
Semi-diameter
 of Moon, 118–19, 121
 of Sun, 117–18
Serson's horizontal top, 94
Settings, achronychal, cosmical and heliacal, 5
Severus, 61
Sexagesimal system, 7, 36
Sextant, 76, 81, 87–92
Shadwell, Capt. C. F. A., 53–4, 158, 256
Short double-altitude method, 177
Sidereal
 day, 32, 33, 36, 264
 hour angle, 317
 year, 15
Sight reduction, 317–43
 cosine-haversine method, 325, 333
 intercept method, 317, 329
 graphical methods, 334–40
 longitude by chronometer, 317, 318, 328

Sight reduction (*contd.*)
  mechanical methods, 340–2
  Ogura's method, 331–3
  Yonemura's method, 333–4
  See also Position-line navigation
Simpson-Baikie, E. B., 307
Slide-rule, 340–1
Smeaton, J., 94
Smith, C., 79–80
Snell, W., 100
Solar
  day, 32, 33–4, 36, 43, 49, 264
  year, 34, 35–6
Solarometer, 342
Solstices, 33, 344
Solstitial points, 33
Souillagouet, 328
Southern Cross, 140–1
Spherical trigonometry, 16, 155, 304, 349–55
  cosine formula, 155, 351–2
  four-parts formula, 352–3
  Napier's rules, 47, 354–5
  sine formula, 349–51
Stars
  altitude curves, 340
  catalogues of
    Brahe, 25
    Flamsteed, 186
    Hipparchus, 13, 14, 18
  culminations, 129
  declination, 140, 141–2
  double-altitude problem, 144–5, 156–9, 161, 292
  hour angles, 244, 268, 317, 347
  meridian altitudes, 129, 139–40
  observation for chronometer error, 258, 263
  parallax, 25, 106, 345
  parallax-in-altitude, 120
  real motion, 345
  right ascension, 14, 48
  timekeeping by, 35, 36, 38

Stella Maris, 37–8
  See also Pole Star
Stereographic projection, 334, 339
Stevens, J., 136
Strabo, 129
Sturmy, S., 335
Sub-solar point, 268
Sub-stellar point, 268
Sullivan, Capt., 275
Sumner, Capt. T. H., 30, 160, 164, 270, 275–84, 290–3, 308, 319, 321, 323
Sun
  altitude-measuring instruments, 60–73, 89, 139
  apparent motion, 5, 15, 36, 343
  chronometer error observations, 257–63
  chronometer rating observations, 264–5
  culmination, 32, 165
  declination, 14, 33, 46, 128, 137, 138, 141–3, 309–10, 345
  eclipses, 183–4, 256
  horizontal parallax, 120
  hour circle, 346
  irradiation, 121–2
  latitude observations
    change in altitude, 179
    double-altitude, 143–56
    equal altitudes, 174
    ex-meridian altitude, 172–3, 175–6, 177
    meridian altitude, 137–9, 162–163
  longitude observations
    chronometer method, 245, 275–6
    sun-lunar distance, 196–7, 205–8
  position-line navigation, 275–84, 290, 293
  rotation, spots, 27
  timekeeping, 41–2

INDEX

Sun-dials, 37
Sunrise and sunset, observation for longitude, 226–7

Tables
  A, B and C, 262, 284–90, 318, 324
  A and K, 333
  Alphonsine, 21, 310
  Andrews, 152
  angular distances between stars, 158, 161
  Aquino, 329–30
  atmospheric refraction, 100–10
  azimuth, 284–90, 312, 323–41, 328
  Bairnson, 176
  Ball, 327–8, 329, 330
  Bertin, 329
  Blackburne, 198–9, 288–9, 328
  Braga, 331
  Brent, Walter and Williams, 306
  Burdwood, 162, 284, 312
  Burton, 284
  Cambridge, 226
  Cassini, 186, 187, 318
  chronometer, 324–6
  Clairaut, 201
  Coleman, 237
  corrections for second difference, 242
  Damoiseau, 189
  Davis, 253, 284, 312, 330, 324–6
  declinations
    of Moon, 142
    of stars, 140–2
    of Sun, 46, 60, 137, 141–3, 309–10
  Delambre, 189
  dip, 114–17
  Douwes, 148, 150
  Duhamel, 285
  Espinasse, 288

Tables (*contd.*)
  ex-meridian, 165, 173, 174–9, 283, 306, 312
  Flamsteed, 186–7
  Galileo, 185
  Goodwin, 331
  Guyon, 329
  Hall, 234
  Halley, 200
  Hansen, 218
  Hariot, 115
  haversines, 152, 168, 252–4, 325
  Heath, 284–90
  Hommey, 319
  horizontal refraction, 234
  Hughes, 334
  Inman, 110, 168, 242, 252
  inspection, 318, 326–7
  Ivory, 110
  Johnson, 176, 274
  Jupiter's satellites, 185–9
  Lalande, 318
  Laverty, 283, 289
  Lecky, 289–90
  Lemonnier, 200
  logarithm, 46, 323, 325
  logarithmic
    differences, 231
    haversines, 252–3, 325, 333
    risings, 173, 175, 247–8
    trigonometric, 46
  longitude factor, 283–4
  lunar distance, 312, 315
  Lynn, 156–7, 318–19, 320, 324
  Mackay, 110, 116
  Margetts, 336
  maritime positions, 254
  Martelli, 323
  Maskelyne, 110, 115, 148, 150, 188, 226, 239–40, 248, 311–312
  Mayer, 201–5
  meridional parts, 46

INDEX

Tables (*contd.*)
  Moon, 28, 29
    declination, 142
    horizontal parallax, 120–1
    lunar distances, 29, 199–205
    Right Ascension, 142, 191
  Moore, 110, 116
  Norie, 110, 116, 118, 237, 242, 284, 333
  Ogura, 331–2
  Pagel, 274
  Pole Star, 134–6, 168–9, 309
  proportional logarithms, 237
  Raper, 110, 116–17, 242, 247, 254
  Riddle, E., 110, 116, 175
  Riddle, J., 110
  Right Ascensions
    of Moon, 142, 191
    of stars, 46, 48, 142
    of Sun, 46, 49
  Rios, 120, 217–18
  Robertson, 116, 118
  Rosser, 173, 289
  Rudolphine, 26, 198
  Rust, 331, 335
  Shadwell, 158
  short-method, 155, 312–13, 317–334
  sight-reduction, 318–34
  Souillagouet, 328
  Stars
    angular distances, 156–9, 161
    declinations, 142
    hour angles, 317
    Right Ascensions, 46, 48, 142
    southing, 48
    total correction, 117
  Sun
    declination, 46, 49, 142–3, 309–11
    double-altitude, 148, 150
    Right Ascension, 48, 49
    total correction, 118
    true bearing, 162, 323–4

Tables (*contd.*)
  tan lat, 265, 266
  Thomson, 321–2
  tides, 46
  Towson, 175–6, 320, 329
  versines, 168, 252–3
    reduced, 176
  Wakeley, 323
  Wargentin, 189
  Weston, 266
  Wright, 115
  Yonemura, 333–4
  Zacuto, 310
Tapp, J., 311
Tate, W. G., 154
Taylor, Mrs. J., 110, 116, 155, 250, 261–2
Telegraphic chains, 256–7
Thales, 7–8, 12
Thomson, Sir W. (Lord Kelvin), 319, 323
Tides, 28, 35, 40
Timocharus, 11
Towson, Capt. J. T., 175, 320
Trigonometrical functions, 44
Triquet um, 58
Trivett, Capt. J. F., 161, 285–7
Troughton, 84, 89, 90
True
  place, 119
  Sun, 34
Two-star diagram, 339

Venus, 27
Vernier scale, 82, 84
Versine, 175, 248, 252–3, 355
  reduced, 176
Vertical circles, 346
Villarceau, Y., 293, 294, 301, 302
Vitello, 99

Wakeley, A., 47, 323

Walker, J. R., 306
Walter, 176, 306
Waltherus of Nuremberg, 21
Wargentin, P., 189
Watches, 43, 143, 194
Weems, Capt. P. V. H., 340
Werner, J., 23, 196, 197
Weston, Comdr., 266
Whiston, W., 192–3
White, J., 178
Wier, Capt. P., 324, 335
Williams, 176, 306
Willis, E. J., 342
Wilson, Dr. J., 93–4
Witchell, G., 208, 226
Wollaston prism, 90

Wood Robinson, T., 335
Wright, E., 42, 44, 46, 109, 111, 115, 133–4, 138, 139, 140–2

Year, 15
Yonemura, A., 333
Young, Prof. J. R., 213, 214–15
  Dr. T., 314–15

Zacuto, A. ben S., 310
Zeiss, C., 136
Zenith distance, 330
Zodiac, 7, 344
  stars, 237

527   Cotter, Charles H.
COT
      A history of
         nautical astronomy

DATE			

DISCARDED

from
INDIAN VALLEY
PUBLIC LIBRARY
Indian Valley Public Library

NO RENEWALS

© THE BAKER & TAYLOR CO.